Introduction to Computability

Fred Hennie
Massachusetts Institute of Technology

ADDISON-WESLEY PUBLISHING COMPANY
Reading, Massachusetts · Menlo Park, California
London · Amsterdam · Don Mills, Ontario · Sydney

This book is in the
ADDISON-WESLEY SERIES IN
COMPUTER SCIENCE AND INFORMATION PROCESSING

Michael A. Harrison
Consulting Editor

ISBN 0-201-02848-4
BCDEFGHIJ-MA-7987

Preface

The rapid growth of the use of digital computers over the last twenty years has led to increased interest in the study of algorithms. This book introduces some of the classical ways of modeling algorithmic processes and develops a characterization of the functions that can be evaluated algorithmically. Its purpose is to explore the fundamental limits of digital computation and, indirectly, the expressive power of programming languages. Although concerned primarily with the ultimate capabilities of algorithms, it also provides an appropriate background for current work on the complexity and efficiency of algorithms.

A fundamental study of algorithmic computation may be undertaken for several reasons. For the mathematician, it introduces the important branch of mathematics known as recursive function theory. For the computer scientist, it helps define the limits of the computing art and lends a new perspective on the nature of computation. In particular, it forces the computer scientist to face the possibility that there may be no algorithmic procedure for dealing with a given computational problem. And for those who wish to explore such topics as computational complexity, formal languages, and design of algorithms, the study of computability provides a useful starting point.

While the informal idea of an algorithm is quite old, careful formulations of this idea are all relatively recent. It was not until the 1930's that the work of Turing, Church, and Post provided satisfactory formal models for the notion of an algorithmic computation and led to the definition of computable functions. These formal models were mathematical in nature, but served to characterize the capabilities of "physical" computations. Then as digital computers and programming languages were developed, it became clear that the computations that can be described by general-purpose programming languages are precisely those represented by the earlier, more abstract, models.

It is this relationship between programming and computability that makes the theory of computability of direct interest to computer scientists. One of the themes of the book is that the equivalence of the programmable functions and the mathematically computable functions stems from certain basic features common to all general-purpose programming languages. This theme is developed by abstracting some of the essential properties of programming languages and exploring the consequences of these properties. Several mathematical models for algorithmic computation are then introduced and shown to satisfy the same basic properties. Finally, it is argued that the properties in question uniquely characterize the set of functions that can be evaluated algorithmically.

The major part of this book has been used as a text in a senior-level elective subject on computability and logic. The whole book provides adequate material for a one-quarter course on computability or, with minimal augmentation, a one-semester course. Shorter introductions to computability can be obtained by skimming or omitting certain sections. In particular, the material on Turing machines (Chapters 2 and 3) lends itself naturally to less detailed treatment. The discussion of general recursive functions (Section 5.3) can be omitted entirely without loss of continuity. The later parts of Sections 6.2 and 6.3 can also be omitted if desired.

Except for a basic familiarity with sets, relations, and functions, very little specific mathematical background is required for an understanding of the material presented here. However, since both abstraction and careful reasoning are involved, a certain amount of "mathematical sophistication" proves helpful. Programming experience is also desirable, since the basic properties introduced in Chapter 1 are derived from the properties of programming languages, but it is by no means essential. Thus both computer science students with some mathematical experience and mathematics students with some programming experience should find themselves well prepared.

The text has been organized to serve both the basic goal of characterizing the functions that can be evaluated algorithmically and the broader goal of providing a foundation for more advanced studies. For this reason certain concepts and techniques are presented in a more comprehensive form than is needed for our immediate purposes. As a result, the reader who wants to investigate recursive function theory, mathematical logic, or formal languages will be prepared to do so without having to go back and reconstruct his or her understanding of computability on a broader base.

No book—certainly no technical book—is the product of its author alone. I am pleased to acknowledge here the contributions of several colleagues who have had a major influence on this book and the course from which it arose. The basic philosophy of the course grew out of discussions with Manuel Blum, whose ideas helped shape the goals and outline of the text. Much of the detailed technical development was influenced by Michael M. Hammer, who faced with me the day-to-day problems of planning and teaching the course during its early years. Donna J. Brown provided valuable advice in the later stages of development and had a major part in the selection and organization of the problems. Thanks are also due Marsha E. Baker and Terry Bayer, who cheerfully and skillfully typed what must have seemed endless versions of the manuscript.

Cambridge, Massachusetts   F. H.
September 1976

Contents

Chapter 1

Properties of Algorithmic Computation

Our basic goal in this book is to answer the question: What is the scope and what are the limitations of discrete computation? In order to achieve this goal we must of course reach an understanding of what is meant by discrete computation and how discrete computations may be carried out.

The distinction between discrete and analog computation can be made in a variety of ways. For our purposes, the essential difference between the two is that discrete computation consists in the evaluation of functions from one countable set to another, whereas analog computation consists in the evaluation of functions from one uncountable set to another. Of particular interest within the realm of discrete computation is the evaluation of functions from the set of natural numbers to the set of natural numbers, and we shall be concerned almost exclusively with such functions.

It is also important at the outset to recognize a second distinction—that between functions and algorithms. A function is simply a relationship between the members of one set and those of another. An algorithm, on the other hand, is a procedure for evaluating a function. That is, an algorithm provides a way of finding, for any given element in the domain of a function, the corresponding value of the function. To oversimplify, a function *is* and an algorithm *does*, or tells how to do. Note that the relationship between functions and algorithms is not one-to-one. Many different algorithms can be used to evaluate the same function, and, as we shall see, there is a practical sense in which some functions cannot be evaluated by any algorithm. Thus our basic question becomes: What functions from one countable set to another can be evaluated algorithmically?

The answer to this question obviously depends on the precise nature of the algorithms permitted. We will eventually investigate several computing models that might be used as the basis for a definition of "algorithmic computation." But such investigations will inevitably involve us in the details of basic operations, sequencing of operations, access to memory, and so on. Since these

1

details can obscure our understanding of the general nature of algorithmic computation, we prefer to avoid them whenever possible. Fortunately, the various classes of general-purpose computing algorithms share several basic properties that by themselves go a long way toward specifying the kinds of functions that can be evaluated.

This chapter will accordingly be devoted to investigating, in abstract form, some of the basic properties of useful classes of algorithms. As a result of this investigation we will obtain an overview of the capabilities and limitations of algorithmic computation without becoming embroiled in the details of practical implementation and, in fact, without having to say precisely what an algorithm is. Subsequent chapters will examine some specific classes of algorithms in order to illustrate the results obtained here and to provide the foundation for a sensible definition of an "algorithmically computable function."

1.1 FUNCTIONS AND ALGORITHMS

Our first job is to review the concept of a function and the basic terminology used to describe functions. We will then discuss the nature of a computing algorithm and develop the notion of a family of algorithms, by which we mean a class of algorithms having a common repertoire of basic instructions and a common descriptive language. In Section 1.2, we will introduce five important properties that are shared by the families of algorithms associated with general-purpose programming languages. We will then show that these properties impose certain fundamental constraints on the functions that can be evaluated by the algorithms in a family.

Functions

Functions are most precisely described in terms of sets, and we rely on the set-theoretic approach for our basic definitions. In practice, however, the set-theoretic approach can become cumbersome, and recourse is usually made to more familiar, albeit less precise, descriptive means. Here we review both the formal and the informal terminology and notation of functions.

A function is basically a set of ordered tuples. To be more precise, suppose that $X_1, \ldots, X_n$ and Y are specified sets and that X denotes the Cartesian product $X_1 \times X_2 \times \cdots \times X_n$. A *partial function* from X to Y is then defined to be a subset f of $X_1 \times \cdots \times X_n \times Y$ such that, for each choice of x_1 in $X_1, \ldots, x_n$ in X_n, there is *at most* one element y in Y for which $(x_1, \ldots, x_n, y)$ belongs to f. A *total function* from X to Y is defined to be a subset f of $X_1 \times \cdots \times X_n \times Y$ such that for each choice of x_1 in $X_1, \ldots, x_n$ in X_n there is *exactly* one element y in Y for which $(x_1, \ldots, x_n, y)$ belongs to f. Inasmuch as X is the product of n component sets, a partial or total function from X to Y is commonly referred to as an *n-variable function* or, synonymously, as a *function of n variables*.

It is clear from the preceding definitions that a total function is just a special type of partial function. We will use the term *function* somewhat loosely to mean either total or partial function, prefixing the appropriate adjective as precision or emphasis requires. It must be recognized, however, that in most branches of mathematics the word function means, by definition, what we have agreed to call a total function. Even in the literature of computability and recursive function theory, the word function, used by itself, usually means total function.

It should be noted that whether a set f is a function from X to Y depends not only on the set f but also on the sets X and Y.

Example 1-1. Suppose that f is the set of ordered triples $\{(0, 0, a), (0, 1, b), (1, 0, c), (1, 1, b)\}$. Then f is certainly a (two-variable) function from $\{0, 1\} \times \{0, 1\}$ to $\{a, b, c\}$. In fact, it is a total function from $\{0, 1\} \times \{0, 1\}$ to $\{a, b, c\}$. The same set f is also a function from $\{0, 1, 2\} \times \{0, 1, 2\}$ to $\{a, b, c\}$, but in this case not a total function since no member of f has (say) 0, 2 as its first two components. The set f is also a (total) function from $\{0, 1\} \times \{0, 1\}$ to $\{a, b, c, d\}$, and a (nontotal) function from $\{0, 1, 2\} \times \{0, 1, 2\}$ to $\{a, b, c, d\}$. However, f is not a function from $\{0, 2\} \times \{0, 2\}$ to $\{a, b, c\}$, since f is not a subset of $\{0, 2\} \times \{0, 2\} \times \{a, b, c\}$.   □

The *domain* of a function f from $X_1 \times \cdots \times X_n$ to Y is the set of all those n-tuples $(x_1, \ldots, x_n)$ for which there exists an element y in Y such that $(x_1, \ldots, x_n, y) \in f$. Thus f is a total function from $X_1 \times \cdots \times X_n$ to Y iff[†] the domain of f is $X_1 \times \cdots \times X_n$. The *range* of a function f from $X_1 \times \cdots \times X_n$ to Y is the set of all those elements y in Y for which there exists some choice of elements $x_1, \ldots, x_n$ such that $(x_1, \ldots, x_n, y) \in f$. The domain and range of the function f are commonly denoted dom f and ran f, respectively.

If f is a function from $X_1 \times \cdots \times X_n$ to Y and if $(x_1, \ldots, x_n) \in$ dom f, we say that f is *defined* for the (combination of) *arguments* $x_1, \ldots, x_n$; if $(x_1, \ldots, x_n) \notin$ dom f, we say that f is *undefined* for the arguments $x_1, \ldots, x_n$. If f is defined for the arguments $x_1, \ldots, x_n$ and if $(x_1, \ldots, x_n, y) \in f$, we say that the *value* of f for the arguments $x_1, \ldots, x_n$ is y, or that f *maps* $(x_1, \ldots, x_n)$ into y. If f is undefined for the arguments $x_1, \ldots, x_n$, we say that the *value* of f for the arguments $x_1, \ldots, x_n$ is also *undefined*. Thus a function from $X_1 \times \cdots \times X_n$ to Y is total iff its value is defined for every n-tuple of argument values in $X_1 \times \cdots \times X_n$.

The value of a function f for the arguments $x_1, \ldots, x_n$ is normally denoted $f(x_1, \ldots, x_n)$, whether that value is defined or not. Thus the statement $f(x_1, \ldots, x_n) = y$ is synonymous with the statement $(x_1, \ldots, x_n, y) \in f$. The statement $f(x_1, \ldots, x_n) \neq y$ means simply that the value of $f(x_1, \ldots, x_n)$ is *not* y.

[†] "iff" is an abbreviation for "if and only if."

This includes the possibility that the value of $f(x_1, \ldots, x_n)$ is undefined as well as the possibility that the value of $f(x_1, \ldots, x_n)$ is defined but not equal to y.

It is often convenient to have a simple way of indicating whether a function value is defined or not. For this purpose we use the symbol $\downarrow$ to mean "is defined" and the symbol $\uparrow$ to mean "is undefined." Thus we write $f(a, b, c)\uparrow$ to indicate that the value of the function f is undefined for the combination of arguments a, b, c.

Example 1-2. Let f be the function $\{(0, 0, b), (1, 2, d), (1, 1, b), (0, 1, c), (2, 0, c)\}$ from $\{0, 1, 2\} \times \{0, 1, 2\}$ to $\{a, b, c, d\}$. Note that the domain of f is $\{(0, 0), (0, 1), (1, 1), (1, 2), (2, 0)\}$ and the range of f is $\{b, c, d\}$. Since dom $f \neq \{0, 1, 2\} \times \{0, 1, 2\}$, f is not a total function. In particular, f is defined for the argument combinations $(0, 0)$, $(0, 1)$, $(1, 1)$, $(1, 2)$, $(2, 0)$ and undefined for the combinations $(0, 2)$, $(1, 0)$, $(2, 1)$, $(2, 2)$. Among the statements that can be made about the values of f are: $f(1, 2) = d$, $f(2, 0) = c$, $f(1, 0)\uparrow$, $f(0, 1)\downarrow$, and so on. $\square$

It is often necessary to distinguish between the name of a function and the value of that function. For this reason it is dangerous to speak of "the function $f(x_1, x_2)$," since when we subsequently write "$f(x_1, x_2)$" it may not be clear whether we are referring to the function itself as a set or to the value of the function for certain arguments x_1, x_2. To avoid this ambiguity, we will normally use single symbols such as f, g, h, F, G, H, etc., as names for functions and reserve the notation $f(x_1, \ldots, x_n)$ to denote the value of the function f for the argument values $x_1, \ldots, x_n$. Thus instead of speaking of "the function $f(x_1, x_2) = x_1 + x_2$," we should say something like "the two-variable function f such that $f(x_1, x_2) = x_1 + x_2$."

We will often have occasion to combine two functions so that the values of one become arguments of the other. This process of combination is called composition, and the new function that results from it is called a composite function. In order to explain precisely what is meant by the composition of partial functions or of functions of more than one variable, we resort to a set theoretic definition. In other words, we define the desired composite function as a set by specifying exactly what ordered tuples it contains.

Let us suppose that h is an m-variable partial function and that $g_1, \ldots, g_m$ are n-variable partial functions. Then the composition of h with $g_1, \ldots, g_m$ is the n-variable function f defined so that $(x_1, \ldots, x_n, z) \in f$ iff there exist elements $y_1, \ldots, y_m$ such that:

$$(x_1, \ldots, x_n, y_1) \in g_1$$

$$\vdots$$

$$(x_1, \ldots, x_n, y_m) \in g_m$$

and

$$(y_1, \ldots, y_m, z) \in h.$$

We indicate that the function f is obtained by composing h with $g_1, \ldots, g_m$ by writing $f = h \circ (g_1, \ldots, g_m)$. It is important to remember that the expression $h \circ (g_1, \ldots, g_m)$ is a *name* for f, and does not denote a value of f.

In most applications of functional composition, $g_1, \ldots, g_m$ are functions from some common set $X_1 \times \cdots \times X_n$ to the respective sets $Y_1, \ldots, Y_m$, while h is a function from $Y_1 \times \cdots \times Y_m$ to Z. In this case, the composite function $h \circ (g_1, \ldots, g_m)$ is a function from $X_1 \times \cdots \times X_n$ to Z. If $g_1, \ldots, g_m$ and h are all total, the composite function is also total. In this case the value of f can be obtained by first evaluating $g_1, \ldots, g_m$ and then using the results as arguments in the evaluation of h. If $g_1, \ldots, g_m$ and h are not all total, it follows from the definition that the value of the function $h \circ (g_1, \ldots, g_m)$ will be undefined for any combination of argument values for which any one of the functions $g_1, \ldots, g_m$ is undefined or for which the values of $g_1, \ldots, g_m$ are defined but the resulting value of h is not. In any event, we will write $f(x_1, \ldots, x_n) = h(g_1(x_1, \ldots, x_n), \ldots, g_m(x_1, \ldots, x_n))$ to indicate how the value of f is related to the values of $g_1, \ldots, g_m$ and h.

Example 1-3. Consider the one-variable functions

$$g_1 = \{(1, a), (3, b), (4, d), (5, b), (6, c)\}$$

$$g_2 = \{(1, d), (2, c), (3, a), (4, c), (6, b)\}$$

from $\{1, 2, 3, 4, 5, 6\}$ to $\{a, b, c, d\}$ and the two-variable function

$$h = \{(a, c, s), (a, d, q), (b, a, p), (b, c, t), (c, c, r), (c, d, t), (d, c, r)\}$$

from $\{a, b, c, d\} \times \{a, b, c, d\}$ to $\{p, q, r, s, t\}$. By definition, the composite function $h \circ (g_1, g_2)$ is the one-variable function

$$f = \{(1, q), (3, p), (4, r)\}$$

from $\{1, 2, 3, 4, 5, 6\}$ to $\{p, q, r, s, t\}$. Note that $f(1) = q$ because $g_1(1) = a$, $g_2(1) = d$, and $h(a, d) = q$. Similarly, $f(3) = h(g_1(3), g_2(3)) = h(b, a) = p$ and $f(4) = h(g_1(4), g_2(4)) = h(d, c) = r$. On the other hand, $f(2)$ is undefined because g_1 is undefined for the argument 2, while $f(5)$ is undefined because g_2 is undefined for the argument 5. The value $f(6)$ is also undefined because, although $g_1(6) = c$ and $g_2(6) = b$, the value $h(c, b)$ is undefined. □

Number-Theoretic Functions

We shall be concerned primarily with functions that map natural numbers or n-tuples of natural numbers into natural numbers. Such functions are called *number-theoretic functions*. We follow the common practice of using N to denote the set of natural numbers. Thus an *n-variable number-theoretic function* is a function from N^n to N. A total number-theoretic function is obviously one that is defined for every possible choice of natural numbers for its arguments. Thus ordinary addition is a total number-theoretic function, whereas subtraction is not.

Two special classes of number-theoretic functions will appear repeatedly in our investigations. One is the class of constant functions. A *constant* number-theoretic function of n variables is one that maps each n-tuple of natural numbers into the same natural number. We will use the symbol $C_m^{(n)}$ to denote the *n*-variable constant function whose value is m. Thus

$$C_m^{(n)}(x_1, \ldots, x_n) = m \qquad \text{for all } x_1, \ldots, x_n \in N.$$

When the context permits, we may omit the superscript (n). Strictly speaking, though, it is necessary to distinguish between constant functions of different numbers of variables because they are in fact different functions—even though they may have the same value.

The second special class consists of the projection functions or, as they are sometimes called, identity functions. For each choice of positive integers n and i, where $i \leq n$, we define the ith *n-variable projection function* to be the function that maps each member of N^n into its ith component. This function will be denoted $P_i^{(n)}$. Thus $P_i^{(n)}$ is the number-theoretic function such that

$$P_i^{(n)}(x_1, \ldots, x_n) = x_i \qquad \text{for all } x_1, \ldots, x_n \in N.$$

As in the case of the constant functions, we may omit the superscript (n) when the context permits.

By composing a given function h with appropriately chosen projection and constant functions, we can in effect permute the arguments of h, substitute constant values for one or more arguments, and expand or contract the set of arguments. In doing this it is important to remember that the functions with which h is to be composed must all have the same number of variables and that the number of such functions must equal the number of arguments of h. The ability to "manipulate" the arguments of a function h by means of projection and constant functions will play an important role in our later studies, and the reader should make sure that he understands how such manipulations are performed.

Example 1-4. Suppose that h is a given two-variable number-theoretic function. Among the functions that can be obtained by composing h with constant and projection functions are the following:

$$
\begin{aligned}
f_1 &= h \circ (P_2^{(2)}, P_1^{(2)}) & f_4 &= h \circ (P_1^{(1)}, P_1^{(1)}) \\
f_2 &= h \circ (P_1^{(2)}, P_1^{(2)}) & f_5 &= h \circ (P_1^{(3)}, P_2^{(3)}) \\
f_3 &= h \circ (P_1^{(2)}, C_0^{(2)}) & f_6 &= h \circ (C_7^{(3)}, C_{15}^{(3)})
\end{aligned}
$$

Here f_1, f_2, and f_3 are all functions of two variables, since the functions $P_1^{(2)}$, $P_2^{(2)}$, and $C_0^{(2)}$ are functions of two variables. The function f_1 is in effect obtained from h by interchanging the arguments of h. More precisely, the values of f_1 are related to those of h by the equation $f_1(x_1, x_2) = h(x_2, x_1)$. Similarly, we see that the values of f_2 and f_3 are related to those of h by the equations $f_2(x_1, x_2) = h(x_1, x_1)$ and $f_3(x_1, x_2) = h(x_1, 0)$. The function f_4 is a function of one variable, since $P_1^{(1)}$ is a function of one variable; its values are specified by the equation $f_4(x) = h(x, x)$. The functions f_5 and f_6 are functions of three variables; their values are specified by the equations $f_5(x_1, x_2, x_3) = h(x_1, x_2)$ and $f_6(x_1, x_2, x_3) = h(7, 15)$. Thus f_6 is the function $C_{h(7, 15)}^{(3)}$.   □

In addition to the constant and projection functions, we will from time to time have occasion to deal with the class of empty functions. By the *empty number-theoretic function of n variables*, we mean simply the n-variable function that is undefined for every choice of argument values. This function will be denoted $\Phi^{(n)}$. Thus

$$\Phi^{(n)}(x_1, \ldots, x_n)\uparrow \qquad \text{for all } x_1, \ldots, x_n \in N.$$

Note that the composition of an empty function with any other function(s) necessarily yields an empty function. Moreover, the empty function $\Phi^{(n)}$ can be obtained by composing *any* nontotal function with appropriate constant functions.

Example 1-5. Let φ be a three-variable number-theoretic function whose value is undefined for the combination of argument values a, b, c. Then

$$\varphi \circ (C_a^{(2)}, C_b^{(2)}, C_c^{(2)})$$

is the two-variable empty function $\Phi^{(2)}$.   □

Algorithms

While functions can be defined quite simply and precisely in terms of sets, a good definition of an algorithm is not easy to arrive at. We shall not even try to make such a definition here. Instead, we shall discuss briefly the characteristics that intuition and common sense suggest an algorithm ought to have. From these informally described characteristics we will extract two features that motivate a formal definition of certain abstract classes of algorithms. It is these classes, rather than specific algorithms, that we will study in the remainder of this chapter.

Basically, an algorithm specifies a routine, mechanistic way of evaluating a function. This evaluation is to be carried out by some sort of computing agent, which may be human, mechanical, electronic, or whatever. Since in practice the characteristics of an algorithm may be influenced by the intended computing agent, we shall, at least temporarily, consider each algorithm as being associated with a particular computing agent. Now, what features ought such a combination of algorithm and computer have if the algorithm is to determine a routine, mechanical computation?

Let us begin from an operational point of view. Associated with each algorithm is a countable set of objects on which the algorithm is designed to operate and among which the results produced by the algorithm are to be included. This set will be called the *domain* of the algorithm in question. A typical domain might be the set of natural numbers (or suitable representations thereof) or the set of strings composed of certain alphanumeric characters. At the beginning of any computation one or more members of the domain are designated as the argument or arguments to which the algorithm is to be applied.

An algorithm itself consists of a collection of instructions, each of which specifies a basic operation that is to be carried out on one or more members of the domain. In order that these instructions be properly executed, they must be presented to the computing agent in a form that the latter can correctly interpret. Thus each algorithm must be represented by a *description* in a suitable language. We shall have more to say about the nature of this description later on.

If a computation is to proceed mechanistically, both the sequence of operations to be carried out and the particular object(s) to which each operation is to be applied must be well defined. This sequence of operations and specification of objects must be determined by the algorithm, in accordance with certain "operating rules" associated with the algorithm and computing agent. For example, the operating rules might simply require that each instruction explicitly indicate the locations of the objects to which it is to be applied and that the instructions making up an algorithm be fed to the computing agent and executed one at a time, without repetition, until the end of the algorithm is reached. More sophisticated operating rules might permit the intermediate results obtained during the course of a computation to determine the instruction to be followed next and the objects to which it should be applied, as in the case of algorithms defined by practical programming languages.

Note that the operating rules of some algorithms may permit the execution of one or more instructions to be repeated over and over again, perhaps indefinitely. Thus the computation prescribed by an algorithm need not terminate for every choice of argument(s). But when a computation does terminate, that fact should be readily apparent and the outcome of the computation should be well defined. Moreover, this outcome should depend only on the particular argument(s) to which the algorithm was applied. In other words, each algorithm should determine the computation of a partial function on its associated domain.

To put these ideas in perspective, let us examine some simple examples. First consider the Euclidean algorithm for finding greatest common divisors, where the computing agent is a human equipped with pencil and paper. Here the domain is the set of natural numbers. The choice of the basic operations depends on the background of the human involved—i.e., on the particular computing agent used. Let us take the basic operations to be, roughly speaking, those of writing numbers down on paper, testing numbers for equality with zero, and computing remainders. The descriptive language is English, in which one possible description is:

"Step 1. Let m and n be given natural numbers. Compute the remainder of n with respect to m. Call this remainder r, and proceed to Step 2.

Step 2. If $r = 0$, terminate the computation with m as the result. If $r \neq 0$, replace n by m and m by r. Now repeat Step 1."

According to the operating rules used here, the algorithm specifies both the operation to be performed next, and the numbers on which it is to be performed, in terms of results obtained so far.

Next consider the algorithm defined by a FORTRAN program for selecting the largest number in a list. Here the computing agent is a particular combination of a computer and a suitable compiler or operating system. The domain is a subset of the integers. The basic operations are those of addition, subtraction, multiplication, storing, etc. The descriptive language comprises the set of legal FORTRAN programs—i.e., a subset of the strings that can be formed from certain alphanumeric characters. Again the sequence of operations and the numbers to which each operation is to be applied are specified by the algorithm in terms of results already obtained.

These examples should make two points clear. First, the familiar types of algorithms do in fact consist of instructions that describe basic operations to be carried out on elements from some domain, and the choice of elements and sequencing of operations are in fact determined by the algorithms in conjunction with certain established "operating rules." Second, the business of pinning down exactly what the basic operations are, how they are to be specified, and how they are to be sequenced may well be complicated and tedious. We prefer to avoid, at least temporarily, becoming involved in these details. Thus we propose to concentrate on what an algorithm does, rather than on how it does it.

But first we must face a question we have so far avoided: should any limit be placed on the size of an algorithm or the length of its description? In order to settle this question, let us adopt the fundamental principle that in practical computation it is the computing agent that is to do the actual evaluation of the function in question, not the person who prepares the algorithm. Thus we must rule out algorithms that amount to "infinite table lookups" in the sense that they require the programmer to determine and include in the algorithm the values of the desired function for an infinite number of argument values.

In fact, from a practical point of view, we must rule out any algorithm that would require an infinite amount of work to prepare. This means, in effect, that we must restrict our attention to algorithms that consist of finite numbers of instructions and therefore can be represented by finite descriptions, where by a finite description we mean a finite collection of symbols from some countable alphabet. Such descriptions usually, but not always, take the form of finite strings of symbols from the alphabet in question. And those that do not can normally be transformed into linear strings in a natural and obvious way. (This is one of the things we accomplish when we convert a two-dimensional flow chart into a one-dimensional program.) Thus we shall suppose that each of the algorithms with which we deal can be described by a finite string of symbols from some countable alphabet.[†]

Classes of Algorithms

Algorithms can be put rather naturally into classes, each of which consists of all the algorithms that share a common domain, a common set of basic operations, a common set of operating rules, and a common descriptive language. One example of such a class is the set of algorithms, described in English, to be implemented on a desk calculator without auxiliary memory. Another is the set of algorithms defined by the programs that can be written in a certain assembly language for execution on a certain digital computer. It is with such classes of algorithms that we will be primarily concerned, rather than individual algorithms. Of course, our definition of a "class" of algorithms is still somewhat vague. This situation will be remedied shortly. In the meantime, a rough intuitive understanding is sufficient to motivate the discussion that follows.

In practice, some algorithms are meant to evaluate functions of fixed numbers of variables. The familiar Euclidean algorithm is a case in point. Other algorithms are meant to evaluate functions of arbitrary numbers of variables. Algorithms that select the largest or smallest number from an arbitrarily long list are examples of this latter type. At first sight there might seem to be a fundamental difference between these two types of algorithms, but such is not really the case.

With an appropriate choice of operating rules, *any* algorithm can be viewed as being of the type that operates on arbitrary numbers of arguments, even though it may have been "designed" for a fixed number of arguments. For it is an easy matter to assign *some* outcome to each situation in which an

---

[†] One may argue that in some cases an infinite algorithm can be generated by a second, finite, algorithm—thereby occasioning only a finite amount of work on the part of the programmer. But in any reasonably sophisticated computing environment the infinite algorithm could then be replaced by the finite one, appropriately modified so as to arrange for the execution of the infinite algorithm as it is generated.

algorithm is applied to the "wrong" number of arguments. Perhaps excess arguments can simply be ignored; perhaps some fixed values can be provided in lieu of missing arguments; perhaps the ensuing computation can be made to yield some fixed value, say 0. At worst, we can always agree that the outcome of the computation is undefined whenever an algorithm is applied in a situation for which it was not designed.

Thus every algorithm can be interpreted as defining some partial function of n variables for every positive integer n. The exact way in which this is done is not important. What is important is that we can, without loss of generality, restrict our attention to classes of algorithms in which each algorithm has associated with it some well-defined function for every possible number of arguments.[†]

Since we are mainly interested in computing number-theoretic functions, we shall concentrate on algorithms whose domains are large enough to include representations of all the natural numbers. Any denumerably infinite domain will serve this purpose, since the members of a denumerably infinite set can always be put in a one-to-one correspondence with the natural numbers. The exact nature of this correspondence is not important. It is not even necessary that every domain element stand for some natural number. All that matters is that every denumerable domain can be thought of as containing representations of the natural numbers. Of course in adopting this point of view we are faced with the nuisance of converting natural numbers into their representations before applying algorithms to them, and later converting the results produced by the algorithms back into natural numbers. But in practice some such coding and decoding must always be done, since real computers in fact operate on representations of numbers rather than numbers themselves.

Thus from now on we will consider classes of algorithms with denumerable domains and assume that in each case a means has been chosen for representing natural numbers within the domain. Every algorithm in such a class can be interpreted as evaluating some partial n-variable number-theoretic function for each positive integer n. This function is defined as follows. If the algorithm in question eventually terminates when applied to a given n-tuple of natural number representations and if the resulting outcome is itself the representation of a natural number, the value of the function is taken to be that natural number. If the algorithm fails to terminate or if it terminates but produces an outcome that does not represent a natural number, the value of the function is taken to be undefined.

One of the important features of general-purpose computers is the fact that programs can be made to modify or create other programs. Indeed, it is possible to write programs—such as compilers—that are capable of constructing

---

[†] Remember that a function is a set, and can be perfectly well defined as a *set* even though its *values* for certain arguments are undefined.

very complicated and sophisticated programs when given appropriate instructions. Of course, the reason this is possible is that programs can be represented by members of the domain on which programs operate. Since we will want to investigate classes of algorithms that have the capability of modifying or creating other algorithms, we must consider classes whose domains include not only representations of the natural numbers, but also representations of all the algorithms in the class.

This requirement is not as serious a constraint as it sounds. Remember that each algorithm is to have a finite description made up of symbols from a countable alphabet. Since the number of such descriptions is itself countable, every description can be represented by a natural number. For example, suppose that every algorithm in a certain class is described by a finite string of symbols from a certain finite alphabet. The descriptions associated with the various members of the class can then be arranged in lexicographic order[†] and assigned the natural numbers 0, 1, 2, 3, ... in sequence. This reduces the problem of modifying or creating algorithm descriptions to that of manipulating natural numbers—i.e., to the problem of evaluating number-theoretic functions. Thus as long as a class of algorithms has a denumerable domain, every member of that class can be thought of as having the ability to modify or create the representations of other algorithms.

The essential features of the preceding discussion can be summarized as follows. We propose to study classes of algorithms that are capable of evaluating number-theoretic functions. To this end, it is sufficient to consider classes having denumerable domains, for the natural numbers can always be represented within such domains. Moreover, we can assume that each algorithm defines some (partial) n-variable number-theoretic function for each positive integer n. Finally, since each algorithm has a finite description, each algorithm can be represented by some natural number—i.e., by some member of the domain.

Abstract Families of Algorithms

In this chapter we will not be especially concerned with the defining characteristics of a class of algorithms—the basic operations, the operating rules, the descriptive language, etc. Instead, we shall be concerned with certain fundamental limitations on the number-theoretic functions that can be computed by the members of a class. We will therefore ignore the details of operations and descriptions and will concentrate on two features common to all classes having denumerable domains: the fact that every algorithm can be viewed as evaluating some n-variable number-theoretic function for each choice of n, and the fact that every algorithm can be represented by a natural number.

[†] A *lexicographic ordering* of a set of strings is one in which the strings are arranged in order of increasing length, strings of the same length being listed "alphabetically" according to some designated "alphabetical order" for the individual symbols.

We now make a formal definition that incorporates these two features. By an *abstract family of algorithms* we mean a denumerable set of otherwise undefined objects called *algorithms*, each of which has associated with it exactly one (partial) *n*-variable function for each positive integer *n*. Each algorithm is assigned a natural number, called the *index* of that algorithm, and every natural number is assigned as an index to some algorithm. The algorithm whose index is *i* will be denoted A_i. The *n*-variable number-theoretic function associated with the algorithm A_i will be referred to as the *n*-variable function *computed* by A_i. This function will be denoted $F_i^{(n)}$, and *i* will be said to be an *index* for it. Since many different algorithms may compute the same function, one function may have many different indices.

It is important to recognize that this definition of a family of algorithms is a generous one. It places no restrictions on the nature of the algorithms in a family, since it says nothing about basic operations, descriptions, etc. It allows the most sophisticated types of operations and operating rules as well as the simplest. It imposes only two requirements: that there be some scheme whereby each algorithm can be interpreted as computing a number-theoretic function of any desired number of variables; and that some correspondence be established between the members of the family and the natural numbers.

Since we are interested in practical as well as theoretical issues, we should note that practical programming languages can be used to define families of algorithms in a rather natural way. We simply view each legal program in a language as representing an algorithm, and define the *n*-variable function computed by that algorithm in the manner suggested earlier. In so doing, we may adopt a pragmatic point of view, defining the values of the function in question to be the results that would be obtained by running the program on an actual computing system. Or we may choose to ignore the limitations imposed by finite word length and memory size, thereby obtaining the function "ideally" computed by the given program. In any event, we may suppose that the family of algorithms is indexed by arranging the legal programs in some convenient order and assigning natural numbers in sequence.

While our definition of a family of algorithms is motivated by consideration of practical programming languages, we shall find it more instructive for the time being to deal with families that are not derived from practical languages. The following example provides two instances of such families. Although quite artificial, these families will be useful in illustrating properties to be discussed in Section 1.2.

Example 1-6

a) Let $\mathcal{Q}_1$ denote the family of algorithms in which the *n*-variable function computed by the *i*th algorithm is defined by the equation

$$F_i^{(n)}(x_1, \ldots, x_n) = x_1 + \cdots + x_n + i.$$

Thus $F_3^{(2)}(7, 6) = 16$, $F_3^{(1)}(5) = 8$, $F_{17}^{(3)}(1, 0, 4) = 22$, etc.

b) Let $\mathcal{Q}_2$ denote the family of algorithms in which:

for even values of i,

$$F_i^{(n)}(x_1, \ldots, x_n) = \begin{cases} 0 & \text{if} & x_1 + \cdots + x_n = i/2 \\ \uparrow & \text{if} & x_1 + \cdots + x_n \neq i/2 \end{cases}$$

for odd values of i,

$$F_i^{(n)}(x_1, \ldots, x_n) = \begin{cases} 0 & \text{if} & x_1 + \cdots + x_n = (i-1)/2 \\ x_1 + \cdots + x_n + 1 & \text{if} & x_1 + \cdots + x_n \neq (i-1)/2 \end{cases}$$

Thus $F_{30}^{(2)}(8, 7) = 0$, $F_{30}^{(2)}(8, 3)\uparrow$, $F_{15}^{(1)}(7) = 0$, $F_{15}^{(1)}(8) = 9$, and so on.   □

In spite of the generality of the notion of a family of algorithms, there is one important constraint imposed by our insistence that algorithms have finite descriptions: a family of algorithms can contain only a countable number of members. Since the number of n-variable number-theoretic functions is uncountable, there are more such functions than there are algorithms in any given family. This observation leads us to our first fundamental limitation on the capabilities of families of algorithms.

Theorem 1-1. *Let $\mathcal{Q}$ be any abstract family of algorithms. Then for each positive integer n, there exists an n-variable number-theoretic function that is not computed by any member of $\mathcal{Q}$.*

A function that is computed by some member of a family $\mathcal{Q}$ will be said to be *algorithmic in $\mathcal{Q}$* or *computable in $\mathcal{Q}$*. Alternatively, we may say that the family $\mathcal{Q}$ *contains* the function in question. A function that is not computed by any member of $\mathcal{Q}$ will be said to be *nonalgorithmic in $\mathcal{Q}$* or *noncomputable in $\mathcal{Q}$*. When the family $\mathcal{Q}$ is understood, we may refer to a given function as being simply algorithmic or nonalgorithmic, computable or noncomputable. In the same context, we may indicate that some member of the family computes a function having certain properties by saying that there exists an algorithmic function having those properties.

Theorem 1-1 tells us that no matter what family $\mathcal{Q}$ we choose, some number-theoretic functions are bound to be nonalgorithmic in $\mathcal{Q}$. (Indeed, an uncountable number of functions must be nonalgorithmic in $\mathcal{Q}$.) Note that the lack of algorithms for evaluating certain functions is not something that can be remedied by more sophisticated programming procedures or computing devices; it is an inevitable consequence of the fact that we require algorithms to have finite descriptions. Of course, it is the particular choice of the family $\mathcal{Q}$—as

	0	1	2	3	$\cdots$
0	$F_0^{(1)}(0)$	$F_0^{(1)}(1)$	$F_0^{(1)}(2)$	$F_0^{(1)}(3)$	$\cdots$
1	$F_1^{(1)}(0)$	$F_1^{(1)}(1)$	$F_1^{(1)}(2)$	$F_1^{(1)}(3)$	$\cdots$
2	$F_2^{(1)}(0)$	$F_2^{(1)}(1)$	$\cdots$		
3	$F_3^{(1)}(0)$	$F_3^{(1)}(1)$	$\cdots$		
$\vdots$	$\vdots$				

Figure 1-1

determined in practice by a choice of basic operations, operating rules, etc.—
that determines which functions fail to be algorithmic.

The reasoning leading to Theorem 1-1 is an example of an existence proof;
it argues for the existence of nonalgorithmic functions without actually exhibiting any of them. We now show how to specify, for any given family of algorithms, a particular function that is not computable by the members of that
family. We do this by means of a common and important technique known as
diagonalization.

Let $\mathcal{C}$ be any given family of algorithms. To specify a one-variable function
g that is not computable in $\mathcal{C}$, we need only make sure that g differs in some
way from each of the functions that are computable in $\mathcal{C}$. A simple way of doing
this is to define g so that, for each natural number i, the value of g differs from
that of the function $F_i^{(1)}$ for the argument value i. In adopting this strategy, we
are defining g in terms of the values found along the main diagonal of an array
like that of Fig. 1-1, where the entry in row j, column k is the value of $F_j^{(1)}(k)$. It
is this tabular interpretation that provides the term "diagonalization."

Specifically, we may define the function g as follows:

$$g(x) = \begin{cases} F_x^{(1)}(x) + 1 & \text{if} \quad F_x^{(1)}(x) \quad \text{is defined} \\ 0 & \text{if} \quad F_x^{(1)}(x) \quad \text{is undefined.} \end{cases}$$

Since for each index i the value $g(i)$ differs from the value $F_i^{(1)}(i)$, it is clear that
the function g differs from each of the functions $F_0^{(1)}$, $F_1^{(1)}$, $F_2^{(1)}$, ... and so cannot be computed by any member of $\mathcal{C}$.

To put the argument more formally, suppose that g were computed by
some member of $\mathcal{C}$. In particular, suppose that g were computed by the algorithm A_w, so that $g = F_w^{(1)}$. Since g is by definition a total function, this would

mean that $F_w^{(1)}(w)$ is defined, and that $F_w^{(1)}(w) = g(w) = F_w^{(1)}(w) + 1$. This contradiction forces us to conclude that g cannot in fact be computed by any member of $\mathcal{C}$.

Example 1-7. As in Example 1-6, let $\mathcal{C}_1$ denote the abstract family of algorithms in which the n-variable function computed by the algorithm A_i is defined by the equation

$$F_i^{(n)}(x_1, \ldots, x_n) = x_1 + \cdots + x_n + i.$$

Applying the diagonalization method described above, we find that the function g, where

$$g(x) = F_x^{(1)}(x) + 1 = 2x + 1$$

is a one-variable function that is not computed by any member of $\mathcal{C}_1$. (There are, of course, many others.) $\qquad\square$

This concludes our introduction to the concept of a family of algorithms. We are now ready to explore some of the fundamental capabilities and limitations of such families.

1.2 FIVE BASIC PROPERTIES

In this section we will specify five properties that families of algorithms often have and will begin to derive some of the consequences of these properties. Although the properties that we shall consider do not hold for all families, they do hold for the families defined by general-purpose programming languages. In fact, our choice of properties is motivated by certain basic features of programming languages. Thus the conclusions that we reach in this section and the next will be relevant to real computing situations.

We begin by noting two simple and obvious properties. If a family of algorithms is to be at all useful for practical computation, there are certain elementary functions that it must be capable of evaluating. Our first property specifies some of these functions.

Property 1. (The Base-Function Property). A family of algorithms satisfies this property if it contains algorithms that compute the following number-theoretic functions:

1. *the constant function $C_m^{(n)}$, for each choice of positive integer n and natural number m;*
2. *the projection function $P_i^{(n)}$, for each choice of positive integers i and n, where $1 \le i \le n$;*
3. *the one-variable successor function S, where $S(x) = x + 1$.*

Another essential feature of practical computation is the ability to use subroutines. By this we mean, basically, the ability to use the results obtained from one or more computations as arguments of a new computation. Thus we are interested in situations in which the algorithms of a family are composable with one another or, to put it more precisely, situations in which the set of functions computed by the members of a family is closed under functional composition. This requirement motivates our second property.

Property 2. (*The Closure Property*) *A family of algorithms satisfies this property if, whenever it contains algorithms that evaluate the m-variable function h and the n-variable functions* $g_1, \ldots, g_m$, *it also contains an algorithm that evaluates the composite function* $h \circ (g_1, \ldots, g_m)$.

By themselves, these two properties tell us relatively little about the nature of algorithmic computation. But in conjunction with some of the other properties to be considered in this section, they allow us to draw some unexpected conclusions concerning both the capabilities and limitations of algorithmic computation.

Universal Functions

Our third property is motivated by the fact that general-purpose programming languages can be used to write interpreters. Suppose that the programs of a certain programming language have been indexed in some natural way, say lexicographically. Then given the index of a program, together with appropriate data, it should be possible to reconstruct the program and apply it to the given data. In fact, if the language in question is really a general-purpose language, it should be possible to write a program that carries out the desired reconstruction and application. Such a program may be thought of as an interpreter for the given language, inasmuch as it is able to reproduce the effect of any designated program in the language on any designated set of data.

In order to capture this feature of programming languages in abstract terms, we must formalize the role of an interpreter. Assume that $\mathcal{C}$ is a family of algorithms in which, as usual, the n-variable function evaluated by the ith algorithm is denoted $F_i^{(n)}$. Then for each positive integer n we define the $(n + 1)$-variable function $U^{(n+1)}$ so that

$$U^{(n+1)}(i, x_1, \ldots, x_n) = F_i^{(n)}(x_1, \ldots, x_n) \qquad (1\text{-}1)$$

for all values of $x_1, \ldots, x_n$ and i. This function $U^{(n+1)}$ will be called the $(n + 1)$-variable *universal function* for the family $\mathcal{C}$, since the result of applying $U^{(n+1)}$ to the index i and the "data" $x_1, \ldots, x_n$ is just the result that would be obtained if the ith algorithm were applied to the arguments $x_1, \ldots, x_n$. From this remark we see that the abstract counterpart of a language in which it is possible to write an interpreter for the language itself is a family whose universal functions are computable by members of the family. We therefore propose:

Property 3. (*The Enumeration Property*) *A family of algorithms satisfies this property if, for each positive integer n, the family contains an algorithm that evaluates the universal function $U^{(n+1)}$, where*

$$U^{(n+1)}(i, x_1, \ldots, x_n) = F_i^{(n)}(x_1, \ldots, x_n).$$

An algorithm that evaluates the universal function $U^{(n+1)}$ will be referred to as a *universal algorithm* for n-variable functions. It may happen that a single algorithm evaluates the universal functions for several, or even all, values of n. But this need not be the case; the Enumeration Property only requires the existence of some universal algorithm for each value of n.

Note that the Enumeration Property is different in nature from the first two properties discussed. Whether Properties 1 and 2 hold for a given family depends only on the set of functions evaluated by that family. But whether the Enumeration Property holds depends not only on the functions that are evaluated, but also on the way in which these functions are indexed. Thus a given collection of algorithms may contain a universal algorithm under one indexing scheme and not under another.

Example 1-8

a) As in Example 1-6, let α_1 denote the family of algorithms in which $F_i^{(n)}(x_1, \ldots, x_n) = x_1 + \cdots + x_n + i$. For this family, the values of the universal function $U^{(n+1)}$ are given by

$$U^{(n+1)}(i, x_1, \ldots, x_n) = F_i^{(n)}(x_1, \ldots, x_n) = x_1 + \cdots + x_n + i.$$

Since $x_1 + \cdots + x_n + i$ is by definition the value of $F_0^{(n+1)}(i, x_1, \ldots, x_n)$, we see that $U^{(n+1)}$ is the function $F_0^{(n+1)}$. Thus α_1 satisfies the Enumeration Property; in fact, the algorithm A_0 is a universal algorithm for every value of n.

b) Now suppose that the family α_1 is reindexed so that what was originally the ith algorithm receives as its new index either $i + 1$ or $i - 1$ according as i is even or odd. Thus A_0 becomes A_1, A_1 becomes A_0, A_2 becomes A_3, A_3 becomes A_2, and so on. It is easy to see that under this new indexing scheme the universal functions are not algorithmic.[†] Note, for instance, that if the new universal function $U^{(2)}$ is to be algorithmic, there must be some index u such that $F_u^{(2)}(i, x) = F_i^{(1)}(x)$ for all i and x. But $F_u^{(2)}(i, x) = x + i + k$, where k is either $u + 1$ or $u - 1$ and is independent of i. On the other hand,

$$F_i^{(1)}(x) = \begin{cases} x + i - 1 & \text{if} \quad i \text{ is odd} \\ x + i + 1 & \text{if} \quad i \text{ is even.} \end{cases}$$

Thus there is no choice of u that will make $F_u^{(2)}(i, x) = F_i^{(1)}(x)$ for all i and x, and $U^{(2)}$ is not algorithmic.   □

[†] It is important to understand that the definition of the universal functions for a given family depends on the way in which the members of that family are indexed. Thus changing the indexing scheme will in general also change the universal functions.

Although the existence of universal algorithms depends on the indexing scheme used, most useful families have universal algorithms under a wide variety of indexing schemes. We note, for example, that as long as it is possible to convert algorithmically from one indexing scheme to another, the existence of universal algorithms for one scheme implies the existence of universal algorithms for the other. (See Problem 1.2.3.) Thus considerable latitude may be exercised in the choice of an indexing scheme without jeopardizing Property 3.

The ability of a family to evaluate its own universal functions has some unexpected consequences. For instance, it is tempting to suppose that a family all of whose algorithms evaluate total functions is more "powerful" than a family in which some algorithms evaluate nontotal functions. But this is not an appropriate point of view. Indeed, if a family contains a universal algorithm, is closed under composition, and contains algorithms for the projection and successor functions, it *cannot* consist entirely of algorithms that compute total functions.

The essence of the argument used to prove this statement can be summarized as follows. Suppose that a family $\mathcal{C}$ satisfies Properties 1, 2, and 3 but that its members compute only total functions. Since it is possible to evaluate the universal function $U^{(2)}$ in $\mathcal{C}$, it must also be possible to evaluate the function f, where

$$f(x) = F_x^{(1)}(x) + 1.$$

In other words, f must be the function $F_w^{(1)}$ for some index w. Applying f to its own index, we obtain a contradiction:

$$f(w) = F_w^{(1)}(w) + 1 = f(w) + 1.$$

Thus in fact the members of $\mathcal{C}$ cannot all compute total functions. Filling in the details of this argument, we have:

Theorem 1-2. *Let $\mathcal{C}$ be a family of algorithms for which Properties 1, 2, and 3 hold. Then not all the members of $\mathcal{C}$ compute total functions; in particular, some member of $\mathcal{C}$ must compute a nontotal one-variable function.*

Proof. Assume that, to the contrary, the one-variable functions computed by the members of $\mathcal{C}$ are all total. By hypothesis, the functions $P_1^{(1)}$, S, and $U^{(2)}$ are all algorithmic in $\mathcal{C}$. The fact that $\mathcal{C}$ is closed under composition then means that the one-variable function

$$f = S \circ (U^{(2)} \circ (P_1^{(1)}, P_1^{(1)}))$$

is also algorithmic in $\mathcal{C}$. Suppose that w is an index of the function f; that is, suppose that $f = F_w^{(1)}$. Then

$$F_w^{(1)}(x) = f(x) = S(U^{(2)}(P_1^{(1)}(x), P_1^{(1)}(x))) = S(U^{(2)}(x, x))$$

$$= \begin{cases} F_x^{(1)}(x) + 1 & \text{if} \quad F_x^{(1)}(x)\downarrow \\ \uparrow & \text{if} \quad F_x^{(1)}(x)\uparrow \end{cases}.$$

Now, we have assumed that the one-variable functions computed by the members of α are all total. Thus $F_w^{(1)}$ must be total and in particular must be defined for the argument value w. Therefore

$$F_w^{(1)}(w) = F_w^{(1)}(w) + 1.$$

From this contradiction we conclude that our original assumption is untenable and that in fact some member of α computes a nontotal one-variable function. □

Another way of stating Theorem 1-2 is that if all the algorithms in a given family do compute total functions, then that family must fail to include universal algorithms, or fail to be closed under composition, or else fail to contain algorithms for the projection or successor functions.

Example 1-9

a) Let α be a family of algorithms whose members are capable of evaluating precisely those functions that can be obtained by means of repeated composition from the constant and projection functions, together with addition, multiplication, exponentiation, and the extraction of (integer) quotients, remainders, and square roots. A straightforward induction argument suffices to show that all the algorithms of α compute total functions. Since α satisfies Properties 1 and 2, it must fail to satisfy Property 3. Thus *no matter how the algorithms of α are indexed*, this family cannot evaluate its own universal functions.

b) Again consider the family α_1 described in Examples 1-6, 1-7, and 1-8. Since this family consists entirely of algorithms that compute total functions and since it does contain universal algorithms, it must fail to satisfy either Property 1 or Property 2. A little thought shows that it does not in fact contain algorithms for the projection functions. □

Once it has been established that one nontotal function is algorithmic in a given family, it is usually possible to show that a variety of other nontotal functions must also be algorithmic in that family. The following extension of Theorem 1-2 provides a useful illustration of this fact.

Corollary 1-2.1. Let the family α satisfy Properties 1, 2, and 3. Then for every positive integer n, the empty function $\Phi^{(n)}$ is algorithmic in α.

Proof. According to Theorem 1-2, some member of α must evaluate a nontotal one-variable function. Suppose that this nontotal function f is undefined for the argument value a. Since α satisfies Properties 1 and 2, the composite function $h = f \circ (C_a^{(n)})$ must also be algorithmic in α. But

$$h(x_1, \ldots, x_n) = f(C_a^{(n)}(x_1, \ldots, x_n)) = f(a)$$

for every choice of $x_1, \ldots, x_n$. Thus h is undefined for all argument values and is in fact the empty function $\Phi^{(n)}$.   ☐

A further illustration is provided by the next example.

Example 1-10. Suppose that the family $\mathcal{C}$ satisfies Properties 1, 2, and 3. Then some member of $\mathcal{C}$ must evaluate a one-variable function that is neither total nor empty. That is, some member of $\mathcal{C}$ must evaluate a one-variable function that is defined for at least one argument value and undefined for at least one argument value. To see this, note that according to Property 1 the function $C_0^{(1)}$ is algorithmic in $\mathcal{C}$, while according to Corollary 1-2.1 the function $\Phi^{(1)}$ is algorithmic in $\mathcal{C}$. Let a be any index for $C_0^{(1)}$ and let b be any index for $\Phi^{(1)}$. Now, the one-variable function $\varphi = U^{(2)} \circ (P_1^{(1)}, P_1^{(1)})$ is certainly algorithmic in $\mathcal{C}$. Moreover,

$$\varphi(a) = U^{(2)}(P_1^{(1)}(a), P_1^{(1)}(a)) = U^{(2)}(a, a) = F_a^{(1)}(a) = C_0^{(1)}(a)$$

and

$$\varphi(b) = U^{(2)}(P_1^{(1)}(b), P_1^{(1)}(b)) = U^{(2)}(b, b) = F_b^{(1)}(b) = \Phi^{(1)}(b).$$

Thus φ is defined for the argument value a and undefined for the argument value b, as required.   ☐

Decision Functions

In the realm of practical computation it is important to be able to make decisions based on given data or intermediate results. Thus in our abstract study of computation we will be interested in algorithms that evaluate various *decision functions*. By this term we mean, generally speaking, a function that assumes one value when its arguments satisfy a specified condition and another value when its arguments do not satisfy that condition. Thus the three-variable function whose value is 5 when the sum of its argument values is 17 and whose value is 8 otherwise is an example of what we will refer to as a decision function.

Among the simplest decision functions are those that in effect determine whether or not two given arguments are equal and produce one of two different values accordingly. Of particular interest is the four-variable function Λ, where

$$\Lambda(x, y, s, t) = \begin{cases} s & \text{if} & x = y \\ t & \text{if} & x \neq y. \end{cases} \tag{1-2}$$

We will refer to this function as the *selection function*. The fact that all general-purpose programming languages are capable of evaluating Λ motivates our next property.

Property 4. (*The Selection Property*) *A family of algorithms satisfies this property if it contains an algorithm that evaluates the selection function Λ defined by Eq. (1-2).*

Although only one particular decision function is required to be algorithmic by Property 4, this property together with Properties 1 and 2 imply that many other decision functions must also be algorithmic. The next example provides a simple illustration of this fact.

Example 1-11. Suppose that $\mathcal{C}$ is a family of algorithms in which Properties 1, 2, and 4 are satisfied. Then the one-variable function f, where

$$f(x) = \begin{cases} 0 & \text{if} & x = 0 \\ 1 & \text{if} & x \neq 0 \end{cases}$$

must be algorithmic in $\mathcal{C}$. To see this, we need only note that f is the composite function $\Lambda \circ (P_1^{(1)}, C_0^{(1)}, C_0^{(1)}, C_1^{(1)})$; the fact that f is algorithmic then follows immediately from Properties 1, 2, and 4. ☐

It is important to remember that the value of a composite function is undefined whenever the value of *any* of the functions from which it is formed is undefined. As the next example shows, this fact must be kept in mind when the selection function is composed with various partial functions.

Example 1-12. Let $\mathcal{C}$ be a family of algorithms in which Properties 1, 2, 3, and 4 all hold, and let f be any one-variable function that is algorithmic in $\mathcal{C}$. We propose to show that the one-variable function $\hat{f}$, where

$$\hat{f}(x) = \begin{cases} 0 & \text{if} & f(x) = 0 \\ \uparrow & \text{if} & f(x) \neq 0 \end{cases}$$

is also algorithmic in $\mathcal{C}$.

According to Corollary 1-2.1, the empty function $\Phi^{(1)}$ is necessarily algorithmic in $\mathcal{C}$. We might therefore try to obtain $\hat{f}$ by composing the selection function Λ with the functions f, $C_0^{(1)}$, $C_0^{(1)}$, and $\Phi^{(1)}$, so that

$$\hat{f}(x) = \Lambda(f(x), C_0^{(1)}(x), C_0^{(1)}(x), \Phi^{(1)}(x))$$
$$= \Lambda(f(x), 0, C_0^{(1)}(x), \Phi^{(1)}(x)).$$

The intent here is simple: If $f(x) = 0$, the value of $\hat{f}(x)$ is to be that of $C_0^{(1)}(x)$, namely 0; while if $f(x) \neq 0$, the value of $\hat{f}(x)$ is to be that of $\Phi^{(1)}(x)$, namely undefined. But there is a catch. Even when $f(x) = 0$, the value of $\Phi^{(1)}(x)$ is undefined, and this causes the value of the composite function to be undefined, rather than 0.

The difficulty may be circumvented by using the selection function to produce an *index* for one of the functions $C_0^{(1)}$ or $\Phi^{(1)}$ and using that index as an argument of the universal function $U^{(2)}$. In particular, suppose that a is an index for the function $C_0^{(1)}$ and b is an index for the function $\Phi^{(1)}$. We may then define $\hat{f}$ by composition as follows:

$$\hat{f} = U^{(2)} \circ (\Lambda \circ (f, C_0^{(1)}, C_a^{(1)}, C_b^{(1)}), P_1^{(1)}).$$

Since $C_0^{(1)}$, $C_a^{(1)}$, and $C_b^{(1)}$ are all total functions, the value of $\Lambda \circ (f, C_0^{(1)}, C_a^{(1)}, C_b^{(1)})$ will be defined whenever the value of f is defined. Thus if $f(x)$ is defined and equal to 0, we have

$$\hat{f}(x) = U^{(2)}(\Lambda(0, 0, a, b), x) = U^{(2)}(a, x) = C_0^{(1)}(x) = 0.$$

If $f(x)$ is defined and equal to k, where $k \neq 0$, we have

$$\hat{f}(x) = U^{(2)}(\Lambda(k, 0, a, b), x) = U^{(2)}(b, x) = \Phi^{(1)}(x)\uparrow.$$

And of course if $f(x)$ is undefined, $\hat{f}(x)$ will also be undefined. Thus the composite function $U^{(2)} \circ (\Lambda \circ (f, C_0^{(1)}, C_a^{(1)}, C_b^{(1)}), P_1^{(1)})$ has the desired properties.   □

The selection function plays a critical role in determining the capabilities and limitations of algorithmic computation. To illustrate this role, consider the problem of determining whether the function associated with a given algorithm is defined for a given combination of arguments. In more familiar terms, this is equivalent to the problem of determining whether a program eventually terminates with a natural number output after being supplied with a given combination of input values. Since we wish to deal with algorithms or programs that may not terminate, and since it is often important to know whether an algorithm or program does terminate, this problem is of considerable practical interest.

Of course, we can always apply the given algorithm to the given argument values and see what happens. If the algorithm terminates, we will eventually discover that fact and so will be able to determine the value of the function being computed. But if the algorithm does not terminate, there may be no obvious way of recognizing that fact, and we might continue the computation indefinitely without ever being sure that the process was not going to terminate eventually. Thus what we would really like to have is an *algorithmic* way of determining whether or not a given function is defined for a given combination of argument values.

To formulate the issue more precisely, let us suppose that $\mathcal{a}$ is some specified family of algorithms. For each positive integer n we define the total $(n + 1)$-variable decision function $D^{(n+1)}$ as follows.

$$D^{(n+1)}(i, x_1, \ldots, x_n) = \begin{cases} 1 & \text{if} \quad F_i^{(n)}(x_1, \ldots, x_n)\downarrow \\ 0 & \text{if} \quad F_i^{(n)}(x_1, \ldots, x_n)\uparrow \end{cases} \qquad (1\text{-}3)$$

The function $D^{(n+1)}$ will be called the *n*-argument *domain function* for the family $\mathcal{a}$, since its value for the arguments $i, x_1, \ldots, x_n$ tells us whether the

n-tuple $(x_1, \ldots, x_n)$ belongs to the domain of the function $F_i^{(n)}$. The question now becomes: how can the various domain functions associated with the family α be evaluated? In particular, can these functions be evaluated by appropriately chosen algorithms within the family itself?

The answer is that most useful families do not contain such algorithms. Indeed, any family α that satisfies Properties 1 through 4 *cannot* contain algorithms that evaluate the domain functions of α. The strategy of the argument used to prove this claim can be described quite simply, using the function $D^{(2)}$ as an example. If the function $D^{(2)}$ is to be algorithmic in a family in which Properties 1 through 4 hold, then the one-variable function $\hat{D}$ must also be algorithmic, where

$$\hat{D}(x) = \begin{cases} 0 & \text{if} & D^{(2)}(x, x) = 0 \\ \uparrow & \text{if} & D^{(2)}(x, x) \neq 0. \end{cases}$$

Let w be an index of this algorithmic function $\hat{D}$. Applying $\hat{D}$ to the argument w and making use of the definition of $D^{(2)}$, we see that $\hat{D}(w)$ will be defined (and equal to 0) iff $D^{(2)}(w, w) = 0$, or equivalently, iff $F_w^{(1)}(w)$ is undefined. But since $F_w^{(1)}$ is the function $\hat{D}$, this means that $\hat{D}(w)$ is defined iff it is undefined. Because of this contradiction, we conclude that $D^{(2)}$ cannot in fact be algorithmic.

The details of this argument are made precise in:

Theorem 1-3. *Let α be any family of algorithms in which Properties 1 through 4 hold. Then for no value of n does α contain an algorithm that evaluates the domain function $D^{(n+1)}$, where*

$$D^{(n+1)}(i, x_1, \ldots, x_n) = \begin{cases} 1 & \text{if} & F_i^{(n)}(x_1, \ldots, x_n)\downarrow \\ 0 & \text{if} & F_i^{(n)}(x_1, \ldots, x_n)\uparrow. \end{cases}$$

Proof. We carry out the proof for the case $n = 1$; extension to the general case is left to the reader. Assume, then, that the function $D^{(2)}$ is algorithmic in α. Note that Property 1 ensures that the function $C_0^{(1)}$ is algorithmic in α, while Corollary 1-2.1 ensures that $\Phi^{(1)}$ is algorithmic in α. Let a be an index of $C_0^{(1)}$ and let b be an index of $\Phi^{(1)}$. Following an approach similar to that of Example 1-12, we now define $\hat{D}$ to be the composite function $U^{(2)} \circ (\Lambda \circ (D^{(2)} \circ (P_1^{(1)}, P_1^{(1)}), C_0^{(1)}, C_a^{(1)}, C_b^{(1)}), P_1^{(1)})$. Thus

$$\hat{D}(x) = U^{(2)}(\Lambda(D^{(2)}(x, x), 0, a, b), x) = \begin{cases} U^{(2)}(a, x) & \text{if} & D^{(2)}(x, x) = 0 \\ U^{(2)}(b, x) & \text{if} & D^{(2)}(x, x) \neq 0 \end{cases}$$

$$= \begin{cases} 0 & \text{if} & D^{(2)}(x, x) = 0 \\ \uparrow & \text{if} & D^{(2)}(x, x) \neq 0. \end{cases}$$

Now, Properties 1 through 4, together with the assumption that $D^{(2)}$ is algorithmic, ensure that $\hat{D}$ is algorithmic. Let w be an index of $\hat{D}$. Applying $\hat{D}$ to w gives

$$\hat{D}(w) = \begin{cases} 0 & \text{if} & D^{(2)}(w, w) = 0 \\ \uparrow & \text{if} & D^{(2)}(w, w) \neq 0 \end{cases}$$

$$= \begin{cases} 0 & \text{if} & F_w^{(1)}(w)\uparrow \\ \uparrow & \text{if} & F_w^{(1)}(w)\downarrow \end{cases}$$

$$= \begin{cases} 0 & \text{if} & \hat{D}(w)\uparrow \\ \uparrow & \text{if} & \hat{D}(w)\downarrow. \end{cases}$$

Thus $\hat{D}(w)$ is defined (and equal to 0) iff $\hat{D}(w)$ is undefined. From this contradiction we conclude that $D^{(2)}$ cannot in fact be algorithmic.   □

There are many restricted versions of the domain functions that are also nonalgorithmic in any family that satisfies Properties 1 through 4. We turn now to what is undoubtedly the most important and interesting of these restricted functions. For each family of algorithms $\mathfrak{C}$, we define the one-variable function d so that

$$d(x) = \begin{cases} 1 & \text{if} & F_x^{(1)}(x)\downarrow \\ 0 & \text{if} & F_x^{(1)}(x)\uparrow. \end{cases} \tag{1-4}$$

This function will be called the *diagonal domain function* for the family $\mathfrak{C}$, inasmuch as it refers to phenomena associated with the main diagonal of an array like that of Fig. 1-1. The next theorem, whose proof is similar to that of Theorem 1-3, shows that even the restricted decision function d is necessarily nonalgorithmic in families of practical interest.

Theorem 1-4. *Let $\mathfrak{C}$ be any family of algorithms in which Properties 1 through 4 hold. Then the diagonal domain function d, defined so that*

$$d(x) = \begin{cases} 1 & \text{if} & F_x^{(1)}(x)\downarrow \\ 0 & \text{if} & F_x^{(1)}(x)\uparrow \end{cases}$$

is not algorithmic in $\mathfrak{C}$.

Proof. Assume that d is algorithmic. Let $\hat{d}$ denote the composite function $U^{(2)} \circ (\Lambda \circ (d, C_0^{(1)}, C_a^{(1)}, C_b^{(1)}), P_1^{(1)})$, where again a is an index for $C_0^{(1)}$ and b is an index for $\Phi^{(1)}$. Thus

$$\hat{d}(x) = U^{(2)}(\Lambda(d(x), 0, a, b), x) = \begin{cases} U^{(2)}(a, x) & \text{if} & d(x) = 0 \\ U^{(2)}(b, x) & \text{if} & d(x) \neq 0 \end{cases}$$

$$= \begin{cases} 0 & \text{if} & d(x) = 0 \\ \uparrow & \text{if} & d(x) \neq 0. \end{cases}$$

Now, if d is algorithmic, $\hat{d}$ must also be algorithmic. Let w be an index for $\hat{d}$. Recalling that $d(x) = 0$ iff $F_x^{(1)}(x)\uparrow$, we have

$$\hat{d}(w) = \begin{cases} 0 & \text{if} \quad d(w) = 0 \\ \uparrow & \text{if} \quad d(w) \neq 0 \end{cases} \quad = \begin{cases} 0 & \text{if} \quad F_w^{(1)}(w)\uparrow \\ \uparrow & \text{if} \quad F_w^{(1)}(w)\downarrow \end{cases} \quad = \begin{cases} 0 & \text{if} \quad \hat{d}(w)\uparrow \\ \uparrow & \text{if} \quad \hat{d}(w)\downarrow. \end{cases}$$

Again we have reached a contradiction, and so must conclude that the function d is not algorithmic. □

Roughly speaking, Theorems 1-3 and 1-4 say that as long as a family of algorithms is capable of evaluating the constant and projection functions, interpreting the members of the family, and making simple decisions, the domain functions for that family are too complicated to be evaluated by members of the family.[†] What are the implications of these results for general-purpose programming languages? Simply that we should not expect to be able to write a program that is capable of determining whether or not an arbitrary program in the same language eventually halts with a natural number output.

Index Computation

One important feature of practical programming languages is the fact that they can be used to write programs that modify or create other programs. For instance, it is certainly possible to write a program that, when presented with programs for evaluating arbitrary functions h and g, produces a program for evaluating the composite function $h \circ (g)$. Similarly, it is possible to write a program that, when presented with an arbitrary number k, produces a program for evaluating the function f, where $f(y) = k + y$. Note that many tasks of this kind are performed by compilers and simple operating systems. Indeed, we may consider one of the major roles of compilers and operating systems to be that of modifying or creating programs or descriptions of programs.

When it comes to expressing this feature of programming languages in abstract terms, we are faced with a problem. Of the many possible program-modifying capabilities of practical languages, which should be incorporated in our formal model and how can they be characterized? Somewhat surprisingly, there turns out to be one simple kind of program-modifying ability from which all the others can be derived. (This claim will be substantiated in Section 6.1.) Let us now examine this special case and try to formalize it.

Suppose that P is a program written in a general-purpose programming language and that P is designed to be supplied with two input variables x and y. Then for each value of x, there must be a new program P_x that yields the same result when applied to the single input y as P does when applied to x and y. The only difference between P and P_x is that all references to the first input of P are

---

[†] As Problem 1.2.7 shows, the full powers of Properties 1 through 4 are not really needed to establish the noncomputability of the diagonal domain function.

changed in P_x to references to the constant x, and all references to the second input of P are changed to references to the (only) input of P_x. Thus the program P is easily converted into the program P_x. In fact, if the language in question is really a general-purpose language, it should be possible to write a program that, given x and a suitable description of P, produces a description of P_x.

Now consider the abstract counterpart of this situation. Suppose that $\mathcal{A}$ is a family of algorithms for which Properties 1 and 2 hold, and that $F_i^{(2)}$ is one of the two-variable functions computed by the members of $\mathcal{A}$. For each natural number x, let f_x denote the one-variable function such that

$$f_x(y) = F_i^{(2)}(x, y)$$

for all y. (Note that the subscript x is not an index of f_x, but only a part of the name of the function.) Since $f_x = F_i^{(2)} \circ (C_x^{(1)}, P_1^{(1)})$, and since Properties 1 and 2 hold, it follows that f_x is an algorithmic function and has one or more indices. The values of these indices obviously depend on the values of i and x. We may therefore think of looking for a function that tells us how to get an index for f_x from any given values of i and x. In other words, we are interested in a total two-variable function s such that

$$F_{s(i,\,x)}^{(1)}(y) = f_x(y) = F_i^{(2)}(x, y)$$

for all values of i and x. Such a function s will be referred to as an *index-computing function* for f_x.

In general there will be many such functions s, since for any given choice of i and x the function f_x will usually have more than one index. The formal counterpart of the programming situation described above is that in which at least one of these index-computing functions is algorithmic. Thus we are interested in families that are capable of evaluating a total function s such that $F_{s(i,\,x)}^{(1)}(y) = F_i^{(2)}(x, y)$ for all choices of i and x. Extending this requirement so as to provide for arbitrary numbers of variables, we propose our last major property.

Property 5. *(The S-M-N Property) A family of algorithms satisfies this property if, for each choice of positive integers m and n, the family contains an algorithm evaluating a total $(m + 1)$-variable function $s_{n,\,m}$ such that*

$$F_{s_{n,m}(i,\,x_1,\,\ldots,\,x_m)}^{(n)}(y_1, \ldots, y_n) = F_i^{(n+m)}(x_1, \ldots, x_m, y_1, \ldots, y_n)$$

for all i, x_1, ..., x_m, and y_1, ..., y_n.

Note that the function $s_{1,\,1}$ is just the index-computing function s discussed above. Whenever the context permits we will omit the subscripts n and m from the function name $s_{n,\,m}$.

Example 1-13. Again consider the family α_1, in which

$$F_i^{(n)}(x_1, \ldots, x_n) = x_1 + \cdots + x_n + i.$$

We now note that this family satisfies the S-M-N Property. In particular, $F_0^{(m+1)}$ serves as the function $s_{n, m}$ for every choice of n. To verify this, we observe first of all that $F_0^{(m+1)}$ is total. Moreover,

$$F_{F_0^{(m+1)}(i, x_1, \ldots, x_m)}^{(n)}(y_1, \ldots, y_n) = F_{x_1 + \cdots + x_m + i}^{(n)}(y_1, \ldots, y_n)$$

$$= y_1 + \cdots + y_n + (x_1 + \cdots + x_m + i)$$

$$= x_1 + \cdots + x_m + y_1 + \cdots + y_n + i$$

$$= F_i^{(m+n)}(x_1, \ldots, x_m, y_1, \ldots, y_n)$$

which is just what the S-M-N Property requires.   □

The S-M-N Property simply requires that certain index-computing functions be algorithmic. By themselves, these index-computing functions may not seem particularly interesting. But when the S-M-N Property holds in conjunction with some of the other properties we have discussed, it can be used to establish the computability of a wide variety of other index-computing functions. We now illustrate this role of the S-M-N Property by means of some simple examples.

Suppose that α is a family of algorithms in which Properties 1, 2, and 5 hold and in which the familiar two-variable multiplication function is algorithmic. Let g denote this multiplication function. For each natural number x, let f_x denote the one-variable function such that

$$f_x(y) = g(x, y) = xy.$$

Thus $f_3(5) = 15, f_7(3) = 21$, etc. Evidently each of the functions f_x is algorithmic in α, since f_x is in fact the composite function $g \circ (C_x^{(1)}, P_1^{(1)})$. Thus each function f_x has one or more indices, the values of which depend on the choice of x. (Again we must emphasize the distinction between an index of the function f_x and the subscript x, which is only part of the name of f_x.)

We now claim that because α satisfies Properties 1, 2, and 5, it must contain an algorithm that computes an index for f_x when applied to the value x. To see this, note that since the multiplication function g is algorithmic in α, it has an index. That is, $g(x, y) = F_w^{(2)}(x, y)$ for some natural number w. The S-M-N Property then guarantees the existence of a two-variable algorithmic function s (officially called $s_{1, 1}$) such that

$$F_{s(w, x)}^{(1)}(y) = F_w^{(2)}(x, y) = g(x, y) = xy = f_x(y).$$

Thus $s(w, x)$ is an index for f_x. Noting that w is a fixed number, we see that this index $s(w, x)$ is really a function of x alone. In particular, if we define the function r so that $r = s \circ (C_w^{(1)}, P_1^{(1)})$, we have

$$F_{r(x)}^{(1)}(y) = F_{s(w,\, x)}^{(1)}(y) = f_x(y).$$

The function r, which is algorithmic by virtue of Properties 1 and 2, is the desired index-computing function.[†]

The preceding example suggests a general approach to the problem of showing that certain index-computing functions are algorithmic. Suppose that $f_0, f_1, f_2, \ldots, f_x, \ldots$ is an infinite series of one-variable functions, each of which is known to be algorithmic in the family $\mathcal{C}$, and that we wish to establish the existence of an algorithmic function that computes an index for f_x from the value of x. We first define a two-variable auxiliary function g such that

$$g(x, y) = f_x(y).$$

If it can be shown that g is algorithmic and hence that g has some index w, the S-M-N Property will guarantee the existence of an algorithmic function s such that

$$F_{s(w,\, x)}^{(1)}(y) = F_w^{(2)}(x, y) = g(x, y) = f_x(y).$$

The desired index-computing function r may then be obtained by defining r to be the composite function $s \circ (C_w^{(1)}, P_1^{(1)})$.

Our next result provides a simple application of this strategy.

Theorem 1-5. *Let $\mathcal{C}$ be a family of algorithms that satisfies Properties 1, 2, and 5. Then there must exist a total, algorithmic one-variable function k such that $F_{k(x)}^{(1)} = C_x^{(1)}$.*

Proof. We first define the auxiliary function g so that

$$g(x, y) = C_x^{(1)}(y) = x.$$

Since $g = P_1^{(2)}$, g is certainly algorithmic in $\mathcal{C}$. Let w be any index for g. The S-M-N Property guarantees the existence of a total algorithmic function s such that

$$F_{s(w,\, x)}^{(1)}(y) = F_w^{(2)}(x, y) = g(x, y) = C_x^{(1)}(y).$$

Defining k to be the function $s \circ (C_w^{(1)}, P_1^{(1)})$ now yields the desired index-computing function, since k is certainly total and algorithmic, and

$$F_{k(x)}^{(1)}(y) = F_{s(w,\, x)}^{(1)}(y) = C_x^{(1)}(y). \qquad \square$$

An analogous strategy can be used when the functions of interest are characterized by more than one parameter. Suppose that $\mathcal{C}$ is a family that satisfies the Closure Property, so that the composition of any two one-variable

---

[†] It is actually possible to establish the computability of r using *only* the S-M-N Property, without invoking Properties 1 and 2. See Problem 1.2.11.

algorithmic functions is also algorithmic. Is it possible to find an algorithmic function that computes an index for this composite function from the indices of the given functions? The answer is yes, provided the family satisfies Properties 1, 3, and 5 as well.

Theorem 1-6. *Let $\mathcal{C}$ be any family in which Properties 1, 2, 3, and 5 hold. Then there exists a total, algorithmic, two-variable function h such that $F^{(1)}_{h(x_1, x_2)} = F^{(1)}_{x_1} \circ (F^{(1)}_{x_2})$.*

Proof. We first define the three-variable auxiliary function g so that

$$g(x_1, x_2, y) = F^{(1)}_{x_1}(F^{(1)}_{x_2}(y)) = U^{(2)}(x_1, U^{(2)}(x_2, y)).$$

Since $g = U^{(2)} \circ (P^{(3)}_1, U^{(2)} \circ (P^{(3)}_2, P^{(3)}_3))$, the function g is algorithmic in $\mathcal{C}$ and has some index w. The S-M-N Property guarantees the existence of a total algorithmic three-variable function s (officially called $s_{1,2}$) such that

$$F^{(1)}_{s(w, x_1, x_2)}(y) = F^{(3)}_w(x_1, x_2, y) = g(x_1, x_2, y) = F^{(1)}_{x_1}(F^{(1)}_{x_2}(y)).$$

Defining h to be the function $s \circ (C^{(2)}_w, P^{(2)}_1, P^{(2)}_2)$ then yields the desired index-computing function. $\square$

The proofs of Theorem 1-5 and 1-6 illustrate the way in which Property 5 can be used to establish that certain index-computing functions are algorithmic. Similar arguments can be used to show that a wide variety of other index-computing functions are also algorithmic. Since computing an index of a function can be viewed as the formal counterpart of producing a description of a program, we see that Property 5 does at least partially reflect the ability of practical programming languages to modify and create descriptions of programs.

1.3 INHERENT CAPABILITIES AND LIMITATIONS

The preceding section introduced five basic properties that families of algorithms might have and illustrated some of the consequences that follow from those properties. In this section we will examine families of algorithms that, like the families derived from programming languages, satisfy all five properties. Our goal is two-fold: to obtain a better understanding of the inherent capabilities and limitations of algorithmic computation, and to develop certain techniques for establishing the computability or noncomputability of functions.

Standard Families of Algorithms

An abstract family of algorithms that satisfies all five of the basic properties discussed in Section 1.2 will be called a *standard family* of algorithms. From this point on, we will consider only standard families, and will attempt to discover some of the features common to all such families. We begin with a brief discussion of the relationship between standard families and programming languages.

The five basic properties were intended to represent common and desirable features of general-purpose programming languages. Yet when implemented in a real-life computing environment, the algorithms defined by programming languages do not quite satisfy these properties. The underlying reason is simply that the storage capacity of real computing systems is limited. As a consequence, real systems cannot be used to do such things as interpret arbitrarily long programs or construct descriptions of arbitrarily complex algorithms (or even multiply arbitrarily large numbers). Thus in practice we cannot really write programs that evaluate the functions required by the basic properties.

But these deficiencies arise from what might be called "economic" constraints, and do not represent limitations on the inherent expressive power of programming languages. If a program fails to yield a particular function value because sufficient storage capacity is lacking, we can in principle obtain the desired value by running the same program on a larger computer. Indeed, we can readily imagine an idealized computer that is like a real computing system in all essential respects except that its word length and memory size are unbounded. Such a computer would be capable of storing and manipulating arbitrarily large numbers and program descriptions and so could be used to simulate the execution of arbitrarily long programs or to produce descriptions of arbitrarily long programs. Thus when implemented on an idealized computer, the family of algorithms defined by a general-purpose programming language would in fact satisfy the basic properties of Section 1.2.

Our concern in this book is with the inherent capabilities and limitations of algorithms for discrete computation, not with the economic limitations imposed by physical computing systems. It is therefore appropriate to imagine that programs are to be executed by an idealized computer, rather than by a real—and therefore limited—computer. And since the use of a general-purpose programming language in the context of an idealized computer determines a standard family of algorithms, it is appropriate to examine the properties of such families. Any limitations we may discover concerning the computing capabilities of standard families must represent limitations on the inherent capabilities of programming languages, and these limitations can in turn be expected to have implications for practical computation.

The consequences that follow from the five basic properties are of two general kinds: that certain functions must be computable in every standard family of algorithms, and that certain functions cannot be computed in any standard family. We have already seen examples of both types of functions, and will now use the concerted power of all five properties to derive additional examples. In the process, we intend not only to improve our understanding of the computational capabilities of standard families, but also to develop basic techniques for establishing the computability or noncomputability of functions.

We begin by examining an important technique known as reduction. This technique makes it possible to show that many functions relating to the behavior of the members of a standard family are necessarily noncomputable in that family. We then introduce a fundamental result known as the Recursion Theorem. This theorem will ultimately enable us to show that a great many functions of practical interest must be computable in any standard family. Finally, we will return to the relationship between abstract families and real-life computation, and begin the job of characterizing the class of functions that can be evaluated by what would be regarded on intuitive grounds as algorithmic means.

Reduction

The usual way of establishing that a function is nonalgorithmic is to assume that it is and show that this assumption leads to a contradiction. There are two general ways of obtaining such a contradiction. The first method, which is the one that we have used so far, consists in obtaining a contradiction by applying an appropriately specified function to its own index. This method is often referred to loosely as *diagonalizing*. The second method consists in showing that if the given function were algorithmic, then some other function, already known not to be algorithmic, would in fact have to be algorithmic. This second method of arriving at a contradiction is known as *reducing* one function to another. Reduction is an important and useful technique. To see how it works, let us examine a typical application.

Consider the problem of determining whether or not the one-variable function computed by a specified algorithm is total. To formalize this problem for a given family of algorithms $\mathcal{C}$, we define the decision function t so that

$$t(x) = \begin{cases} 1 & \text{if} & F_x^{(1)} \text{ is total} \\ 0 & \text{if} & F_x^{(1)} \text{ is not total.} \end{cases} \tag{1-5}$$

We will refer to t as the (one-variable) *totality function* for the family $\mathcal{C}$. It is now natural to ask whether this totality function can be evaluated by some member of $\mathcal{C}$. As we shall show, the answer is no, as long as the family satisfies Properties 1 through 5.

As the first step in establishing this claim, we define a set of auxiliary functions $f_0, f_1, \ldots, f_x, \ldots$ so that

$$f_x(y) = \begin{cases} 0 & \text{if} & F_x^{(1)}(x)\downarrow \\ \uparrow & \text{if} & F_x^{(1)}(x)\uparrow. \end{cases}$$

In other words, we define f_x to be the constant function $C_0^{(1)}$ if $F_x^{(1)}(x)$ is defined and the empty function $\Phi^{(1)}$ if $F_x^{(1)}(x)$ is undefined. Now, if the family in question satisfies all five properties, the function f_x is certainly algorithmic, for each natural number x. Moreover, as the next result shows, there is an algorithm that computes an index for f_x given the value of x.

Lemma 1-1. *Let $\mathcal{C}$ be any standard family of algorithms. Then there exists a total, algorithmic one-variable function q such that $F^{(1)}_{q(x)}$ is the function f_x, where*

$$f_x = \begin{cases} C_0^{(1)} & \text{if} & F_x^{(1)}(x)\downarrow \\ \Phi^{(1)} & \text{if} & F_x^{(1)}(x)\uparrow. \end{cases}$$

Proof. First define the two-variable function g so that

$$g(x,\, y) = \begin{cases} 0 & \text{if} & F_x^{(1)}(x)\downarrow \\ \uparrow & \text{if} & F_x^{(1)}(x)\uparrow. \end{cases}$$

Since $g = C_0^{(1)} \circ (U^{(2)} \circ (P_1^{(2)},\, P_1^{(2)}))$, it follows that g is algorithmic in $\mathcal{C}$. Suppose that w is an index of g. The S-M-N Property then guarantees the existence of a total algorithmic function s such that

$$F^{(1)}_{s(w,\, x)}(y) = F^{(2)}_w(x,\, y) = g(x,\, y) = \begin{cases} 0 & \text{if} & F_x^{(1)}(x)\downarrow \\ \uparrow & \text{if} & F_x^{(1)}(x)\uparrow. \end{cases}$$

Defining q to be the function $s \circ (C_w^{(1)},\, P_1^{(1)})$, we have

$$F^{(1)}_{q(x)}(y) = \begin{cases} 0 & \text{if} & F_x^{(1)}(x)\downarrow \\ \uparrow & \text{if} & F_x^{(1)}(x)\uparrow. \end{cases}$$

Hence

$$F^{(1)}_{q(x)} = \begin{cases} C_0^{(1)} & \text{if} & F_x^{(1)}(x)\downarrow \\ \Phi^{(1)} & \text{if} & F_x^{(1)}(x)\uparrow \end{cases}$$

as required.   □

Now note that the function f_x is total if $F_x^{(1)}(x)$ is defined and nontotal if $F_x^{(1)}(x)$ is undefined. This fact gives us a way of showing that the function t is not algorithmic. Suppose that, to the contrary, the family $\mathcal{C}$ contained an algorithm for evaluating t. We could then use that algorithm to evaluate the diagonal domain function for $\mathcal{C}$, as follows. Given a value of x for which we wish to determine whether $F_x^{(1)}(x)$ is defined, first apply the algorithm specified in Lemma 1-1 so as to obtain an index of the function f_x. Then apply the assumed algorithm for evaluating the function t to the index in question. If the assumed algorithm reveals that f_x is total, we conclude that $F_x^{(1)}(x)$ is defined; if the assumed algorithm reveals that f_x is not total, we conclude that $F_x^{(1)}(x)$ is undefined. Thus the existence of an algorithm for evaluating the function t would imply the existence of an algorithm for evaluating the diagonal domain function d. Since we know that the latter function cannot be algorithmic in any family that satisfies Properties 1 through 5, we must conclude that t cannot be algorithmic in such a family either.

The preceding argument is made precise in our next theorem.

Theorem 1-7. *Let $\mathcal{C}$ be any standard family of algorithms. Then $\mathcal{C}$ does not contain an algorithm that evaluates the totality function t, where*

$$t(x) = \begin{cases} 1 & \text{if} & F_x^{(1)} \text{ is total} \\ 0 & \text{if} & F_x^{(1)} \text{ is not total.} \end{cases}$$

Proof. Assume that t is algorithmic in $\mathcal{C}$. Let q be the index-computing function specified in Lemma 1-1, and define h to be the composite function $t \circ (q)$. Since $\mathcal{C}$ is closed under composition, h must be algorithmic. According to the definition of t,

$$h(x) = t(q(x)) = \begin{cases} 1 & \text{if} & F_{q(x)}^{(1)} \text{ is total} \\ 0 & \text{if} & F_{q(x)}^{(1)} \text{ is not total.} \end{cases}$$

But according to the definition of q, the function $F_{q(x)}^{(1)}$ is total (and identical to $C_0^{(1)}$) iff $F_x^{(1)}(x)$ is defined. Therefore

$$h(x) = \begin{cases} 1 & \text{if} & F_x^{(1)}(x)\downarrow \\ 0 & \text{if} & F_x^{(1)}(x)\uparrow, \end{cases}$$

which means that h is in fact the diagonal domain function for $\mathcal{C}$. Since Theorem 1-4 ensures that the diagonal domain function is *not* algorithmic, our original assumption is untenable and t cannot be algorithmic either.   □

It is important to understand the strategy used in the proof of Theorem 1-7. Put in its simplest form, it is this: We showed, by means of a suitable constructive argument, that if the totality function were algorithmic, then the diagonal domain function would also have to be algorithmic. Since the diagonal domain function is known not to be algorithmic, we must conclude that the totality function cannot be algorithmic. We describe this approach by saying that we have *reduced* the computation of the diagonal domain function to that of the totality function. Note carefully the direction of the argument. We do not try to show that if the diagonal domain function were algorithmic then the totality function would have to be algorithmic, for in doing so we would not obtain a contradiction.

The heart of a reduction argument, then, consists in showing that the computability of a given function φ would imply the computability of some function θ that is known not to be computable. Two problems are involved here: choosing the noncomputable function θ, and establishing the necessary relationship between φ and θ. Unfortunately, these two problems are interdependent, and there are no firm guidelines for dealing with them.

The proof of Theorem 1-7 illustrates one of the most common forms of reduction arguments, wherein the given function φ (whose role is played by t in Theorem 1-7) is a decision function whose arguments are indices of algorithms. In this case it is usually appropriate to choose θ (whose role is played by d in Theorem 1-7) to be a function of the same type. To establish the necessary relationship between φ and θ, we attempt to define a series of auxiliary functions $f_0, f_1, \ldots, f_x, \ldots$ so that the value of φ when applied to an index of f_x is the same as the value of θ when applied to x. We then try to demonstrate the existence of an algorithmic function r that maps each natural number x into an index of f_x. If this can be done, we will have succeeded in reducing the computability of θ to that of φ, for the composite function $\varphi \circ (r)$ will in fact be the function θ.

As a second example of this approach, consider the problem of determining whether a given algorithm evaluates a specified function, say the projection function $P_1^{(1)}$. Again assuming that $\mathcal{C}$ is a family in which all five properties hold, let us define the decision function p so that

$$p(x) = \begin{cases} 1 & \text{if} & F_x^{(1)} = P_1^{(1)} \\ 0 & \text{if} & F_x^{(1)} \neq P_1^{(1)}. \end{cases}$$

In order to show that p is not algorithmic in $\mathcal{C}$, we shall show that the computability of the totality function t—which we already know to be nonalgorithmic—can be reduced to that of p. To this end, we shall try to define a series of auxiliary functions $f_0, f_1, \ldots, f_x, \ldots$ such that f_x is the projection function $P_1^{(1)}$ iff $F_x^{(1)}$ is total. If this can be done, the application of an assumed algorithm for p to an *index* of f_x would determine whether $F_x^{(1)}$ is total, and we will have achieved the desired contradiction.

Perhaps the simplest way of choosing the auxiliary functions $f_0, f_1, \ldots, f_x, \ldots$ is to let f_x be the composite function $P_1^{(2)} \circ (P_1^{(1)}, F_x^{(1)})$, so that

$$f_x(y) = P_1^{(2)}(P_1^{(1)}(y), F_x^{(1)}(y)) = P_1^{(2)}(y, F_x^{(1)}(y)).$$

Then if $F_x^{(1)}$ is total, f_x will also be total and will in fact be identical to $P_1^{(1)}$. However, if $F_x^{(1)}$ is undefined for some argument value, f_x will be undefined for the same argument value and hence will not be identical to $P_1^{(1)}$.

It is clear that each of the functions f_x is algorithmic in $\mathcal{C}$. To show that there exists an algorithm that computes an index for f_x given x, we follow the usual procedure. First define the two-variable function g so that $g(x, y) = f_x(y)$. Since

$$g = P_1^{(2)} \circ (P_2^{(2)}, U^{(2)} \circ (P_1^{(2)}, P_2^{(2)})),$$

we know that g is algorithmic and has some index w. The S-M-N Property guarantees the existence of a total algorithmic function s such that

$$F_{s(w, x)}^{(1)}(y) = F_w^{(2)}(x, y) = g(x, y) = f_x(y).$$

The composite function $r = s \circ (C_w^{(1)}, P_1^{(1)})$ is therefore a total algorithmic function such that $F_{r(x)}^{(1)} = f_x$.

At this point we are ready to argue formally that the function p is not algorithmic in any standard family.

Lemma 1-2. *Let $\mathcal{C}$ be a family of algorithms in which Properties 1 through 5 hold. Then $\mathcal{C}$ does not contain an algorithm that evaluates the function p, where*

$$p(x) = \begin{cases} 1 & \text{if} & F_x^{(1)} = P_1^{(1)} \\ 0 & \text{if} & F_x^{(1)} \neq P_1^{(1)}. \end{cases}$$

Proof. Assume that the function p is algorithmic in $\mathcal{C}$. Then the composite function $h = p \circ (r)$ must also be algorithmic, where r is the index-computing function described above. According to the definition of p,

$$h(x) = p(r(x)) = \begin{cases} 1 & \text{if} & F_{r(x)}^{(1)} = P_1^{(1)} \\ 0 & \text{if} & F_{r(x)}^{(1)} \neq P_1^{(1)}. \end{cases}$$

But according to the definition of r, $F_{r(x)}^{(1)} = P_1^{(1)}$ iff $F_x^{(1)}$ is total. This means that the function h is in fact the totality function t. Since t is known not to be algorithmic in $\mathcal{C}$, we must conclude that p is also not algorithmic in $\mathcal{C}$. ☐

Although the argument used to establish Lemma 1-2 is typical, some reduction arguments assume an especially simple form. Consider the problem of determining whether or not two algorithms evaluate the same function. In particular, let us define the two-variable *equivalence function e* so that

$$e(x, y) = \begin{cases} 1 & \text{if} & F_x^{(1)} = F_y^{(1)} \\ 0 & \text{if} & F_x^{(1)} \neq F_y^{(1)}. \end{cases}$$

Showing that e is not algorithmic in any standard family is quite easy, since the computability of the nonalgorithmic function p reduces almost immediately to that of e. In fact, the evaluation of p is just a special case of the evaluation of e.

Theorem 1-8. *Let $\mathcal{C}$ be any standard family of algorithms. Then no member of $\mathcal{C}$ evaluates the equivalence function e.*

Proof. If e were algorithmic in $\mathcal{C}$, the function $e \circ (P_1^{(1)}, C_w^{(1)})$ would also have to be algorithmic, where w is any index for the projection function $P_1^{(1)}$. But

$$e(P_1^{(1)}(x), C_w^{(1)}(x)) = e(x, w) = \begin{cases} 1 & \text{if} & F_x^{(1)} = P_1^{(1)} \\ 0 & \text{if} & F_x^{(1)} \neq P_1^{(1)} \end{cases}$$

so that in fact $e \circ (P_1^{(1)}, C_w^{(1)})$ is the function p defined in Lemma 1-2. Since p is not algorithmic, it follows that e cannot be algorithmic. ☐

It must be emphasized that the direction of the argument used in Theorem 1-8 is important. The fact that p is a "special case" of the function e means that the computability of e implies the computability of p, hence that the noncomputability of p implies the noncomputability of e. But the argument does not work the other way around. If it had been the function e that was known to be noncomputable, we could *not* have used that fact to deduce the noncomputability of p. Indeed, there are many situations in which a restricted version of a function is computable even though the function itself is not. Thus attention must be given to ensuring that the desired conclusion follows logically from the argument presented.

The Recursion Theorem

We now turn to what must be considered one of the most important consequences of the five basic properties. This result, known as the Recursion Theorem, has a number of far-reaching and unexpected consequences of its own. We will eventually use this theorem to help us show that all practical families in which the five properties hold must compute the same set of functions. But before we can use the Recursion Theorem for such purposes, we must understand what it says.

Suppose that f is a total one-variable function that is algorithmic in a certain family α. This function may be thought of as defining a mapping from α into α whereby the algorithm A_i is mapped into the algorithm $A_{f(i)}$. By extension, we can also think of f as defining a mapping on the one-variable functions evaluated by α, whereby $F_i^{(1)}$ is mapped into $F_{f(i)}^{(1)}$. Note that this latter "mapping" is not really a function, since there is no guarantee that $F_{f(i)}^{(1)}$ and $F_{f(j)}^{(1)}$ will be the same function whenever $F_i^{(1)}$ and $F_j^{(1)}$ are the same function. It is nevertheless convenient to think of f as providing a way of transforming the function $F_i^{(1)}$ into the function $F_{f(i)}^{(1)}$ for each choice of i.

Of special interest is the situation in which $F_i^{(1)}$ and $F_{f(i)}^{(1)}$ turn out to be the same function. This will obviously be the case if $f(i) = i$, but since one function can have many different indices, it can also happen even though $f(i) \neq i$. A value of i such that $F_i^{(1)} = F_{f(i)}^{(1)}$ is called a *fixed point* for the function f.[†] It is natural to suppose that the existence or nonexistence of such a fixed point depends on the particular function f and on the manner in which the members of α are indexed. The surprising result is that as long as α is indexed in such a way that the five basic properties hold, *every* total algorithmic one-variable function f is guaranteed to have at least one fixed point. This is the content of the Recursion Theorem, which we now set out to prove.

Since the argument is somewhat complicated, let us begin with an informal version phrased in programming terms. Suppose that the legal programs of a certain language have been indexed in some appropriate way. Let us denote these programs P_0, P_1, P_2, etc., and write $P(x)$ to represent the result that is obtained when program P is applied to the number x. Next suppose that f is a total, one-variable function computed by some program in the language. Then

---

[†] This choice of terminology is unfortunate, since it suggests that $f(i) = i$, which need not be true.

we can think of f as operating on indices so as to map each program P_i into a corresponding program $P_{f(i)}$, which we will denote P'_i for short. What we wish to argue is that there must be some program P such that P and P' compute the same one-variable function.

If the language in question is in fact a general-purpose language, it should be possible to design a program R that behaves as follows. When given the number u as an argument, R does the following two things:

1. constructs a program Q_u that, when applied to the number x, first simulates the computation performed by program P_u on input u and, if that computation terminates, proceeds to act like the program $P_{P_u(u)}$ for input x;

2. uses the given function f to compute an index of the program Q'_u.

Let v be the index of this program R; that is, assume that R is the program P_v.

Now consider the program Q_v—that is, the program that R constructs when given the number v. By definition, when Q_v is applied to the input x, it will produce the same result as the program $P_{P_v(v)}$ would produce when applied to x. But since P_v is the program R, and since $R(v)$ is by definition the index of the program Q'_v, it follows that $P_{P_v(v)}$ is the program Q'_v. Thus Q_v produces the same result when applied to x as does the program Q'_v, which means that Q_v is the desired "fixed-point" program.

The first step in formalizing this argument is to incorporate the role of the various programs Q_u in our abstract model. Suppose that $\mathcal{C}$ is a family of algorithms for which Properties 1 through 5 hold. For each natural number u, we define θ_u to be the one-variable function such that[†]

$$\theta_u(x) = \begin{cases} F_{F_u(u)}(x) & \text{if} & F_u(u)\!\downarrow \\ \uparrow & \text{if} & F_u(u)\!\uparrow. \end{cases}$$

Noting that $\theta_u(x) = U^{(2)}(U^{(2)}(u, u), x)$, we see that θ_u is in fact the composite function $U^{(2)} \circ (U^{(2)} \circ (C_u^{(1)}, C_u^{(1)}), P_1^{(1)})$. Since $\mathcal{C}$ satisfies Properties 1, 2, and 3, and since the value of u is fixed, it follows that the function θ_u must be algorithmic in $\mathcal{C}$.

Of course, just which algorithm in $\mathcal{C}$ computes the function θ_u depends on the value of u. We indicate this dependence by specifying the one-variable function ρ so that

$$F_{\rho(u)}(x) = \begin{cases} F_{F_u(u)}(x) & \text{if} & F_u(u)\!\downarrow \\ \uparrow & \text{if} & F_u(u)\!\uparrow. \end{cases}$$

In other words, for each natural number u, the function ρ is to yield the index of an algorithm that computes the function θ_u. (Note that since the value of $\rho(u)$ is

---

[†] For typographical convenience, we will omit the superscript (1) from expressions of the form $F_x^{(1)}(y)$ in the remainder of this section.

to be an index, $\rho(u)$ is always defined, even when $F_u(u)$ is not.) Our next job is to show that the family $\mathcal{C}$ is capable of evaluating such a function ρ.

The argument follows the usual lines. First define g to be the composite function $U^{(2)} \circ (U^{(2)} \circ (P_1^{(2)}, P_1^{(2)}), P_2^{(2)})$, so that

$$g(u, x) = U^{(2)}(U^{(2)}(u, u), x) = \begin{cases} F_{F_u(u)}(x) & \text{if} & F_u(u){\downarrow} \\ \uparrow & \text{if} & F_u(u){\uparrow}. \end{cases}$$

Obviously g is algorithmic in $\mathcal{C}$ and has some index w. Moreover, there exists a total two-variable algorithmic function s such that

$$F_{s(w, u)}(x) = F_w^{(2)}(u, x) = g(u, x).$$

Now define ρ to be the function $s \circ (C_w^{(1)}, P_1^{(1)})$. Since s, $C_w^{(1)}$, and $P_1^{(1)}$ are all total and algorithmic, ρ is also total and algorithmic. And, by construction

$$F_{\rho(u)}(x) = F_{s(w, u)}(x) = g(u, x) = \begin{cases} F_{F_u(u)}(x) & \text{if} & F_u(u){\downarrow} \\ \uparrow & \text{if} & F_u(u){\uparrow}. \end{cases}$$

Thus ρ is the desired index-computing function.

Although we have so far confined our attention to functions of one variable, only a minor notational change is required to adapt the preceding arguments to functions of several variables. Thus we have:

Lemma 1-3. *Let $\mathcal{C}$ be any standard family of algorithms and let n be any positive integer. Then there exists a total algorithmic function ρ such that*

$$F_{\rho(u)}^{(n)}(x_1, \ldots, x_n) = \begin{cases} F_{F_u(u)}^{(n)}(x_1, \ldots, x_n) & \text{if} & F_u(u){\downarrow} \\ \uparrow & \text{if} & F_u(u){\uparrow}. \end{cases}$$

In terms of our informal programming argument, the function ρ performs the first of the two tasks required of the program R: it computes an index of the functional counterpart of the program Q_u. The remainder of the formal argument is now fairly straightforward.

Theorem 1-9. *(The Recursion Theorem) Let $\mathcal{C}$ be any standard family of algorithms, let f be any total one-variable function that is algorithmic in $\mathcal{C}$, and let n be any positive integer. Then there exists a natural number m such that $F_m^{(n)} = F_{f(m)}^{(n)}$.*

Proof. Let ρ denote the index-computing function described in Lemma 1-3. Then the composite function $f \circ (\rho)$ is a total algorithmic function and has some index v. If we choose $m = \rho(v)$, we have:

$$F_m^{(n)}(x_1, \ldots, x_n) = F_{\rho(v)}^{(n)}(x_1, \ldots, x_n) = \begin{cases} F_{F_v(v)}^{(n)}(x_1, \ldots, x_n) & \text{if} & F_v(v){\downarrow} \\ \uparrow & \text{if} & F_v(v){\uparrow}. \end{cases}$$

Since $F_v^{(1)} = f \circ (\rho)$ is a total function, $F_v(v)$ must be defined. Therefore $F_m^{(n)}(x_1, \ldots, x_n) = F_{F_v(v)}^{(n)}(x_1, \ldots, x_n)$. But

$$F_{F_v(v)}^{(n)}(x_1, \ldots, x_n) = F_{f(\rho(v))}^{(n)}(x_1, \ldots, x_n) = F_{f(m)}^{(n)}(x_1, \ldots, x_n)$$

so that in fact $F_m^{(n)} = F_{f(m)}^{(n)}$.   □

Although many important applications of the Recursion Theorem lie beyond the scope of this book, we can nevertheless illustrate a few relatively simple and unobvious consequences of this fundamental result. In each case we suppose that α is a standard family, hence a family for which the Recursion Theorem holds.

First consider the case in which the function f is the successor function. Since this function is total and algorithmic, the Recursion Theorem guarantees that there must be at least one value of m for which $F_m^{(1)} = F_{f(m)}^{(1)} = F_{m+1}^{(1)}$. In other words, if the members of the family α are listed according to their indices, there must be some pair of consecutive algorithms that compute the same one-variable function. (Of course, this function need not be total, or even nonempty.) In general, if we take f to be the function obtained by composing the successor function with itself k times, we find that there must be some value of m for which $F_m^{(1)} = F_{m+k}^{(1)}$. Thus for each choice of k there must be two algorithms that are located exactly k positions apart in the listing and that compute the same one-variable function.

From these remarks we see that in any family for which the Recursion Theorem holds, some one-variable function must have more than one index. Making the obvious extension to n-variable functions, we have:

Theorem 1-10. *If α is a standard family of algorithms, then for each positive integer n there exists an algorithmic n-variable function that has more than one index.*

Or, to phrase the theorem the other way around, if α is a family in which no two algorithms compute the same function, α must fail to satisfy one or more of Properties 1 through 5. (The families α_1 and α_2, first defined in Example 1-6, illustrate this point.)

For a second example, let us consider the one-variable constant functions $C_0^{(1)}, C_1^{(1)}, C_2^{(1)}, \ldots$, each of which may have several indices in the family α. There is, of course, no reason to suppose that the indices of $C_i^{(1)}$ are related in any simple way to the constant value i computed by $C_i^{(1)}$. In particular, there is no reason to suppose that $C_i^{(1)}$ has the number i as one of its indices. But as long as we are dealing with a standard family of algorithms, there must be at least one natural number m that is an index for the constant function $C_m^{(1)}$. For we know from Theorem 1-5 that whenever Properties 1, 2, and 5 hold, there exists a total algorithmic function k such that

$$F_{k(x)}^{(1)} = C_x^{(1)}.$$

Applying the Recursion Theorem to the function k, we find that there must be some natural number m such that

$$F_m^{(1)} = F_{k(m)}^{(1)} = C_m^{(1)}.$$

Thus m is in fact an index of $C_m^{(1)}$.

We conclude with a more sophisticated application of the Recursion Theorem. Again assume we are dealing with a standard family of algorithms $\mathcal{A}$. First note that for each natural number i, the two-variable function f_i whose values are given by $f_i(x, y) = F_i^{(2)}(x, y + 1) + 1$ is certainly algorithmic in $\mathcal{A}$. Moreover, it is by now a routine matter to show that $\mathcal{A}$ must contain an algorithm that computes an index for the function f_i when given the value i. In other words, there must be a total algorithmic function β such that

$$F_{\beta(i)}^{(2)}(x, y) = F_i^{(2)}(x, y + 1) + 1.$$

The details of the derivation of β are left to the reader.

Next assume that the number a is one of the indices of the projection function $P_1^{(2)}$. Then for each natural number i, the two-variable function g_i such that $g_i(x, y) = U^{(3)}(\Lambda(x, y, a, \beta(i)), x, y)$ must be algorithmic in $\mathcal{A}$. And as usual, there must be some algorithm in $\mathcal{A}$ that computes an index for g_i when given the value i. That is, there must be a total algorithmic function r such that

$$F_{r(i)}^{(2)}(x, y) = U^{(3)}(\Lambda(x, y, a, \beta(i)), x, y).$$

Again the argument is routine and the details are omitted.

If we now apply the Recursion Theorem to the function r, we conclude that there must exist some natural number m such that

$$F_m^{(2)}(x, y) = F_{r(m)}^{(2)}(x, y) = U^{(3)}(\Lambda(x, y, a, \beta(m)), x, y).$$

Since β is a total function, the value of $\Lambda(x, y, a, \beta(m))$ is always defined and we may write

$$F_m^{(2)}(x, y) = \begin{cases} U^{(3)}(a, x, y) & = F_a^{(2)}(x, y) & \text{if} & x = y \\ U^{(3)}(\beta(m), x, y) & = F_{\beta(m)}^{(2)}(x, y) & \text{if} & x \neq y. \end{cases}$$

According to the definitions of a and β, this means that

$$F_m^{(2)}(x, y) = \begin{cases} x & \text{if} & x = y \\ F_m^{(2)}(x, y + 1) + 1 & \text{if} & x \neq y. \end{cases}$$

This last equation allows us to deduce some of the values of $F_m^{(2)}$. Certainly $F_m^{(2)}(x, x) = x$, for all x. Next consider the value $F_m^{(2)}(x, x - 1)$, where $x > 0$. Since $x \neq x - 1$, we have

$$F_m^{(2)}(x, x - 1) = F_m^{(2)}(x, (x - 1) + 1) + 1 = F_m^{(2)}(x, x) + 1 = x + 1.$$

Using similar reasoning, we find that

$$F_m^{(2)}(x, x - 2) = F_m^{(2)}(x, x - 1) + 1 = x + 2,$$
$$F_m^{(2)}(x, x - 3) = F_m^{(2)}(x, x - 2) + 1 = x + 3,$$

and in general

$$F_m^{(2)}(x, x - k) = x + k \qquad \text{for} \quad 0 \le k \le x.$$

If we now define h to be the function $F_m^{(2)} \circ (P_1^{(1)}, C_0^{(1)})$, we see that h is a one-variable algorithmic function such that

$$h(x) = F_m^{(2)}(x, 0) = x + x = 2x.$$

What does all this mean? Simply that if the family $\mathcal{C}$ satisfies the five basic properties, then in addition to containing algorithms for the constant, projection, and successor functions, it must also contain an algorithm that evaluates the "doubling function" h. Similar arguments can be used to show that $\mathcal{C}$ must contain algorithms for evaluating many other arithmetic functions as well, including the familiar addition, multiplication, and exponentiation functions. We will return in Chapter 6 to the question of just which functions of this type can be shown to be algorithmic in $\mathcal{C}$. Right now, we wish only to suggest the power of the Recursion Theorem.

Families of Algorithms and Effective Computability

Since the definition of a standard family of algorithms was intended to reflect certain basic features of programming languages, conclusions concerning the capabilities of standard families may be expected to have implications for practical programming. It is therefore appropriate at this point to summarize some of the important characteristics of standard families.

1. Every standard family is capable of evaluating a variety of decision functions based on simple argument comparisons.
2. Every standard family is capable of evaluating a variety of index-computing functions.
3. Every standard family satisfies the Recursion Theorem.
4. No standard family is capable of evaluating its own domain functions.
5. No standard family is capable of evaluating its own totality function.
6. No standard family is capable of evaluating its own equivalence function.

The first three characteristics may be viewed as "positive" attributes: they specify things that can be done in any standard family. Indeed, as we will see in Chapter 6, these characteristics completely determine the set of functions that can be evaluated in every standard family. The last three characteristics may be viewed as "negative" attributes: they specify things that cannot be done in any standard family. Since the significance of these negative attributes is sometimes misunderstood, a few comments about them may be in order.

First, remember that algorithms are assumed to be specified by their indices. Thus when we ask for a means of determining whether a given algorithm computes a total function, we are looking for a procedure that can be applied to the index of the algorithm in question. The negative results of characteristics 4 to 6 imply that certain interesting questions about the properties of algorithms cannot be resolved on the basis of indices alone. In fact—as we will discover in Chapter 6—as long as a family is indexed in such a way that Properties 3 and 5 are satisfied, knowledge of indices provides almost no useful information about the properties of the functions bearing those indices.

On the other hand, characteristics 4 to 6 should not be interpreted as saying that one cannot determine whether a *particular* algorithm eventually yields a defined value, or computes a total function, or whether two *particular* algorithms compute the same function. There may in fact be many particular algorithms for which these questions can be settled. What the negative results do say is that there can be no one algorithm in a standard family that correctly answers one of these questions for every member of the family. Thus in the case of the diagonal domain function, we may be able to find an index n such that $F_n(x) = 1$ for some values of x for which $F_x(x)$ is defined and $F_n(x) = 0$ for some values of x for which $F_x(x)$ is undefined. But no matter which index n we choose, there will always be some value of x for which either $F_n(x)$ is undefined or else $F_n(x)$ is defined but disagrees with the corresponding value of the diagonal domain function.

These conclusions translate directly to the case of programs written in general-purpose languages for execution on an idealized computer. There can be no such program that, when applied to a suitable description of an arbitrary program in the language, correctly determines whether that second program terminates with a defined value, or computes a total function, or computes the same function as some other program. Again, it may be possible to settle these questions for certain specific programs or classes of programs, but not when the program to be tested may be chosen freely from among all the legal programs in the language.

In practical terms, then, we should not expect to be able to write general-purpose programs that test for such properties as termination, totality, or equivalence.[†] Similarly, we should not expect to be able to write general-purpose debugging programs, since the operation of debugging is basically that

---

[†] If programs are to be executed on a computer whose storage capacity is strictly bounded, the number of steps that might be executed prior to termination is also bounded. In this case it is possible, through exhaustive simulations on a larger computer, to determine whether a given program halts, or computes a total function, or whether two programs compute the same function. But because of the extraordinarily large numbers involved, such an exhaustive approach, although possible in principle, is completely unrealistic in practice.

of determining whether a given program evaluates a specified function. Thus if we are concerned with the problem of determining whether programs behave correctly—and this is a problem of considerable current interest—we should be content to deal with appropriate special cases and important subclasses of programs.

Although the domain, totality, and equivalence functions associated with a given standard family $\mathcal{C}_0$ cannot be evaluated within that family, it may be possible to evaluate them within a larger family $\mathcal{C}_1$. To form such a family $\mathcal{C}_1$, we begin with the set of functions computed by the members of $\mathcal{C}_0$ and add the desired domain, totality, and equivalence functions, together with whatever additional functions may be needed to ensure closure under composition. Since the resulting set of functions is necessarily countable, we may assign a new indexed set of algorithms to its members and so arrive at a new family $\mathcal{C}_1$. In fact—although we cannot provide the details here—it is always possible to arrange the new indexing so that $\mathcal{C}_1$ satisfies Properties 1 through 5. Thus every standard family $\mathcal{C}_0$ can be embedded in a larger standard family $\mathcal{C}_1$ in which the domain, totality, and equivalence functions of $\mathcal{C}_0$ are algorithmic.

Of course, the family $\mathcal{C}_1$ has its own domain, totality, and equivalence functions, distinct from those of $\mathcal{C}_0$. And since $\mathcal{C}_1$ is a standard family, it does not contain algorithms for evaluating its own domain, totality, and equivalence functions. In order to provide algorithms for these functions, we must define a still larger family $\mathcal{C}_2$. The family $\mathcal{C}_2$ will in turn have its domain, totality, and equivalence functions, which require a new family $\mathcal{C}_3$, and so on. Thus there exists an infinite hierarchy of standard families, each of which computes functions not computed by any of its predecessors.

The existence of such a hierarchy raises an important practical issue. So far we have treated algorithms as abstract objects, without concern for the manner in which they evaluate their functions. But does it really make sense to speak of augmenting a family of algorithms by adding, say, an algorithm that computes the diagonal domain function for that family? What assurance do we have that there exists a procedure for evaluating this function? Suppose, for example, that the family $\mathcal{C}$ is formed by the programs of a powerful, general-purpose programming language. Can there really exist a more powerful set of algorithms capable of computing functions that are not programmable in the given language?

In order to deal with these questions, we must begin to consider the matter of realizability. While the concept of an abstract algorithm is useful for deducing certain general characteristics of discrete computation, we must not suppose that every function for which we postulate an abstract algorithm can in reality be evaluated in an "algorithmic way." Thus if we wish to draw further conclusions about practical computation, we must restrict our attention to families whose algorithms can in fact be executed in some well-defined, mechanistic way. Such algorithms are called *effective algorithms* and the functions they evaluate are called *effectively computable functions*.

The notion of effective computability is not an easy one to define precisely, and we shall not attempt to do so now. Instead, let us review briefly some of the characteristics that experience and common sense suggest an effective algorithm ought to have.

To begin with, an effective algorithm must provide a finite specification of a sequence of basic operations to be performed. Each basic operation should be well defined, and should be susceptible to implementation in a deterministic, mechanistic manner. At the start of any computation, only a finite amount of information concerning the function to be evaluated is to be provided to the agent executing the algorithm. At each step in the ensuing computation, the particular operation to be performed, the quantities on which it is to be performed, and the disposition of the result must all be explicitly determined by the algorithm. Similarly, the situations in which a computation is to terminate, as well as the outcomes that are to result, must also be explicitly determined by the algorithm. Thus an effective algorithm must specify a routine, mechanistic procedure for evaluating a function.

In the chapters that follow, we will introduce several important classes of functions that can be evaluated by effective algorithms. Study of these classes, their interrelationships, and their possible extensions will lead us to propose a formal counterpart of the notion of effective computability. And since the functions that satisfy this formal definition of computability can be obtained from a standard family of algorithms, all the results of this chapter must apply to them.

1.4 SUMMARY

The purpose of this book is to discover some of the fundamental characteristics of algorithmic computation. In practice, algorithms may be grouped into classes according to the elements on which they operate, the basic operations they use, and the languages in which they are described. But because algorithms are required to have finite descriptions, every such class will contain only a countable number of members and so will be incapable of evaluating all possible number-theoretic functions. What, then, are the basic limitations on the functions that can be evaluated by a given class of algorithms?

The concept of an abstract family of algorithms provides one useful way of addressing this question. An abstract family of algorithms is simply an indexed set of objects, each of which has associated with it some n-variable number-theoretic function for each positive integer n. Such an abstract family may be thought of as a model for a class of real algorithms that are capable of dealing with representations of natural numbers and with representations of algorithms as well. In particular, abstract families can be used to model the classes of algorithms defined by general-purpose programming languages.

Certain fundamental characteristics are shared by all useful programming languages. These include the ability to evaluate various elementary functions, to make decisions, to use subroutines, to write interpreters, and to write programs that generate descriptions of other programs. When expressed in abstract terms, these characteristics yield five basic properties that a family of algorithms might have. And if these properties are satisfied by a given family, certain general consequences can be shown to follow. On the one hand, the members of the family must be capable of evaluating a variety of decision and index-computing functions. On the other hand, the members of the family can not resolve such questions as whether an algorithm yields a defined value for a given argument, whether an algorithm evaluates a total function, or whether two algorithms evaluate the same function.

Two techniques play an especially important role in establishing the computability or noncomputability of functions in a given family of algorithms. The method of reduction represents the most common way of proving that a function is not computable. In using this method, we show that the ability to evaluate the function in question would make it possible to evaluate some other function already known not to be computable; as a consequence, we must conclude that the given function is also not computable. The Recursion Theorem, which holds whenever the five basic properties are satisfied, provides one of the most powerful ways of showing that certain functions are computable. The essence of this theorem is that if f is a computable function mapping algorithms into algorithms, there must be at least one algorithm that evaluates the same function as its image under f. The Recursion Theorem will be used in Chapter 6 to show that a great many recursively defined functions must be computable in any family in which the five basic properties hold.

Because many fundamental characteristics of programming languages can be represented in abstract families of algorithms, the study of such families can provide insight into the inherent capabilities and limitations of programming languages. The advantage of the abstract model for this purpose is that it eliminates the mass of operational detail associated with practical computation. But in eliminating this detail, the abstract approach avoids the issue of whether the functions that it deals with can be evaluated in an effective way. In order to determine the limits of effective computation, we must now turn to models that provide routine mechanistic ways of evaluating functions.

1.5 PROBLEMS

Section 1.1

1.1.1 Let g and h be one-variable number-theoretic functions, and let f be the composite function $h \circ (g)$.

   a) Show that f is total whenever both g and h are total.

   b) Suppose that f is known to be total. Under what conditions must g be total? Under what conditions must h be total?

1.1.2   Consider the number-theoretic functions g_1, g_2, and h whose values are specified as follows:

$$g_1(x) = \begin{cases} x/3 & \text{if 3 divides } x \\ \uparrow & \text{otherwise} \end{cases} \qquad g_2(x) = \begin{cases} x/5 & \text{if 5 divides } x \\ \uparrow & \text{otherwise} \end{cases}$$

def only if x is even multiple of 15 = x2

$$h(y_1, y_2) = \begin{cases} 15 y_1 y_2 & \text{if } y_1 \text{ and } y_2 \text{ are both even} \\ y_1 + y_2 & \text{if } y_1 \text{ is even and } y_2 \text{ is odd, or vice versa} \\ \uparrow & \text{if } y_1 \text{ and } y_2 \text{ are both odd} \end{cases}$$

Describe the values of the composite function $f = h \circ (g_1, g_2)$ as concisely as possible.

1.1.3   Let h be a specified, total, three-variable, number-theoretic function.

a) The functions f_1, f_2, and f_3 are defined as follows:

   i) $f_1 = h \circ (C_3^{(2)}, P_1^{(2)}, P_2^{(2)})$

   ii) $f_2 = C_5^{(1)} \circ (h \circ (C_0^{(3)}, C_1^{(3)}, C_2^{(3)}))$

   iii) $f_3 = h \circ (P_3^{(3)}, h \circ (C_0^{(3)}, P_1^{(3)}, P_2^{(3)}), C_2^{(3)})$

Express the values of f_1, f_2, and f_3 in terms of appropriate values of h.

b) The functions f_4, f_5, and f_6 are to be defined so that:

   i) $f_4(x, y)$ $= h(y, 1, x)$

   ii) $f_5(x, y, z)$ $= h(h(z, z, 2), 0, z)$

   iii) $f_6(w, x, y, z) = h(x, h(x, y, z), h(w, x, y))$

Give formal definitions of f_4, f_5, and f_6 by expressing them as compositions of h with appropriate constant and projection functions.

1.1.4   The family $\mathcal{C}$ consists of all the algorithms that can be represented by flow diagrams made up of the four types of building blocks shown below. Here $x_0, x_1, \ldots, x_i, \ldots$ denote variables, c denotes an arbitrary constant, and

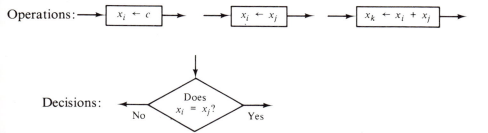

Operations: → $x_i \leftarrow c$ →   → $x_i \leftarrow x_j$ →   → $x_k \leftarrow x_i + x_j$ →

Decisions: ← No   Does $x_i = x_j$?   Yes →

"$x_i \leftarrow a$" means "replace the current value of x_i by the value of a." Building blocks can be interconnected in any way whatever, as long as the resulting diagram specifies a well-defined sequence of operations.

a) Carefully explain how each member of $\mathcal{C}$ can be thought of as defining a (partial) n-variable number-theoretic function for each positive integer n.

b) Do any of the members of $\mathcal{C}$ define nontotal functions? Justify your answer.

c) Show that the two-variable multiplication function is algorithmic in $\mathcal{C}$. In particular, give the flow diagram for a member of $\mathcal{C}$ that evaluates the multiplication function. What three-variable number-theoretic function does this algorithm evaluate?

d) Let $\mathcal{B}$ be the family of algorithms that can be represented by flow diagrams made up of the four types of building blocks shown below.

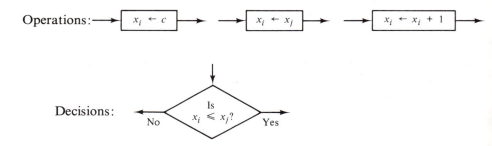

Show that $\mathcal{B}$ is equivalent to the family $\mathcal{C}$. In other words, show that a number-theoretic function is algorithmic in one family iff it is algorithmic in the other.

1.1.5 Recall that every positive integer i has a unique representation as a product of prime factors of the form $i = 2^{e_0}3^{e_1}5^{e_2}7^{e_3}\cdots$, where each exponent e_j is nonnegative and only a finite number of exponents are nonzero. Now consider the abstract family of algorithms $\mathcal{C}$ in which

$$F_i^{(n)}(x_1, \ldots, x_n) = \begin{cases} e_0 + e_1 x_1 + \cdots + e_n x_n & \text{if} \quad i > 0 \\ 0 & \text{if} \quad i = 0 \end{cases}$$

where $e_0, e_1, \ldots, e_n$ are the first $n+1$ exponents in the expression

$$i = 2^{e_0}3^{e_1}5^{e_2}\cdots.$$

a) Show that every function that is algorithmic in $\mathcal{C}$ has an infinite number of indices.

b) What is the smallest index of the function $F^{(2)}_{150} \circ (F^{(2)}_6, F^{(2)}_{45})$?

c) Are all the constant and projection functions algorithmic in $\mathcal{A}$?

1.1.6   A total one-variable number-theoretic function f is said to be *monotone increasing* if $f(n+1) > f(n)$ for all n. Use a diagonalization argument to prove that the number of monotone-increasing one-variable functions is uncountable.

1.1.7   Let $\mathcal{A}$ be any family of algorithms. Define the one-variable function f so that

$$f(x) = \begin{cases} 0 & \text{if } F^{(1)}_x(x) = 1 \\ 1 & \text{otherwise} \end{cases}$$

where $F^{(1)}_x$ denotes the one-variable function computed by the xth member of $\mathcal{A}$. Prove that f is not algorithmic in $\mathcal{A}$.

Section 1.2

1.2.1   For each of the families specified below, determine which of Properties 1 through 3 hold:

a) The family in which $F^{(n)}_i(x_1, \ldots, x_n) = \max (i, x_1, \ldots, x_n)$
b) The family defined in Problem 1.1.5

In each case in which a given property does hold, clearly indicate the algorithm or algorithms that are responsible.

1.2.2   Show that the functions specified below are not algorithmic in any family of algorithms for which Properties 1 and 2 hold.

a) The function f, where $f(x) = \begin{cases} 5 & \text{if } F^{(1)}_x(x) = 3 \\ 1 & \text{otherwise} \end{cases}$

b) The function g, where $g(x, y, z) = \begin{cases} 1 & \text{if } F^{(1)}_x(y) = z \\ 0 & \text{otherwise} \end{cases}$

1.2.3   Let $\mathcal{A}$ be any family of algorithms in which Properties 1, 2, and 3 hold. Suppose that the members of $\mathcal{A}$ are reindexed by means of a certain total algorithmic function f so that the index i is now assigned to what was originally the $f(i)$th algorithm. Show that Properties 1 through 3 are also satisfied under the new indexing scheme.

1.2.4   Let α be a family of algorithms in which Properties 1, 2, and 3 hold.

a) Suppose that a certain total one-variable function r is algorithmic in α. Show that the function f must also be algorithmic, where

$$f(x) = F^{(1)}_{r(x)}(x) + 1.$$

b) Prove that there is no total algorithmic function r such that the values $r(0), r(1), r(2), \ldots$ enumerate the indices of all the total one-variable functions of α. In other words, prove that there is no total algorithmic function r such that ran $r = \{i \mid F^{(1)}_i$ is total$\}$.

1.2.5   The family of algorithms α satisfies Properties 1, 2, and 4. Show that each of the functions whose values are specified below is algorithmic in α. In each case, try to minimize the number of times the selection function is used.

a) $f_1(x) = \begin{cases} 4 - x & \text{if} & x \le 4 \\ 0 & \text{if} & x > 4 \end{cases}$ (4 applications of Λ suffice)

b) $f_2(x, y) = \begin{cases} 0 & \text{if} & x = y = 0 \\ 1 & \text{if} & x = 0 \text{ and } y > 0 \\ 2 & \text{if} & x > 0 \text{ and } y = 1 \\ 3 & \text{if} & x > 0 \text{ and } y \ne 1 \end{cases}$ (3 applications of Λ suffice)

c) $f_3(x, y) = \begin{cases} 0 & \text{if} & x = 3 \text{ and } y = 0 \\ & & \text{or } x = 5 \text{ and } y > 0 \\ 1 & \text{otherwise} \end{cases}$ (2 applications of Λ suffice)

d) $f_4(x, y, z) = \begin{cases} 1 & \text{if an odd number of the} \\ & \text{variables } x, y, z \text{ equal } 0 \\ 0 & \text{otherwise} \end{cases}$ (5 applications of Λ suffice)

1.2.6   A certain family α satisfies Properties 1 through 4. It is known that $F^{(1)}_{17}(x)$ is defined iff x is even and that $F^{(1)}_{241}(x)$ is defined iff x is odd.

a) Describe the values assumed by the three-variable function g, where

$$g = \Lambda \circ (P^{(3)}_2, \quad P^{(3)}_3, \quad F^{(1)}_{17} \circ (P^{(3)}_1), \quad F^{(1)}_{241} \circ (P^{(3)}_1))$$

b) Show that the three-variable function h is algorithmic, where

$$h(x, y, z) = \begin{cases} F^{(1)}_{17}(x) & \text{if} & y = z \\ F^{(1)}_{241}(x) & \text{if} & y \ne z. \end{cases}$$

1.2.7   Consider any family of algorithms for which Properties 2, 3, and 4 hold and in which all the constant and projection functions are algorithmic.

a) Show that there is a total, algorithmic one-variable function θ such that, for all x, $\theta(x) \ne x$.

b) Prove that not all the members of α evaluate total functions.

1.2.8　A family of algorithms is said to satisfy the *Selective Looping Property* if whenever the n-variable function f is algorithmic so is the n-variable function $\hat{f}$, where

$$\hat{f}(x_1, \ldots, x_n) = \begin{cases} 0 & \text{if } f(x_1, \ldots, x_n) = 0 \\ \uparrow & \text{otherwise} \end{cases}$$

　　　a) Show that the family $\mathcal{A}_2$ of Example 1-6 satisfies the Selective Looping Property.

　　　b) Prove that any family for which Property 2 holds and in which the function $\hat{P}_1^{(1)}$ is algorithmic must satisfy the Selective Looping Property.

　　　c) Prove that if a family satisfies the Selective Looping Property, then the diagonal domain function for that family is not algorithmic.

1.2.9　Consider any family of algorithms $\mathcal{A}$ in which Properties 1 through 4 hold. Prove that there are no two natural numbers a and b such that the function h, where

$$h(x) = \begin{cases} a & \text{if} & F_x^{(1)}(x)\!\downarrow \\ b & \text{if} & F_x^{(1)}(x)\!\uparrow, \end{cases}$$

is algorithmic in $\mathcal{A}$.

1.2.10　Does Property 5 hold for the family of Problem 1.2.1(a), in which $F_i^{(n)}(x_1, \ldots, x_n) = \max(i, x_1, \ldots, x_n)$? Justify your answer.

1.2.11　Consider any family of algorithms that satisfies Property 5, but not necessarily Property 1. Show that for each natural number w there exists an algorithmic function r such that

$$F_{r(x)}^{(1)}(y) = F_w^{(2)}(x, y)$$

for all x and y. *Hint:* Use Property 5 twice.

1.2.12　Consider any family of algorithms in which Properties 1, 2, 3, and 5 hold. For each natural number n, let g_n denote the one-variable function such that $g_n(x) = x + n$.

　　　a) Prove that each of the functions $g_0, g_1, \ldots, g_n, \ldots$ is algorithmic.

　　　b) Now consider the two-variable function f, where $f(x, y) = x + y$. Prove that f is algorithmic iff there exists a total algorithmic function r such that $F_{r(n)}^{(1)} = g_n$.

Thus the ability to compute the sum of two arbitrary numbers does not follow from the ability to add any designated constant to a number; it is also necessary to be able to determine the index of a function that adds the designated constant.

1.2.13   A certain family of algorithms satisfies Properties 1 through 5. For each of parts (a) through (e) below, prove that there exists a total algorithmic index-computing function r having the specified property.

a) $F_{r(x)}(y, z) = F_x(z, y)$

b) $F_{r(x)}(y, z) = F_x(y, z + 1) + 1$

c) $F_{r(x)}(y) = \begin{cases} 1 & \text{if} & y = x \\ 0 & \text{if} & y \neq x \end{cases}$

d) $F_{r(x_1, x_2)}(y) = \begin{cases} F_{x_1}(y) & \text{if} & y = x_1 \\ F_{x_2}(y) & \text{if} & y \neq x_1 \end{cases}$

e) $F_{r(x)}(y, z) = \begin{cases} 0 & \text{if} & z = y \\ F_x(y, z + 1) + 1 & \text{if} & z \neq y \end{cases}$

1.2.14   Let $\mathcal{C}$ be any family of algorithms in which Properties 1, 2, and 5 hold, and in which the function $U^{(2)}$ is algorithmic. Show that $\mathcal{C}$ also satisfies Property 3.

1.2.15   In this problem we show that, as long as Properties 1, 2, and 3 hold, Property 5 is equivalent to the ability to compute indices for constant and composite functions.

a) Consider any family of algorithms in which Properties 1, 2, 3, and 5 hold.

   i) Prove that for each positive integer n there exists a total algorithmic function k_n such that $F_{k_n(x)}^{(n)} = C_x^{(n)}$.

   ii) Prove that for each positive integer n there exist total algorithmic functions $h_{n, 1}, h_{n, 2}, \ldots, h_{n, m}, \ldots$ such that

$$F_{h_{n,m}(w, x_1, \ldots, x_m)}^{(n)} = F_w^{(m)} \circ (F_{x_1}^{(n)}, \ldots, F_{x_m}^{(n)}).$$

b) Consider any family of algorithms in which Properties 1 and 2 hold. (Property 3 is not needed here.) Assume that for each positive integer n there exist total algorithmic functions k_n and $h_{n, 1}, h_{n, 2}, \ldots, h_{n, m}, \ldots$ such that

$$F_{k_n(x)}^{(n)} = C_x^{(n)}$$

and

$$F_{h_{n,m}(w, x_1, \ldots, x_m)}^{(n)} = F_w^{(m)} \circ (F_{x_1}^{(n)}, \ldots, F_{x_m}^{(n)}).$$

Prove that Property 5 is satisfied.

Section 1.3

1.3.1 Let $\mathcal{C}$ be a family of algorithms in which Properties 1, 2, 3, and 5 (but not Property 4) are satisfied, and in which the familiar two-variable addition and multiplication functions are algorithmic. Suppose it is known that the function f is not algorithmic in $\mathcal{C}$, where

$$f(x) = \begin{cases} 13 & \text{if} & F_x^{(1)} \text{ is total} \\ 7 & \text{if} & F_x^{(1)} \text{ is not total}. \end{cases}$$

Which of the following functions g_1 and g_2 can be shown not to be algorithmic in $\mathcal{C}$? Justify your answer.

$$g_1(x) = \begin{cases} 5 & \text{if} & F_x^{(1)} \text{ is total} \\ 2 & \text{if} & F_x^{(1)} \text{ is not total} \end{cases}$$

$$g_2(x) = \begin{cases} 29 & \text{if} & F_x^{(1)} \text{ is total} \\ 17 & \text{if} & F_x^{(1)} \text{ is not total} \end{cases}$$

1.3.2 Consider any family of algorithms in which Properties 1 through 5 are satisfied.

a) Show that there exists a total algorithmic function r such that

$$F_{r(x)}(y) = \begin{cases} y & \text{if} & F_x(x)\!\downarrow \\ \uparrow & \text{if} & F_x(x)\!\uparrow. \end{cases}$$

b) Use the result of part (a) to show that none of the functions defined below are algorithmic.

i) $f_1(x) = \begin{cases} 1 & \text{if the range of } F_x^{(1)} \text{ is infinite} \\ 0 & \text{otherwise} \end{cases}$

ii) $f_2(x) = \begin{cases} 1 & \text{if the domain of } F_x^{(1)} \text{ is nonempty} \\ 0 & \text{otherwise} \end{cases}$

iii) $f_3(x, y) = \begin{cases} 1 & \text{if } F_x^{(1)} = F_y^{(1)} \\ 0 & \text{otherwise} \end{cases}$

iv) $f_4(x, y, z) = \begin{cases} 1 & \text{if } F_x(y) = z \\ 0 & \text{otherwise} \end{cases}$

1.3.3 Let $\mathcal{C}$ be a family of algorithms in which Properties 1 through 5 are satisfied, and let f be any specified total, one-variable function known to be algorithmic in $\mathcal{C}$. Let h denote the one-variable function such that:

$$h(x) = \begin{cases} 1 & \text{if} & F_x^{(1)} \text{ is the function } f \\ 0 & \text{if} & F_x^{(1)} \text{ is not the function } f. \end{cases}$$

Prove that h is not algorithmic in $\mathcal{C}$ by:

    a) reduction from the diagonal domain function d;

    b) reduction from the totality function t;

    c) reduction from the general domain function $D^{(2)}$;

    d) reduction from the function p, where $p(x) = \begin{cases} 1 & \text{if} \quad F_x^{(1)} \text{ is } P_1^{(1)} \\ 0 & \text{otherwise.} \end{cases}$

Which of the preceding proofs are also valid for the case in which f is an arbitrary nontotal (but nonempty) function?

1.3.4   Assume that Properties 1 through 5 hold for a certain family of algorithms. Let m denote the one-variable function such that $m(x)$ is the smallest index of $F_x^{(1)}$. Prove that m is not algorithmic.

1.3.5   Prove that neither of the functions defined below is algorithmic in any family of algorithms in which Properties 1 through 5 hold.

    a) $f(x) = \begin{cases} 1 & \text{if} \quad \text{dom } F_x^{(1)} = \{2, 3\} \\ 0 & \text{otherwise} \end{cases}$

    b) $g(x, y) = \begin{cases} 1 & \text{if} \quad \text{dom } F_x^{(1)} = \text{dom } F_y^{(1)} \\ 0 & \text{otherwise.} \end{cases}$

1.3.6   Consider any family of algorithms in which Properties 1 through 5 hold. In each of parts (a) through (e) below, use the Recursion Theorem to establish the existence of a natural number m that satisfies the given conditions. Assume h to be a designated algorithmic function.

    a) $F_m^{(1)}(y) = h(m, y)$ b) $F_m^{(1)}(y) = F_m^{(2)}(m, y)$

    c) $F_m^{(1)}(y) = F_{m+1}^{(1)}(F_{m+2}^{(1)}(y))$ d) $F_m^{(1)}(y) = \begin{cases} 1 & \text{if} \quad y = m \\ 0 & \text{if} \quad y \neq m \end{cases}$

    e) $U^{(2)}(F_m^{(1)}(y), z) = F_m^{(2)}(y, z)$

1.3.7   Consider any family of algorithms in which Properties 1 through 5 hold.

    a) Prove that there exists a total algorithmic function f such that

$$F_{f(x)}(y) = F_y(x) \qquad \text{for all } x \text{ and } y.$$

    b) Use the result of part (a) to show that there exist distinct natural numbers m and n such that $F_n(m)$ is defined and equal to $F_m(n)$.

1.3.8   A certain family of algorithms satisfies Properties 1 through 5 under each of two indexing schemes. Let $F_i^{(n)}$ denote the n-variable function evaluated by the ith algorithm under the first indexing scheme, and let $G_i^{(n)}$ denote the n-variable function evaluated by the ith algorithm under the second scheme. Use the Recursion Theorem to show that there must be some one-variable function that has the same index in both indexing schemes.

1.3.9   Assume that Properties 1 through 5 hold for a certain family of algorithms. Show that there exists an algorithmic function Y that computes the fixed point of any given index-mapping function f. Specifically, show that there exists a total algorithmic function Y such that, whenever i is the index of a total algorithmic function f, $Y(i)$ satisfies the condition $F_{Y(i)}^{(1)} = F_{f(Y(i))}^{(1)}$.

1.3.10   Consider any family of algorithms for which Properties 1 through 5 hold. Show that the Recursion Theorem does not hold when restricted to total functions. Specifically, show that there exists a total algorithmic function f such that:

     i) $F_{f(x)}^{(1)}$ is total whenever $F_x^{(1)}$ is total;

and

     ii) there is no choice of m for which $F_{f(m)}^{(1)}$ and $F_m^{(1)}$ are the same total function.

1.3.11   Let $\mathcal{C}$ be any family of algorithms in which Properties 1 through 5 are satisfied.

    a) Show that there exists a total algorithmic function f such that

$$F_{f(i)}^{(3)}(x,\ y,\ z) = \begin{cases} x & \text{if} & z = y \\ F_i^{(3)}(x,\ y,\ z+1) + 1 & \text{if} & z \neq y. \end{cases}$$

    b) Show that the two-variable addition function is algorithmic in $\mathcal{C}$.

1.3.12   This problem shows that it is possible for an extension of a family of algorithms to be indexed so as to preserve some of the important properties of the original family. Suppose that $\mathcal{C} = \{A_0,\ A_1,\ A_2,\ \ldots\}$ is an abstract family of algorithms whose associated n-variable functions are denoted $F_0^{(n)}, F_1^{(n)}, F_2^{(n)}, \ldots$ in the usual way. Assume that $\mathcal{C}$ satisfies Properties 1 through 5, and that the one-variable function m, where

$$m(x) = \begin{cases} 0 & \text{if} & x = 0 \\ x - 1 & \text{if} & x > 0 \end{cases}$$

is algorithmic in $\mathcal{A}$. Now consider the new family $\mathcal{B} = \{B_0, B_1, B_2, \ldots\}$ whose associated n-variable functions $G_0^{(n)}, G_1^{(n)}, G_2^{(n)}, \ldots$ are defined as follows.

$$G_0^{(1)}(x) = \begin{cases} 1 & \text{if} & F_x^{(1)}(x)\downarrow \\ 0 & \text{if} & F_x^{(1)}(x)\uparrow \end{cases}$$

$$G_0^{(n)} = C_1^{(n)} \qquad \text{for} \quad n > 1$$

$$G_i^{(n)} = F_{i-1}^{(n)} \qquad \text{for} \quad i > 0, n \geq 1$$

a) Explain why $\mathcal{B}$ satisfies Properties 1 and 4. (It does not satisfy Property 2.)

b) Let $V^{(2)}, V^{(3)}, \ldots$ denote the various universal functions of $\mathcal{B}$. Prove that $V^{(n+1)}$ is algorithmic in $\mathcal{B}$ for all $n > 1$. Hint: Express $V^{(n+1)}$ in terms of members of $\mathcal{A}$.

c) Prove that $\mathcal{B}$ satisfies Property 5. Hint: If $\sigma_{n,m}$ denotes an S-M-N function for $\mathcal{B}$, express $\sigma_{n,m}$ in terms of $s_{n,m+1}$ and an index for $V^{(m+n+1)}$.

Chapter 2
Turing Machines

The one essential feature of an effective algorithm is that it can be evaluated in a routine way by a deterministic, mechanistic computing device. This chapter introduces a very elementary model of a computing device that might be used as a basis for defining effective algorithms and effectively computable functions. This model was first proposed by A. M. Turing in 1936, and computing devices patterned after it are accordingly called Turing machines. Turing machines are so simple in structure and so mechanistic in operation that there can be no doubt of their performing effective computations. Yet as we shall see, Turing machines are capable of evaluating extraordinarily complex functions.

2.1 INTRODUCTION

In this section we describe the basic Turing machine model, present a convenient means for describing the internal structure of a Turing machine, and illustrate some of the algorithms that can be executed by Turing machines. Later in the chapter we will consider certain variations on the basic model and will examine ways in which one Turing machine can simulate the behavior of another. In Chapter 3 we will investigate the use of Turing machines for the evaluation of number-theoretic functions.

Basic Definitions

A *Turing machine* consists of an indefinitely long tape coupled to a finite control unit, as indicated in Fig. 2-1. The tape, which acts as the machine's memory, is ruled off into squares. Each square may be inscribed with a single symbol from a designated finite alphabet, or it may be blank. The various tape symbols will usually be represented by numerals or capital letters, with the restriction that the numeral "0" will be reserved to denote a blank square. The

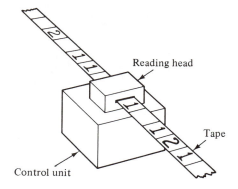

Reading head

Tape

Control unit

Figure 2-1

control unit can shift the tape back and forth through the reading head, but at any given time the reading head is able to examine, or *scan*, just one tape square.

The control unit of a Turing machine is capable of assuming any one of a fixed, finite number of configurations, or *states*, usually denoted q_0, q_1, q_2, etc. The state of the control unit at a given time, together with the tape symbol being scanned by the reading head, uniquely determines how the machine will behave at that time. The actions available to a Turing machine are quite limited; it may either *halt*, thereby terminating its operation, or it may carry out a basic *move*. Each move consists in writing a symbol in the currently scanned tape square, shifting the tape one square to the left or right, and causing the control unit to enter a new state. Of course, the symbol that the machine writes on its tape need not differ from the symbol that is already there, and the new state need not differ from the current state.

The operation of a Turing machine is characterized by the following sequence of events. The machine is initially supplied with a tape in which some finite number of squares are inscribed with symbols and the rest are left blank. Some predesignated square of the tape is placed under the reading head, and the control unit is made to assume some predesignated *starting state*. The machine then goes through a *computation* consisting of a sequence of basic moves. This computation may continue indefinitely, or it may terminate after some finite number of moves. If it does terminate, the pattern of symbols left on the tape is taken to be the outcome of the computation.

The action that a Turing machine takes at a given step of a computation depends on the specific combination of internal state and scanned symbol that it experiences at that step, and on the way that particular machine has been designed. The design specifications of a Turing machine can be compactly described in a *state table*. Figure 2-2 shows the state table for a simple Turing machine having four internal states q_0, q_1, q_2, q_3 and designed to operate on

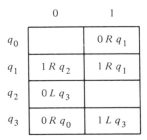

Figure 2-2

tapes in which each square is either blank or else contains the symbol 1. Note that the states are listed down the side of the table, while the tape symbols—including the symbol 0 for a blank square—are listed across the top. Unless otherwise specified, the starting state will be first state in the list; thus in the present example q_0 is the starting state.

Each entry within the table corresponds to a particular combination of current state and scanned symbol. If this combination is one for which the machine is to halt, the entry is left blank. If it is one for which the machine is to make a move, then the entry lists the new tape symbol that the machine is to write, the direction in which the tape is to be shifted, and the new state that the machine is to assume, in that order. Because it is usually most convenient to think of the tape as remaining fixed and the reading head moving back and forth on it, the directional entries in the table indicate the manner in which the *reading head* is to move. Thus the entry "$1 R q_2$" requires the machine to write the symbol 1, move its reading head one square to the right (i.e., shift its tape one square to the left), and enter state q_2.

Once the nature of a state-table description is understood, it is a routine matter to use such a table to trace out the operation of a Turing machine. In doing so, it is convenient to represent the status of a computation by exhibiting the current tape pattern and underlining the currently scanned symbol. Unless otherwise specified, all squares to the left and right of the indicated tape segment are assumed to be blank.

Example 2-1. Suppose that the Turing machine described in Fig. 2-2 is presented with a tape containing the pattern 1 1 1 0 0 0 1 1, preceded and followed by blanks. Assume that the machine is initially in state q_0, and that the tape is positioned so that the left-most 1 is under the reading head, as indicated in line (a) of Fig. 2-3. What computation will ensue?

According to the entry in row q_0, column 1 of the state table, the machine's first move consists in erasing the 1 in the scanned square, shifting the reading head one square to the right, and entering state q_1. Since it now scans another

(a) $\cdots 0\ 0\ \underline{1}\ 1\ 1\ 0\ 0\ 0\ 1\ 1\ 0\ 0\ \cdots$

(b) $\cdots 0\ 0\ 0\ 1\ 1\ 1\ \underline{0}\ 0\ 1\ 1\ 0\ 0\ \cdots$

(c) $\cdots 0\ 0\ 0\ \underline{1}\ 1\ 1\ 0\ 0\ 1\ 1\ 0\ 0\ \cdots$

(d) $\cdots 0\ 0\ 0\ 0\ \underline{1}\ 1\ 1\ 0\ 1\ 1\ 0\ 0\ \cdots$

(e) $\cdots 0\ 0\ 0\ 0\ 0\ 1\ 1\ 1\ \underline{1}\ 1\ 0\ 0\ \cdots$

Figure 2-3

1, its next move is governed by the entry in row q_1, column 1. According to that entry, the machine leaves the currently scanned 1 unchanged, shifts one more square to the right, and remains in state q_1. This operation is repeated at the next step, so that the machine is still in state q_1 when it reaches the first blank square following the initial block of 1's. According to the entry in row q_1, column 0, the machine writes a 1 in this square, shifts one more square to the right, and enters state q_2. The tape now has the appearance shown in line (b) of Fig. 2-3.

Since the machine is now scanning a blank square, it enters state q_3 and moves one square to the left, at which point it scans a 1. As long as it continues to scan 1's, it remains in state q_3 and continues to move to the left, leaving the 1's that it passes over unchanged. When it reaches the blank square at the left end of the block of 1's, it moves one square back to the right and enters state q_0. The tape now has the appearance shown in line (c) of Fig. 2-3.

What the machine has accomplished so far is to shift the left-hand block of 1's one square to the right. Since it is now in essentially the same situation it was in at the beginning of its operation, it will repeat the actions described above. That is, it will erase the first 1 in the left-hand block, add a new 1 at the right end of the block, and shift one more square to the right. Upon finding that next square to be blank, it will return to the left, leaving the tape pattern shown in line (d) of Fig. 2-3. Once more it will erase the first 1 in the left-hand block, add a new 1 at the right end, and shift one square to the right. Now, however, the tape will have the appearance shown in line (e) of Fig. 2-3, the machine will be in state q_2, and it will be scanning a 1. Since the entry in row q_2, column 1 of the state table is blank, the machine now halts.

Once the sequence of moves illustrated in Fig. 2-3 is understood, it is easy to give a general description of the behavior of the machine of Fig. 2-2. When presented with a tape pattern containing two or more blocks of 1's, and when started in state q_0 with the left-most 1 under its reading head, the machine will shift the left-hand block of 1's to the right until it abuts the second block of 1's. When presented with a tape pattern containing a single block of 1's and made to scan the left-most member of that block, the machine will continue indefinitely to shift that block to the right, never halting. What the machine will do when started on a blank square, or somewhere within a block of 1's, is left to the reader to determine.   □

It should be clear from the preceding example that a Turing machine can be thought of as embodying an algorithm for converting one string of symbols into another. The given string is initially inscribed on the machine's tape, and the tape is positioned appropriately in the reading head. The machine is then allowed to carry out its computation, as dictated by its fixed internal structure. If this computation terminates, the outcome of the algorithm is the resulting tape pattern; if the computation does not terminate, the outcome of the algorithm is undefined.

State Diagrams

Although state tables provide compact descriptions of Turing machines, they do not always provide the most convenient vehicle for tracing out a machine's computation. For this purpose a graphical description is usually preferred. We now present a type of graphical description very much like the state diagrams used to represent finite-state machines.

Almost all the Turing machines that we will study have the property that entry into any given state is always accompanied by motion in the same direction. The machine of Fig. 2-2, for example, has this property. Consider state q_1. This state can be entered either from state q_0 or from state q_1 itself, but in each case the reading head moves to the right. Similarly, entry into state q_0 or state q_2 is always accompanied by motion to the right, while entry into state q_3 is always accompanied by motion to the left.

Of course, not all Turing machines have this property. But a machine that does not can always be converted into one that does by adding one or more new states. Specifically, a state that can be entered with motion either to the left or right can be replaced by two equivalent states, one of which is entered whenever motion to the left is required, the other when motion to the right is required. Thus in the discussion to follow we will assume that we are dealing with machines in which each state has associated with it a single direction of motion.

Such a machine can be described by a graph in which states are represented by nodes, or circles, and basic machine moves are represented by arrows from one node to another. To indicate the direction of motion associated with a given state, we write the symbol L or R inside the corresponding node. To indicate that the combination of current state q and scanned tape symbol s causes the machine to write the new symbol s' and to enter state q', we draw an arrow from node q to node q' and label it s/s'. As an illustration, we note that the machine of Fig. 2-2 can be described by the graph of Fig. 2-4(a). Consider, for example, the arrow from node q_1 to node q_2. Since this arrow is labeled

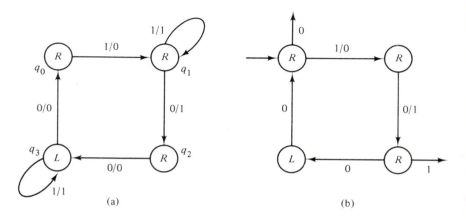

Figure 2-4

"0/1," we infer that when the machine is in state q_1 and scans a blank square it will write a 1 in that square and enter state q_2. Since the directional letter associated with node q_2 is R, the machine will shift to the right at this step in its computation.

In practice, a diagram like that of Fig. 2-4(a) can be simplified in several ways. First, as long as the starting state is properly identified, it is not usually necessary to "name" the various state nodes. As shown in Fig. 2-4(b), we will usually indicate the starting state by means of a short, unlabeled, incoming arrow, and omit the state names q_0, q_1, etc. Second, when the new tape symbol associated with a given arrow is the same as the old tape symbol, we will omit the new symbol. Thus we will write "0" instead of "0/0." (Again, see Fig. 2-4b.) Third, whenever a basic move results in neither a change of tape symbol nor a change of state, we will omit the corresponding arrow entirely. Thus in Fig. 2-4(b) we have deleted the "self-loops" originally found at nodes q_1 and q_3. Of course, once we do this, it becomes necessary to indicate explicitly the situations in which the machine is to halt. This is accomplished by providing a short outgoing arrow, labeled with the appropriate scanned tape symbol, for each situation in which the machine is to halt.

Diagrams like that of Fig. 2-4(b) are called *state diagrams*. As will become apparent, state diagrams provide convenient shorthand descriptions of Turing machines and are useful in both the analysis and design of Turing machines.

Example 2-2. Consider the Turing machine described by the state diagram of Fig. 2-5, where state names have been retained to facilitate the discussion. Suppose that this machine is supplied with a tape containing the pattern 111 (preceded and followed by blanks) and made to scan the left-most 1, as shown in

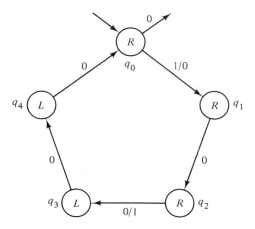

Figure 2-5

Fig. 2-6(a). According to the state diagram, the machine starts in state q_0. Since it initially scans a 1, and since there is an arrow labeled " 1/0 " leading from node q_0 to node q_1, the machine's first move consists of erasing the scanned 1, entering state q_1, and shifting to the right. (Remember that this right shift represents the direction associated with the *new* state q_1, not the old state q_0.) The machine now scans another 1. Since there is only one arrow leaving node q_1, and since that arrow is labeled "0," we see that the machine will remain in state q_1 and continue to move to the right (without changing any tape symbols) until it reaches a blank square. Since the arrow from q_1 to q_2 does not indicate a new tape symbol, the machine does not write anything in that blank square but simply moves one more square to the right and enters state q_2. The tape now has the appearance shown in Fig. 2-6(b). The machine next writes a 1 and shifts to the left, entering state q_3. Without changing any further tape symbols, it continues to move to the left via states q_3 and q_4 until it has encountered a total of two more blank squares, at which point it enters state q_0 and moves one

(a) $\cdots 0\ 0\ \underline{1}\ 1\ 1\ 0\ 0\ 0\ 0\ 0 \cdots$

(b) $\cdots 0\ 0\ 0\ 1\ 1\ 0\ \underline{0}\ 0\ 0\ 0 \cdots$

(c) $\cdots 0\ 0\ 0\ \underline{1}\ 1\ 0\ 1\ 0\ 0\ 0 \cdots$

$\cdots$

(d) $\cdots 0\ 0\ 0\ 0\ 0\ \underline{0}\ 1\ 1\ 0 \cdots$

Figure 2-6

square to the right. (See Fig. 2-6c.) The reader may trace out the remainder of the computation and verify that the machine halts in state q_0 with its tape in the configuration shown in Fig. 2-6(d).   □

Using state diagrams as our basic descriptive device, we will now consider several examples of the ways in which Turing machines can be used to process simple strings of symbols. These examples will help develop some feeling for the kinds of computations that Turing machines can perform and will provide building blocks with which more complicated Turing machines can be designed. When describing string-processing problems, it is common to use a^n as an abbreviation for a block of n consecutive appearances of the symbol a. Although we do not need this notation here, it appears frequently in the problems.

Copying Machines

As a first example, suppose we wish to design a Turing machine with the following characteristics. The machine is to be presented with a tape that contains one or more blocks of consecutive 1's, adjacent blocks being separated by one or more blank squares. One of the blocks is *marked* by the symbol A inscribed in the square immediately to its left. The machine is to begin its computation by scanning this A. Somewhere to the right of the A is a second marking symbol B, and the machine is required to form a copy of the block marked by the A immediately to the right of the B. Thus if the machine is presented with the tape pattern:

$$\ldots 0 \underline{A} 1 1 1 0 1 1 0 0 1 \ldots 0 1 1 1 0 B 0 0 0 0 0 1 1 \ldots$$

it should ultimately produce the pattern:

$$\ldots 0 A 1 1 1 0 1 1 0 0 1 \ldots 0 1 1 1 0 B 1 1 1 0 0 1 1 \ldots$$

We assume that there is sufficient blank space to the right of the B to contain the desired block. The machine is required to halt scanning the symbol A.

Note that such a machine cannot simply scan through the marked block, counting up the number of 1's as it goes, and then move over to the B and write down the desired new block. To do this, it would have to do its "counting" by progressing from one internal state to another. Since there is no limit on the number of 1's to be found in the marked block, such an approach would require the machine to have an infinite number of internal states.

What the machine must do instead is copy the marked block a little bit at a time. Indeed, the simplest strategy is for the machine to copy just one symbol at a time, "checking off" each symbol in the marked block as it is copied. There

(a) 0 $\underline{A}$ 1 1 1 0 1 1 0 0 1 $\cdots$ 0 1 1 1 0 B 0 0 0 0 0 1 1 $\cdots$

(b) 0 A X $\underline{1}$ 1 0 1 1 0 0 1 $\cdots$ 0 1 1 1 0 B 1 0 0 0 0 1 1 $\cdots$

(c) 0 A X X $\underline{1}$ 0 1 1 0 0 1 $\cdots$ 0 1 1 1 0 B 1 1 0 0 0 1 1 $\cdots$

(d) 0 A X X X $\underline{0}$ 1 1 0 0 1 $\cdots$ 0 1 1 1 0 B 1 1 1 0 0 1 1 $\cdots$

(e) 0 $\underline{\underline{A}}$ 1 1 1 0 1 1 0 0 1 $\cdots$ 0 1 1 1 0 B 1 1 1 0 0 1 1 $\cdots$

Figure 2-7

are several slightly different ways of doing this. We more or less arbitrarily choose the following.

1. Mark the first 1 in the block to be copied by replacing it with the symbol X.
2. Move toward the right to the first blank square beyond the symbol B, and write a 1 in that square.
3. Return to the left until the symbol X is encountered, then shift one square to the right.
4. If this square contains a 1, replace that 1 by an X and continue with step (2). If this square is blank, continue with step (5).
5. Move toward the left, changing X's into 1's, until the symbol A is reached, at which point halt.

Figure 2-7 shows how this strategy works for a typical tape pattern. Line (a) shows the initial pattern. Line (b) shows the appearance of the tape after steps (1) through (3) have been carried out. Since the machine is now scanning a 1, it starts a new "pass" during which it replaces the 1 by an X and adds one more 1 to the new block. Lines (c) and (d) show the appearance of the tape after step (3) in each of the next two passes. After the third pass, the machine will be scanning a blank square, which indicates that the basic copying operation has been completed. The machine therefore moves to the left, converting X's to 1's, until it reaches the symbol A and halts, as shown in line (e).[†]
In trying to design a machine that implements this strategy, it is convenient to begin by considering what the machine must do during a typical pass. As shown in Fig. 2-7, the machine begins a typical pass by scanning a 1. It must replace this 1 by an X, move toward the right until it finds the symbol B, move further to the right until it finds the first blank square beyond the B, and then write a 1 in that square. This sequence of events can be realized by the fragmentary state diagram shown in Fig. 2-8(a). The machine must next move to the left

---

[†] Alternatively, the machine might be designed to restore each 1 at the end of the pass in which that 1 is copied.

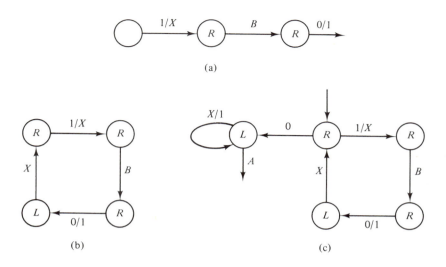

Figure 2-8

until it reaches an X, then move one square to the right, at which point it should be ready to begin the next pass. To accomplish these additional steps, we add a left-moving state to Fig. 2-8(a), arrange for the tape symbol X to cause a transition back to the initial state, and make that state a right-moving state, as shown in Fig. 2-8(b).

Two things remain to be done: make sure that the first pass gets started correctly, and make sure that the computation terminates correctly after the last pass. The former is easy; all that is necessary is to choose the state in the upper left-hand corner of Fig. 2-8(b) to be the starting state. To accomplish the latter, we note that the end of the last pass is signaled when the machine returns to the starting state and scans a blank. (See Fig. 2-7d.) The machine must then shift to the left, changing X's to 1's as it goes, until it reaches the symbol A, at which point it must halt. These last steps are implemented by the added transitions shown in Fig. 2-8(c).

The reader should study the completed diagram of Fig. 2-8(c) until he is convinced that the machine behaves as described. It is also instructive to try to figure out what the machine will do when presented with tape patterns that do not have the specified format of a single A, a single B to the right of the A, and sufficient blank space beyond the B in which to write the new block.

The machine of Fig. 2-8 represents only one of many copying operations that Turing machines can perform. In Section 2.3 we will need a machine that copies a given block into a suitably marked location to the *left* of that block, rather than to the right, as was done above. The given block is to be marked with the symbol B, and the new location is to be marked with the symbol A. Thus a typical initial tape pattern might be

$$\ldots A\ 0\ 0\ 0\ 0\ 0\ 0\ 1\ 1\ 0\ 1\ 1\ 1\ 0 \ldots 0\ 1\ 0\ 0\ B\ 1\ 1\ 1\ 1\ 0\ 0 \ldots .$$

The machine is to begin by scanning the symbol B and is to copy the block of 1's found immediately to the right of the B into the space immediately to the right of the A, so as to produce the pattern

$$\ldots A\ 1\ 1\ 1\ 1\ 0\ 0\ 1\ 1\ 0\ 1\ 1\ 1\ 0 \ldots 0\ 1\ 0\ 0\ B\ 1\ 1\ 1\ 1\ 0\ 0 \ldots .$$

The machine is then to halt scanning the symbol A.

A machine that performs this "right-to-left" copying may be made to operate in much the same way as the copying machine of Fig. 2-8. That is, the machine may be made to mark the first 1 to the right of the B, add a 1 to the right of the A, mark the next 1 to the right of the B, add another 1 to the right of the A, and so on until the copying is completed. Figure 2-9 shows the state diagram of a machine that implements this strategy, as the reader may verify.

Still other types of copying operations are possible. For example, it may be desired to design a machine that copies a pattern made up of several different symbols, rather than a block made up entirely of 1's. Again the step-by-step, one-symbol-at-a-time approach is appropriate. Only two new features must be added. First, between the marking of a symbol in the old pattern and the writing of a symbol in the new pattern, the machine must remember—by means of an appropriate internal state—just which symbol it is to add to the new pattern. Second, some means must be provided for correctly restoring the marked squares to their original condition. Perhaps the easiest way of doing this is to use a different marking symbol for each of the different symbols that might appear in the pattern to be copied. The details of these more general copying operations are left as problems for the reader.

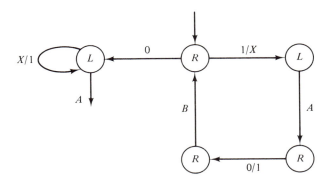

Figure 2-9

Matching Machines

Suppose we need to design a machine that is to be presented with a tape pattern containing two marked blocks of 1's and that is to take different courses of action according as the lengths of the two blocks match or differ. Evidently the critical part of the whole computation is the comparison of the two marked blocks. As our next example, we consider the design of a *matching machine* that performs just this comparison. Once such a matching machine has been realized, it can be augmented with additional states that carry out whatever subsequent operations may be desired.

Our matching machine is to be provided with a tape pattern of the form

$$\ldots \underline{A}\ 1\ 1\ 1\ 1\ 0\ \ldots\ B\ 1\ 1\ 1\ 1\ 0\ \ldots ,$$

with the symbol A under the reading head. The machine is eventually to halt with the same symbol under the reading head and with exactly the same tape pattern with which it started. However, the final state of the machine is to indicate whether or not the lengths of the blocks of 1's following the A and B are equal. In other words, the machine is to halt in one predesignated state if the lengths of the blocks are equal and in another predesignated state if the lengths of the blocks are not equal.

The strategy to be followed by this machine closely resembles that of the copying machines discussed earlier. The machine makes a series of passes across its tape, on each pass marking one symbol in the A block and one symbol in the B block. As usual, this marking is done by replacing the symbols in question with X's. If the members of the two blocks are exhausted during the same pass, the two blocks must have the same length; if one block is exhausted before the other, the blocks must have different lengths.

Figure 2-10 shows the effects of successive passes for a typical initial tape pattern, while Fig. 2-11 shows a portion of a state diagram that will execute these passes. Note that each pass begins with the machine scanning the symbol A. The machine then moves to the right until it encounters the first 1, which it changes to an X. It keeps on moving to the right until it passes the B and

(a)	Initial tape	$\underline{A}\ 1\ 1\ 1\ 1\ 0\ \cdots\ B\ 1\ 1\ 1\ 1\ 0\ \cdots$
(b)	After first pass	$\underline{A}\ X\ 1\ 1\ 1\ 0\ \cdots\ B\ X\ 1\ 1\ 1\ 0\ \cdots$
(c)	After second pass	$\underline{A}\ X\ X\ 1\ 1\ 0\ \cdots\ B\ X\ X\ 1\ 1\ 0\ \cdots$
(d)	After third pass	$\underline{A}\ X\ X\ X\ 1\ 0\ \cdots\ B\ X\ X\ X\ 1\ 0\ \cdots$
(e)	After fourth pass	$\underline{A}\ X\ X\ X\ X\ 0\ \cdots\ B\ X\ X\ X\ X\ 0\ \cdots$

Figure 2-10

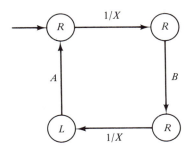

Figure 2-11

reaches the first 1 beyond the *B*. This 1 it also changes to an *X*, at which point it returns to the symbol *A*.

There are two situations in which the series of passes should be interrupted and the computation brought to a close. If the *B* block is shorter than the *A* block, then in the course of some pass the machine will be unable to find an unmarked 1 in the *B* block. In other words, the machine will be moving right in the state represented by the lower right-hand node in Fig. 2-11, and will encounter a blank square rather than a 1. In this event, the machine should return to the left, changing *X*'s back to 1's, until it reaches the symbol *A*, at which point it should halt in a state that indicates that the lengths of the two blocks do not match.

If, on the other hand, the *B* block is at least as long as the *A* block, then at the beginning of some pass the machine will be unable to find an unmarked 1 in the *A* block. (This is the situation shown in line (e) of Fig. 2-10). In other words, the machine will be moving right in the state represented by the upper left-hand node in Fig. 2-11 and will encounter a blank square rather than a 1. When this happens, the machine must determine whether there are any unmarked 1's left in the *B* block. If there are, the blocks do not match; if there are not, the blocks do match. In either case the machine should change *X*'s back to 1's, move left to the symbol *A*, and halt in the appropriate state.

Figure 2-12 shows how the embryonic diagram of Fig. 2-11 can be completed to provide for the proper termination of the matching process and the restoration of the tape pattern.

More elaborate matching operations are also possible. For example, one might wish to design a machine that determines whether or not two patterns, each made up of several different kinds of symbols, are identical. Again a symbol-by-symbol comparison is appropriate, although additional internal states will be needed to "remember" what symbol is currently being matched and it will be convenient to use a different marking symbol for each symbol that might appear in one of the given patterns. We will not pursue the details of such matching machines here.

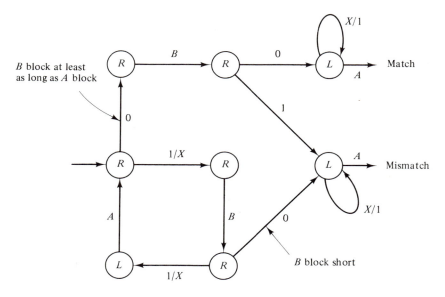

Figure 2-12

Block Substitution

It is often convenient to represent data on a Turing machine tape in the form of blocks of 1's separated by single blanks. For example, the pattern 0 0 1 1 1 0 1 0 1 1 0 1 1 1 1 0 0 might be used to represent the series of numbers 3, 1, 2, 4. The problem we consider next is that of modifying such a sequence of blocks by replacing one of its members by some new block. Of course, if the new block is longer than the one it replaces, the pattern will have to be expanded to make room for it; and if it is shorter than the block it replaces, the pattern will have to be contracted to preserve the single-square separation between blocks.

To define the problem more precisely, let us suppose we are dealing with tape patterns having the following features. At the left end of the pattern is a block of 1's that is marked by being preceded by the symbol B. Somewhere further to the right (after an arbitrary intervening pattern) is a series of blocks of 1's, with single blank squares between adjacent blocks. One of these blocks is marked by the presence of the symbol C immediately to its left. We wish to design a Turing machine that replaces the block marked by the C with a copy of the block marked by the B. In the process, it is to shift the blocks originally lying to the right of the C block as needed to make room for the new block or to preserve the single-blank spacing. Thus if the machine is applied to the tape pattern

... B 1 1 1 1 0 ... 0 0 1 1 1 C 1 1 1 0 1 0 1 1 1 0 1 1 1 1 0 0 ...

it should produce the pattern

  ...$\underline{B}$ 1 1 1 1 1 0 ... 0 0 1 1 1 C 1 1 1 1 1 0 1 0 1 1 1 0 1 1 1 1 1 0 0 ....

The machine is to begin and end its computation by scanning the symbol B.

   Evidently the major problem is that of appropriately shifting the portion of the pattern lying to the right of the C block. One possibility is to design a machine that begins by comparing the lengths of the B and C blocks, thereby determining in which direction and how far to shift the blocks in question. Another, perhaps simpler, approach is to divide the shifting operation into two stages. In the first, or *collapsing*, stage the machine removes the C block entirely, shifting the following blocks to the left until just one blank square remains between the C and the next block of 1's. In the second, or *insertion*, stage the machine forms a copy of the B block immediately to the right of the C, pushing the following blocks just far enough to the right to make room for the new block. Thus when presented with the initial tape pattern

  ...B 1 1 1 1 1 0 ... 0 1 1 1 C 1 1 1 0 1 0 1 1 1 0 1 1 1 1 1 0 0 ...,

the machine would first produce the collapsed pattern

  ...B 1 1 1 1 1 0 ... 0 1 1 1 C 0 1 0 1 1 1 0 1 1 1 1 1 0 0 ...

and then insert a copy of the B block so as to obtain the final pattern

  ...B 1 1 1 1 1 0 ... 0 1 1 1 C 1 1 1 1 1 0 1 0 1 1 1 0 1 1 1 1 1 0 0 ....

Although roundabout, this two-stage process is easy to implement, and represents the approach that we shall adopt. We will first design a separate machine for each stage and then combine them to form a single machine that performs the complete computation.

   The collapsing process can be carried out in a number of passes, during each of which the machine shifts the entire pattern lying to the right of the symbol C one square to the left. Figure 2-13 shows the effects of successive passes on a typical tape pattern. The machine begins each pass in its starting state and scanning the symbol C. It first moves to the right until it has come to the end of the series of blocks of 1's, a situation it recognizes by the appearance of two consecutive blank squares. It then starts to move back to the left. When

(a)	Initial pattern	$\cdots \underline{C}$ 1 1 1 0 1 0 1 1 1 0 1 1 1 1 1 0 0 $\cdots$
(b)	After first pass	$\cdots \underline{C}$ 1 1 0 1 0 1 1 1 0 1 1 1 1 1 0 0 $\cdots$
(c)	After second pass	$\cdots \underline{C}$ 1 0 1 0 1 1 1 0 1 1 1 1 1 0 0 $\cdots$
(d)	After third pass	$\cdots \underline{C}$ 0 1 0 1 1 1 0 1 1 1 1 1 0 0 $\cdots$

Figure 2-13

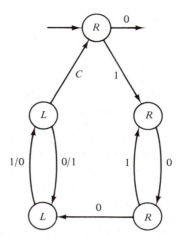

Figure 2-14

it reaches the first 1, it erases that 1, "remembers" (by entering an appropriate internal state) that it was a 1 that was erased, and shifts one square to the left. Here it writes the remembered 1, remembers in turn the symbol displaced by that 1, and shifts to the left again. It continues to work its way to the left, shifting each symbol in the tape pattern one square to the left, until it reaches the C. If the square immediately to the right of the C is now blank, the block that was originally marked by the C has been completely removed, and the

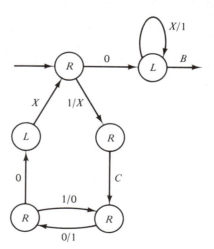

Figure 2-15

series of passes should be terminated. Figure 2-14 shows the state diagram for a machine that accomplishes the collapsing process in the manner just described.

The insertion process can also be carried out in a series of passes. During each pass, the machine marks one new member of the B block with the symbol X, adds one more 1 to the new block being formed next to the C, and shifts the remaining blocks one square to the right. (This shifting is accomplished in the same way as the shifting that took place in the collapsing stage; only the direction is different.) When all the members of the B block have been marked, the series of passes is terminated and the machine moves back to the symbol B, restoring X's to 1's as it goes. Figure 2-15 shows the state diagram for a machine that accomplishes the.insertion process, assuming that the machine begins its operation by scanning the B.

All that remains is to incorporate the machines that perform the collapsing and insertion operations into a single machine that carries out the whole computation. In doing this, we must provide appropriate transitions between the starting point of the overall computation and the starting point of the collapsing operation, as well as between the end of the collapsing operation and the starting point of the insertion operation. Figure 2-16 shows the state diagram for the combined machine. Note that this machine begins its computation by scanning the symbol B. It then moves to the right until it reaches the symbol C, at which point it enters the collapsing stage. Upon completion of the collapsing operation, the machine moves left until it returns to the symbol B, at which point it enters the insertion stage. Since this stage ends with the reading head in the desired final position, no further action is required.

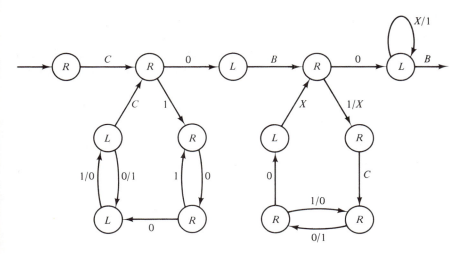

Figure 2-16

Composite Machines

As the preceding example illustrates, the design of a complicated Turing machine is often facilitated by breaking up the desired operation into several parts, each of which can be realized by a relatively simple machine. These "submachines" can then be linked together, with appropriate intermediate states and transitions as needed, to form a machine that performs the desired overall computation. Although this process of combining submachines to form composite machines does not necessarily (or even usually) lead to the most compact realizations, it is often the easiest way to demonstrate that it is possible to design a Turing machine that performs a certain task. For our purposes, establishing the existence of a machine that carries out a given computation is more important than finding the simplest or most elegant machine for the job.

The role of a submachine in a composite machine is much like that of a subroutine in a larger program. Just as a program can call one subroutine several times, so a composite machine can make use of one submachine several times. We illustrate this important possibility with an example.

Consider first the design of a *doubling machine*, which is to behave as follows. When presented with a tape pattern containing a single block of 1's and made to scan the first member of that block, the machine is to extend the block to double its original length. Thus if presented with the pattern

$$\dots 0\ 0\ \underline{1}\ 1\ 1\ 1\ 0\ 0\ 0\ 0\ 0\ 0 \dots$$

the machine is to produce the pattern

$$\dots 0\ \underline{0}\ 1\ 1\ 1\ 1\ 1\ 1\ 1\ 1\ 0\ 0 \dots$$

and is to halt as shown.

The desired doubling operation can be viewed as a special form of copying, in which a copy of the given block of 1's is to be formed immediately to the right of that block, without any intervening blank squares. Thus doubling can be accomplished by a series of passes, as illustrated in Fig. 2-17. During each

(a)	Initial pattern	$\cdots 0\ 0\ \underline{1}\ 1\ 1\ 1\ 0\ 0\ 0\ 0\ 0\ 0 \cdots$
(b)	First pass	$\cdots 0\ 0\ X\ 1\ 1\ 1\ Y\ 0\ 0\ 0\ 0\ 0 \cdots$
(c)	Second pass	$\cdots 0\ 0\ X\ X\ 1\ 1\ Y\ Y\ 0\ 0\ 0\ 0 \cdots$
(d)	Third pass	$\cdots 0\ 0\ X\ X\ X\ 1\ Y\ Y\ Y\ 0\ 0\ 0 \cdots$
(e)	Fourth pass	$\cdots 0\ 0\ X\ X\ X\ X\ Y\ Y\ Y\ Y\ 0\ 0 \cdots$
(f)	Final pattern	$\cdots 0\ 0\ \underline{0}\ 1\ 1\ 1\ 1\ 1\ 1\ 1\ 1\ 0\ 0 \cdots$

Figure 2-17

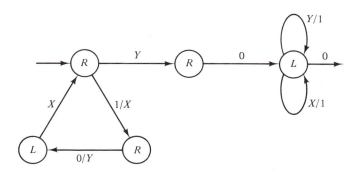

Figure 2-18

pass, the machine marks one additional member of the given block with an X and adds one additional Y at the right end of the block. After all the original 1's have been marked with X's, the series of passes is terminated, and the machine converts all the X's and Y's into 1's and halts. (Note that the use of the intermediate symbol Y is essential. If 1's were added to the end of the original block, the machine would have no way of determining when the original block had been exhausted.)

Figure 2-18 shows the state diagram for a machine that carries out the doubling process. In the discussion that follows, it will be important to understand three things about the behavior of this doubling machine: (1) the machine performs its doubling by extending the given block of 1's to the right; (2) the machine begins by scanning the first member of the block and ends by scanning the blank square immediately to the left of the block; and (3) at no time during its computation does the machine move to the left of that blank square.

We now undertake a somewhat more complex task: the design of an *exponentiating machine*. When presented with a tape pattern consisting of a single block of n 1's, this machine is to erase that block and produce a new block containing exactly 2^n 1's. The machine is to begin its computation by scanning the blank square immediately to the left of the given block and is to end its computation by scanning the blank square immediately to the left of the new block. The position of the new block with respect to the given block is not important; in particular, the new block need not begin in the same square as the given block. Thus if the exponentiating machine is presented with the tape pattern

$$\ldots 0\,\underline{0}\,1\,1\,1\,0\,0\,0\,0\,0\,0\,0\,0\,0\,0\,0\,\ldots ,$$

its final tape pattern might be

$$\ldots 0\,0\,0\,0\,0\,\underline{0}\,1\,1\,1\,1\,1\,1\,1\,1\,0\,0\,0\,\ldots .$$

(a) Original pattern $\cdots$ 0̲ 1 1 1 0 0 0 0 0 0 0 0 0 0 0 $\cdots$

(b) After initializing $\cdots$ 0 1̲ 1 1 0 1 0 0 0 0 0 0 0 0 $\cdots$

(c) After first doubling $\cdots$ 0 0 1̲ 1 0 1 1 0 0 0 0 0 0 0 $\cdots$

(d) After second doubling $\cdots$ 0 0 0 1̲ 0 1 1 1 1 0 0 0 0 0 $\cdots$

(e) After third doubling $\cdots$ 0 0 0 0 0̲ 1 1 1 1 1 1 1 1 0 0 $\cdots$

Figure 2-19

Note that exponentiation can be accomplished by repeated doubling: if a block of length one is successively doubled n times, the result will be a block of length 2^n. Figure 2-19 shows how this fact can be used to design an exponentiation machine. Suppose that the machine is presented with a tape pattern containing a single block of n 1's, as shown in line (a). The machine first "initializes" its tape by writing a single 1 just beyond the original block, as shown in line (b). This 1 is to form the nucleus of a new block of length 2^n. The machine now proceeds to erase the members of the original block one at a time, doubling the length of the new block after each erasure, as shown in lines (c), (d), and (e). When the original block has been completely erased, the new block will consist of 2^n 1's, at which point the machine halts.

The initializing part of the operation is easily implemented. The basic cycle of erasing and doubling is also easy to implement if we make use of the doubling machine of Fig. 2-18 as a submachine. In order to do this, of course, we must make sure that the reading head is properly positioned before the doubling operation begins.

The state diagram for a machine that carries out the entire exponentiation process is shown in Fig. 2-20, where the box labeled "Doubling machine" represents a copy of the machine of Fig. 2-18. The chain of arrows leading from the starting state to the state marked by the asterisk accomplishes the initializing operation and returns the machine to the first 1 in the original block. (See line (a) and (b) in Fig. 2-19). The machine then erases that 1 and moves to the right until it encounters a blank square. At this point the machine is in the correct position to begin doubling the new block, and so should enter the starting state of the doubling submachine.[†] When the doubling has been accomplished, the reading head will be scanning the blank square that precedes the new block. To complete the cycle, the composite machine moves left to the end of the original block and reenters the state marked by the asterisk, ready to erase the next 1 and initiate a new doubling. This process

---

[†] Entry into the doubling machine is facilitated by the fact that the direction of motion assigned to the starting state of that machine coincides with the direction in which the machine moves when initially made to scan a 0.

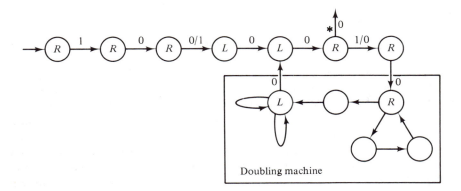

Figure 2-20

is repeated until the original block has been completely erased, at which point the machine halts in the state marked by the asterisk.

This completes our collection of simple Turing machine examples. Although they perform rather mundane operations, these machines illustrate some of the techniques that permit the evaluation of exceedingly complex functions. The following sections will frequently refer back to these simple examples and to the techniques that they embody.

2.2 VARIATIONS ON THE TURING MACHINE MODEL

In this section we will examine several ways in which the basic Turing machine model might be modified. The modifications to be considered include both generalizations—such as providing a machine with more than one tape—and restrictions—such as limiting the number of tape symbols that a machine may use. The primary reason for studying such modifications is to determine how they affect the computing capabilities of the Turing machine model. Ignoring questions of efficiency, we will see that none of the apparent generalizations increases the computing power of the basic model and that many of the apparent restrictions fail to reduce the computing power of the basic model. We are therefore free to employ whichever variant of the basic model is most convenient for a given application.

In showing that certain modifications do not affect the ultimate capabilities of Turing machines, we will be making repeated use of the technique of simulation. By this term we mean the process whereby one type of machine imitates, in a step-by-step way, the computations performed by a second type of machine. Simulation provides the most common way of showing that one computational model is as powerful as another, and we will be using it again in later chapters.

Multitrack Tapes

Let us begin by considering a simple generalization involving the format of the tapes used by Turing machines. In many cases it is convenient to work with machines whose tapes are divided lengthwise into two layers, or *tracks*, as shown in Fig. 2-21. Every square on such a tape consists of two halves, each of which may be independently inscribed with a single symbol. We assume that the machine's reading head is able to scan both halves of a given tape square simultaneously. A basic machine move consists in examining the symbols found in the two halves of the scanned square, writing a new symbol in each half, shifting the tape one square to the left or right, and entering a new state. Machines that operate on such *two-track tapes* will be referred to as *two-track machines*.

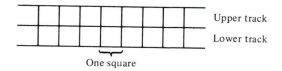

One square

Figure 2-21

In order to describe the behavior of a two-track machine, it is necessary to specify exactly what move the machine will make for every possible combination of the machine's current state, the symbol found in the upper half of the scanned square, and the symbol found in the lower half of the scanned square. As a simple example, Fig. 2-22(a) shows the state-table description of a two-track machine in which each half of a square may either contain the symbol 1 or else be blank. Here the notation $\frac{0}{0}$ represents the situation in which both halves of the scanned square are blank, $\frac{1}{0}$ represents the situation in which the top half of the scanned square contains a 1 and the bottom half is blank, and so on. In general, if a machine operates on a two-track tape in which the top half of each square may contain any one of n (nonblank) symbols and the bottom half may contain any one of m (nonblank) symbols, the state-table description of that machine will require $(n + 1)(m + 1)$ columns.

It should be clear from the preceding remarks and from Fig. 2-22(a) that the use of a two-track tape is only an artifice. Any two-track machine can be viewed as a one-track machine whose tape symbols are ordered pairs, one member of each pair being an upper-track symbol and the other member being a lower-track symbol. If each ordered pair is replaced by an individual symbol, the resulting machine description cannot be distinguished from that of a one-track machine. For instance, replacing the symbol pairs $\frac{0}{0}, \frac{1}{0}, \frac{0}{1}, \frac{1}{1}$ by the respective symbols 0, 1, 2, 3 converts the "two-track machine" of Fig. 2-22(a) into the "one-track machine" of Fig. 2-22(b). Thus whether a given Turing machine is

	$\frac{0}{0}$	$\frac{1}{0}$	$\frac{0}{1}$	$\frac{1}{1}$
q_0	$\frac{0}{0} R\, q_0$	$\frac{1}{1} R\, q_1$	$\frac{0}{1} R\, q_3$	$\frac{1}{1} R\, q_0$
q_1	$\frac{0}{1} L\, q_2$	$\frac{1}{0} R\, q_1$	$\frac{0}{1} R\, q_1$	$\frac{1}{1} R\, q_1$
q_2	$\frac{0}{0} L\, q_2$	$\frac{1}{0} L\, q_2$	$\frac{0}{1} L\, q_2$	$\frac{1}{1} R\, q_0$
q_3		$\frac{1}{0} R\, q_3$	$\frac{1}{1} R\, q_3$	$\frac{1}{1} R\, q_3$

(a)

	0	1	2	3
q_0	$0\, R\, q_0$	$3\, R\, q_1$	$2\, R\, q_3$	$3\, R\, q_0$
q_1	$2\, L\, q_2$	$1\, R\, q_1$	$2\, R\, q_1$	$3\, R\, q_1$
q_2	$0\, L\, q_2$	$1\, L\, q_2$	$2\, L\, q_2$	$3\, R\, q_0$
q_3		$1\, R\, q_3$	$3\, R\, q_3$	$3\, R\, q_3$

(b)

Figure 2-22

to be considered a one-track machine or a two-track machine depends on the interpretation assigned to its tape symbols.

Although the distinction between a one-track machine and a two-track machine lies mainly in the mind of the user, that distinction can greatly simplify the description of certain computations. In effect, the use of a two-track tape permits two different patterns to be superimposed in the same region of the tape without damage to the integrity of either. The following example illustrates the usefulness of the two-track interpretation.

Example 2-3. Let M_1 and M_2 be Turing machines with identical tape alphabets. Assume that both machines have been designed so they eventually halt when presented with tape patterns of a certain class and that, when they do halt, their tapes contain just one block of 1's. Suppose that it is desired to design a new machine **T** which is to behave as follows. When presented with one of the input patterns for which M_1 and M_2 were designed, **T** is to carry out—in sequence—the computations that M_1 and M_2 would perform on that pattern, compare the resulting blocks of 1's, and select the longer block to be its own output pattern.

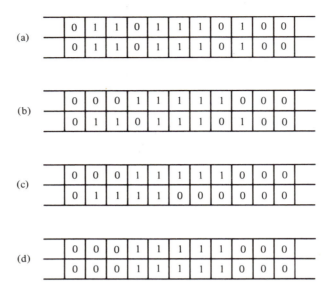

Figure 2-23

If **T** is to do its job successfully, it must save a copy of the input pattern while it is carrying out **M₁**'s computation, so that the input pattern will be available for **M₂**'s computation. There are several ways in which this copy might be saved, such as putting it to one side and trying to keep it out of the way of **M₁**'s computation. But the simplest approach—at least conceptually— is to think of **T** as operating on a two-track tape.

Assume that at the beginning of **T**'s computation, the initial tape pattern is written in both tape tracks, as shown in Fig. 2-23(a). For this purpose it is appropriate to identify the symbol pair $_0^0$ with the single symbol 0 and the pair $_1^1$ with the symbol 1. During the first part of its computation, **T** ignores the symbols in the lower track and carries out the computation of machine **M₁** on the pattern in the upper track, as indicated in Fig. 2-23(b). The machine **T** then proceeds to carry out the computation of **M₂** on the pattern in the lower track, this time ignoring the symbols in the upper track. (See Fig. 2-23c.) When both computations have been completed, **T** compares the resulting blocks of 1's, erases the unwanted block, and leaves the symbols of the desired block written in both tracks, as shown in Fig. 2-23(d). Because the pair $_0^0$ is identified with the single symbol 0 and the pair $_1^1$ with 1, this final tape pattern meets the stated design requirements for the machine **T**.   □

This example gives a brief indication of the conceptual advantages of two-track machines. As might be expected, similar advantages often accrue from dividing tapes into three, four, or more tracks. As long as the number of tracks is finite, a multitrack machine is nothing more than a special representation of a one-track machine. Yet there are many cases in which it is much easier to understand and describe the behavior of a multitrack machine than it is to understand and describe the behavior of the equivalent one-track machine. We shall see several such cases in the remainder of this section.

Multitape Machines

Another way to extend the basic Turing machine model is to provide it with more than one tape. Although the resulting multitape machines are often more convenient to work with than the ordinary one-tape variety, the availability of extra tapes does not result in an increase in ultimate computing power. We will establish this fact for the case of two-tape machines, the generalization to more than two tapes being straightforward.

A *two-tape Turing machine* consists of a finite-state control unit coupled to two different tapes by means of two different reading heads, as shown in Fig. 2-24. A basic move of such a machine consists in writing a new symbol in the currently scanned square of each tape, shifting either or both tapes one square to the left or right, and entering a new internal state. (The two tapes need not be shifted in the same direction.) The particular action that the machine takes at any given step is determined by the machine's current state and the symbols appearing in the currently scanned squares of the two tapes. Thus the structure of a two-tape machine can be specified by a table which lists the machine response associated with each possible combination of internal state and scanned tape symbols.

We now propose to show that given any two-tape Turing machine **M**, it is possible to design an ordinary one-tape machine **T** that simulates the computations of **M**. In order to do this, we must first choose some means of representing both of **M**'s tape patterns on **T**'s single tape. There are many possible representation schemes. For example, **M**'s two patterns might simply be written one after the other on **T**'s tape, with a suitable marking symbol between them. Alternatively, the two patterns might be interleaved, one being written on the

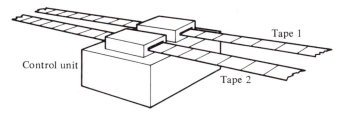

Figure 2-24

	First tape	0	1	1	1	0	1	0	0	

(a) M's tapes:

	Second tape	1	1	0	0	1	1	1	0	

	First pattern	0	1	1	1	0	1	0	0
(b) T's tape:	First marker	0	1	0	0	0	0	0	0
	Second pattern	1	1	0	0	1	1	1	0
	Second marker	0	0	0	0	1	0	0	0

Figure 2-25

even-numbered squares of T's tape, the other on the odd-numbered squares.
Still another approach—the one that we will follow—is to write M's patterns in
separate tracks of T's tape.

Figure 2-25 shows how this is done. We think of T's tape as being divided
into four tracks. The first and third tracks are used to store the current tape
patterns of M's first and second tapes, respectively. These patterns are written
exactly as they are found on M's tapes. The second and fourth tracks of T's tape
are used to record the positions of M's reading heads on their respective tapes.
Each of these tracks is to contain a single 1, located directly under the symbol
currently being scanned on the appropriate tape of M. Thus if M's tapes have
the appearance shown in Fig. 2-25(a), where the underscores indicate the
squares being scanned, then T's tape should have the appearance shown in
Fig. 2-25(b).

Note that we do not try to keep M's tape patterns "lined up" on T's tape
so that the currently scanned symbols are in the same square. Instead, we think
of M's tapes as remaining fixed in space and suppose that M's reading heads
can move back and forth on their tapes independently of one another. Of
course, this is not the only way of accomplishing the desired simulation, but it
is as convenient as any. Investigation of other possible approaches is left to the
reader.

Having decided on a method of representing M's tape patterns on T's tape,
we now turn to the process whereby T simulates M's computations. This
process simply amounts to updating the representation of M's tape patterns in
accordance with the successive moves that M makes. In general, of course, T
will require many moves to reproduce the effect of one of M's moves. We
therefore think of T's computation as being divided into steps, each of which
consists in modifying T's tape record so as to reflect one of M's moves.

In order to update **M**'s tape patterns properly, **T** needs certain auxiliary information. First, it must keep track of **M**'s current state. We assume that **T** carries this information in its own internal state. Thus if **M** has a total of k internal states, **T** will be composed of k almost identical submachines, the particular submachine that **T** is " in " at a given time indicating the current state of **M**. Second, **T** needs to know whether the marking symbol in the fourth track of its tape lies to the right or left of the marking symbol in the second track. This information will also be supplied by **T**'s internal state.

At the beginning of each step in its simulation, **T**'s tape is to contain an up-to-date record of **M**'s patterns, positioned so that the marker symbol in the second track is under **T**'s reading head. Thus at this point **T** knows what symbol **M** is currently scanning on its first tape and, by virtue of its own internal state, what state **M** is currently in. In order to determine what move **M** will make, **T** must first determine what symbol is currently being scanned on **M**'s second tape. **T** accordingly moves its reading head to the square containing the marker in the fourth track. Throughout the course of this motion it must remember, again by means of its internal state, what the scanned symbol on **M**'s first tape is, as well as what internal state **M** is in.

When **T** reaches the second marker, it has enough information to determine exactly what move **M** will make. It can therefore modify the marked symbol of the second tape pattern and shift the marker appropriately. It then returns to the square containing the marker in the second track, modifies the corresponding symbol in the first track, and shifts the marker. Having thus updated the representation of **M**'s tape patterns, **T** enters an internal state that is indicative of **M**'s next state and is ready for the next step in its simulation.

The succeeding steps in the simulation proceed in exactly the same way. If **M** eventually halts, **T** halts at the corresponding step of its simulation, leaving the appropriate representation of **M**'s final tape patterns as its own final tape pattern. Thus as long as **T** is provided with the correctly encoded form of **M**'s initial tape patterns, it will indeed duplicate **M**'s computation.

Of course, the detailed structure of **T** depends on the structure of the machine **M** that is being simulated. And unless **M** is very simple, the state-table description of **T** will be quite complicated. But by this time the reader should be sufficiently experienced in the ways of Turing machines to be able to convince him or herself that, for each two-tape machine **M**, there exists a corresponding machine **T** that simulates the behavior of **M** in the manner outlined above.

Turing machines having more than two tapes can be defined in the obvious way. It should be clear from the preceding discussion that the computations of any such multitape machine can always be simulated by an appropriately designed one-tape machine. Thus while the use of extra tapes may speed up the operation of Turing machines, it does not increase their basic computing capabilities.

Other Generalizations

Several other extensions of the basic Turing machine model can be shown not to result in any change in computing capability. For instance, we might provide an ordinary Turing machine with two or more reading heads operating on the same tape. We think of these reading heads as being attached to the control unit by indefinitely long cables, so they can move back and forth independently of one another. Moreover, we suppose that two heads can scan the same square if necessary, and that some convention exists for determining which head has writing priority. A little thought shows that such a multihead machine can be simulated by an ordinary single-head machine in much the same way that a multitape machine is simulated. Thus the use of more than one head on the same tape does not lead to an increase in computing power. Analogous results hold for machines that can move their reading heads several squares at a time and for machines whose reading heads are allowed to jump in one step to some predesignated square.

Another possible generalization involves the use of multidimensional tapes. Note that a two-dimensional tape may be thought of as an infinite sheet of paper that has been ruled off into squares, like graph paper. In using such a "tape," we suppose that at each step in its computation a Turing machine is able to move its reading head one square up, down, left, or right. As the interested reader may verify, there are several ways in which two-dimensional patterns of symbols can be represented on one-dimensional tapes and kept up to date by an appropriately designed machine. Thus the behavior of a "two-dimensional" Turing machine can always be simulated by some "one-dimensional" Turing machine, from which we conclude that the use of two-dimensional tapes does not provide increased computing power. The same is true of three- or higher dimensional tapes.

In summary, we see that none of the obvious generalizations of the basic Turing machine model lead to an increase in computing power. In each case the equivalence between the basic and modified models can be established by showing that each machine of the more general type can be simulated by a suitably designed machine of the basic type. To be sure, the simulating machine is generally slower and more complicated than the machine that it simulates, and it is generally necessary to encode the tape pattern(s) of the simulated machine in a form that the simulating machine can work with; but because the encodings are in all cases simple to implement, it is still fair to say that the fundamental computing capabilities of the generalizations considered here do not exceed those of the basic Turing machine model introduced in Section 2.1.

Singly Infinite Tapes

The basic Turing machine model can also be restricted in several ways that do not decrease its fundamental computing capability. For our purposes, the most important of these restrictions concerns the unbounded nature of a Turing machine tape.

So far we have assumed that a Turing machine's tape extends indefinitely both to the left and to the right. Such tapes are called *doubly infinite tapes*. In many situations it is more convenient to work with tapes that extend indefinitely far to the right (say), but have a well-defined left end. Such tapes are said to be *singly infinite to the right*. Tapes that are singly infinite to the left can be defined in the obvious way. But we will confine our attention to tapes that extend to the right and will use the term *singly infinite tape* to mean a tape that is singly infinite to the right.

A machine that operates on a doubly infinite tape will be referred to as a *doubly infinite machine*, and one that operates on a singly infinite tape will be referred to as a *singly infinite machine*.[†] If a singly infinite machine is required to move to the left when it is already scanning the left-most square of its tape, it simply runs off the tape, thereby terminating its computation. Since a machine has no built-in mechanism for determining when it is about to run off the end of its tape, it might seem that singly infinite machines are less powerful than doubly infinite machines. But this is not the case; any computation that can be carried out by a doubly infinite machine can also be carried out by a suitably designed singly infinite machine.

To see that this is true, suppose that **M** is an arbitrary Turing machine designed to operate on a (one-track) doubly infinite tape. We wish to design a new machine **T** that operates on a singly infinite tape and yet is able to simulate the computations of **M**. The first step, of course, is to choose a way of representing **M**'s doubly infinite tape patterns on **T**'s singly infinite tape. Of the many possible representation schemes, we more or less arbitrarily adopt the following.

Let the squares of **M**'s doubly infinite tape be numbered as shown in Fig. 2-26(a), where square zero is the initially scanned square. If we think of **T**'s singly infinite tape as being divided into two tracks, the squares of **M**'s tape can be represented by half squares of **T**'s tape in the manner indicated in Fig. 2-26(b). In particular, square zero is represented by the upper half of **T**'s left-most square, the positively numbered squares are represented by the half squares in the upper track of **T**'s tape, and the negatively numbered squares are represented by the half squares in the lower track of **T**'s tape. The lower half of **T**'s left-most square is to contain the special marking symbol $, which will enable **T** to recognize when it has reached the end of its tape.

The machine **T** is to begin its computation with **M**'s initial tape pattern written in the appropriate half squares of its upper and lower tracks and with

---

[†] Although convenient, this terminology is illogical; one cannot tell by looking at a state table or diagram whether the machine it describes is to be thought of as a singly infinite machine or a doubly infinite machine.

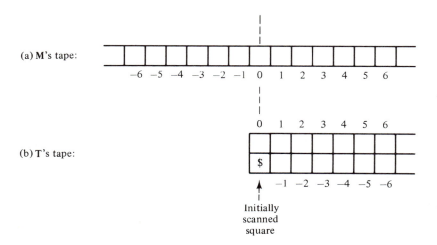

Figure 2-26

the left-most square of its tape under its reading head. Thus if **M** begins a certain computation with the tape pattern shown in Fig. 2-27(a), **T** should begin the corresponding computation with its tape inscribed as shown in Fig. 2-27(b). In effect, **M**'s tape pattern is "folded" at the initially scanned square so that it can be accommodated on **T**'s singly infinite tape.

The machine **T** now proceeds to duplicate the computation performed by **M**. As long as **M** does not move to the left of its initially scanned square, **T** carries out this computation in the upper track of its tape, ignoring the symbols found in the lower track. If and when **M** ventures to the left of its initially scanned square, **T** turns back to the right and begins to operate on the lower track of its tape. As long as **M** remains to the left of the initially scanned square, **T** continues to work in its lower track, shifting right when **M** shifts left, and vice versa. Once **M** returns to the initially scanned square, **T** returns to the upper track. It is, of course, the presence of the special symbol $ that enables **T** to switch back and forth between tracks at just the right times.

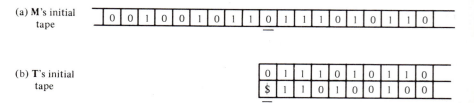

Figure 2-27

The detailed design of **T** obviously depends on the structure of **M**. Basically, however, **T** consists of two "copies" of **M**. One copy executes the portion of the computation that takes place in the upper track. Each state of this copy is assigned the same direction of motion as the corresponding state of **M**, and transitions among states are arranged so that lower-track symbols are ignored while upper-track symbols are operated upon in the same way that **M** operates on its tape symbols. The second copy executes the portion of the computation that takes place in the lower track. Each state is assigned the direction of motion opposite to that of its counterpart in **M**, and transitions are arranged so that upper-track symbols are ignored. Transitions from one copy to the other are triggered by the appearance of the symbol $ in the lower half of the scanned square.

Example 2-4. Suppose that the machine **M** described in Fig. 2-28(a) is designed to operate on a doubly infinite tape. Then the machine **T** described in Fig. 2-28(b) will duplicate the computations of **M** on a singly infinite tape.

The top three states in Fig. 2-28(b) represent the "upper-track copy" of **M**. Solid arrows connecting the states of this copy duplicate the transitions of **M**. To simplify the diagram, we have used the symbol X to stand for either

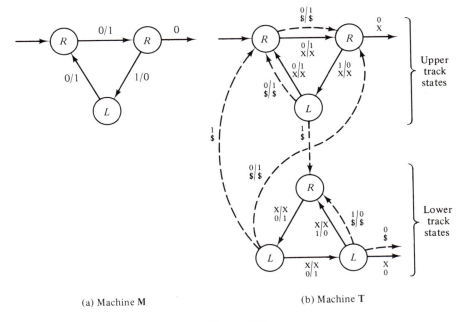

(a) Machine **M**   (b) Machine **T**

Figure 2-28

a 0 or a 1. Thus the label "$\frac{0}{x}/\frac{1}{x}$" indicates that when **T** sees a blank in the upper half of its scanned square and *either* a blank or a 1 in the lower half, it writes a 1 in the upper half and leaves the lower half unchanged. In a similar way, the bottom three states represent the "lower-track copy" of **M**. For convenience, this copy has been drawn upside down; note that the right-moving state of this copy corresponds to the left-moving state of **M**.

The broken arrows in Fig. 2-28(b) represent the transitions occasioned by the symbol $.[†] Note that it is the upper-track symbol appearing in conjunction with the $ that determines what action the machine should take. If this symbol is one that would ordinarily cause **T** to shift right, **T** should simply ignore the $, make the appropriate change in the upper-track symbol, and continue operating in the same track as before. If, however, the symbol appearing in conjunction with the $ is one that would ordinarily cause **T** to shift left, **T** must change tracks and shift right instead. Detailed verification of the diagram of Fig. 2-28(b) is left to the interested reader. □

Since the construction outlined above is a general one that can be applied to any given machine **M**, we conclude that any computation that can be carried out by a doubly infinite machine can also be carried out by a suitably designed singly infinite machine. Thus no loss of computing capability is entailed by restricting our attention to machines that operate on singly infinite tapes. This fact will prove helpful in Section 2.3, where we consider the design of a "universal" Turing machine, and in Chapter 3, where we consider the use of Turing machines for the evaluation of number-theoretic functions.

Of course, the simulation of a doubly infinite computation by a singly infinite machine requires that tape patterns be properly encoded for representation on a singly infinite tape. But this encoding is quite simple and can often be accomplished automatically. Of special importance is the case in which the doubly infinite machine being simulated begins and ends its computation with nothing but blank tape to the left of the initially scanned square. In this case it is appropriate to identify the single symbols 0, 1, 2, ... with the symbol pairs $\frac{0}{0}$, $\frac{1}{0}$, $\frac{2}{0}$, ..., respectively. When this is done, the right-hand half of the initial doubly infinite tape will, except for the missing $, be exactly the pattern that must be supplied to the singly infinite machine. Similarly, the final pattern produced by the singly infinite machine will, except for the extra $, be exactly the right-hand half of the pattern produced by the doubly infinite machine. Thus in this special case the only encoding and decoding required is the insertion and deletion of the special symbol $, both of which are readily accomplished by the simulating machine.

[†] Since the symbol $ appears only in the left-most tape square, **T** can scan this symbol only when it is in its initial state or when it has just moved to the left.

Other Restrictions

Several other restrictions of the basic Turing machine model can be shown not to result in any decrease in computing capability. For example, we might modify the basic model so that at each step in its computation a machine may either write a new symbol or shift its tape, but not both. It is easy to see that, with the use of extra states, a restricted machine of this kind can be made to imitate the behavior of a normal machine. Each step of the normal machine's computation simply requires two steps in the restricted machine's computation—one in which it makes the necessary symbol change and enters an appropriate intermediate state, and one in which it makes the necessary tape shift. Details are left to the reader; see Problem 2.2.6.

Somewhat more interesting is the case in which restrictions are placed on the number of tape symbols that a Turing machine can use. It is natural to suppose that machines having large tape alphabets can perform more complex computations than machines having small tape alphabets. But this is not so. In fact, the computations of any given Turing machine can be simulated by an appropriately designed machine that uses only one nonblank tape symbol.

As usual, the central problem is that of representation. To make matters specific, suppose that the machine **M** uses the tape symbols $0, s_1, \ldots, s_k$ and that we wish to design a machine **T** that simulates **M** using only the tape symbols 0 and 1. Then we need a way of encoding arbitrary strings formed from the symbols $0, s_1, \ldots, s_k$ into strings of 0's and 1's. For this purpose it is convenient to think of the squares of **T**'s tape as being grouped into blocks of length k and to use each block to store an encoded representation of one symbol from **M**'s tape. In particular, we may simply assign the patterns $00 \ldots 0, 10 \ldots 0$, $110 \ldots 0, 1110 \ldots 0, \ldots, 111 \ldots 1$ to represent the symbols $0, s_1, s_2, s_3, \ldots, s_k$, respectively.

Once such an encoding scheme has been adopted, it is easy to see how to design a machine **T** that simulates the behavior of the given machine **M**. Assume that when it begins to imitate a given step in **M**'s computation, **T** scans the left-most square in the block that corresponds to **M**'s currently scanned square. The machine **T** may then move to the right through the block in question, using its internal states to count up the number of 1's that it finds. When it reaches the end of the block, it will have complete information about the symbol currently scanned by **M** and so may go back and write in the representation of **M**'s new symbol, shift one block to the left or right, and enter the appropriate next state, at which point it will be ready to imitate **M**'s next move.

Thus one nonblank tape symbol suffices for the simulation of any Turing machine computation. Of course, the price that must be paid for reducing the number of tape symbols used by a Turing machine is an increase in the number of internal states and the necessity of encoding and decoding tape patterns.

It is interesting to note that the encoding and decoding of patterns consisting solely of 0's and 1's can be accomplished by Turing machines that use only the symbols 0 and 1. (See Problem 2.2.8.) To illustrate the significance of this fact, suppose that the machine **M** has been designed to perform a certain transformation of strings of 0's and 1's, but that in the process it uses various auxiliary tape symbols in addition to 0 and 1. Then it must be possible to design a new machine **T** that encodes the initial pattern of 0's and 1's presented to **M**, simulates the computation performed by **M**, and decodes the result, all without using any symbols other than 0 and 1.[†] Thus the use of auxiliary symbols in the course of Turing machine computations is a matter of convenience, not logical necessity.

It is also possible to reduce the number of internal states used by a Turing machine at the expense of increasing the number of tape symbols. In particular, the computations of any given Turing machine can be simulated by a suitably designed Turing machine having only two internal states. The simulation process is rather complicated, and we will not attempt to describe it here.

2.3 A UNIVERSAL TURING MACHINE

In the preceding section we saw how a suitably chosen machine of one kind could be made to simulate a particular machine of a second kind. We now consider the possibility of designing a single machine of a specified kind that is capable of simulating any other machine of that same kind. A machine having this capability is called a *universal machine*. Since no loss of computing power is entailed, we will restrict our attention to machines that operate on one singly infinite tape. Thus we wish to design a one-tape, singly infinite machine that is able to simulate the computations of any other one-tape, singly infinite machine.

If a fixed machine is to imitate the behavior of an arbitrarily specified machine, it must be supplied with two things: a description of the machine that it is to simulate, and a description of that machine's initial tape pattern. We therefore begin by showing how such machine and tape descriptions can be represented on the tape of a universal machine. Once the format of these descriptions has been agreed upon, we can turn to the design of the universal machine itself.

Machine and Tape Descriptions

A Turing machine can be characterized by three functions: one that maps combinations of current states and scanned symbols into new tape symbols, one that maps combinations of current states and scanned symbols into direc-

---

[†] Since **T** may not be able to determine the limits of the nonblank portion of the tape with which it is presented, it cannot always encode the entire tape pattern before beginning the simulation. However, it can always do the encoding bit by bit as needed to continue the simulation.

	0	1
q_0		$0\,R\,q_1$
q_1	$1\,R\,q_2$	$1\,R\,q_1$
q_2	$0\,L\,q_3$	
q_3	$0\,R\,q_0$	$1\,L\,q_3$

Figure 2-29

tions of motion, and one that maps combinations of current states and scanned symbols into new states. These functions can be compactly described by a single quinary relation, namely the relation that holds among the elements q, s, s', d, q' iff the combination of current state q and scanned symbol s causes the machine to write the symbol s', move in the direction denoted by d (where d is either L or R), and enter state q'. This relation can in turn be described by listing the quintuples for which it holds. Thus each Turing machine can be represented by a list of quintuples.

As a simple example, consider the Turing machine whose state table is given in Fig. 2-29. This machine is represented by six quintuples, one for each nonblank entry in the table:

$$(q_0, 1, 0, R, q_1), (q_1, 0, 1, R, q_2), (q_1, 1, 1, R, q_1),$$
$$(q_2, 0, 0, L, q_3), (q_3, 0, 0, R, q_0), (q_3, 1, 1, L, q_3)$$

The quintuple $(q_0, 1, 0, R, q_1)$, for instance, corresponds to the entry in row q_0, column 1 of the table. It specifies that when the machine is in state q_0 and scans a 1, it will erase that 1, move right, and enter state q_1. The other quintuples are interpreted in a similar way. Note that no quintuples are provided for the situations in which the machine is to halt.

Quintuple descriptions can be simplified through the use of natural numbers. We suppose that the tape symbols of each Turing machine are formally denoted 0, 1, 2, etc. (Of course, we may still use other symbols as informal abbreviations.) We will also assign the number 0 to denote the state q_0, 1 to denote q_1, and so on. Finally, we will use the numbers 0 and 1 to denote the directions L and R, respectively. With these substitutions, the quintuple description of the machine of Fig. 2-29 becomes:

$$(0, 1, 0, 1, 1), (1, 0, 1, 1, 2), (1, 1, 1, 1, 1),$$
$$(2, 0, 0, 0, 3), (3, 0, 0, 1, 0), (3, 1, 1, 0, 3)$$

Thus every Turing machine can be described by a set of quintuples of natural numbers.

Next we must choose a way of representing natural numbers and k-tuples of natural numbers as patterns on a Turing machine tape. Of the many possible ways of doing this, we will use the so-called *unary notation*. In this notation, the natural number n is represented by a block of $n + 1$ consecutive 1's. Thus the unary representation of zero is 1, that of one is 1 1, that of two is 1 1 1, and so on. To represent the k-tuple $(n_1, \ldots, n_k)$, we simply string together the unary representations of $n_1, \ldots, n_k$, leaving a single blank square between adjacent blocks of 1's. Thus the pattern 1 1 1 1 0 1 1 1 0 1 0 1 1 1 1 1 1 0 1 1 is the unary representation of the quintuple (3, 2, 0, 5, 1).

A complete Turing machine description can be provided by stringing together the unary representations of the quintuples corresponding to that machine. To make it easy to distinguish one quintuple from the next, we will require that *two* blank squares be left between adjacent quintuple representations. For the most part, the order in which the various quintuples are listed is not important. However, we shall adopt the convention that the *first* quintuple will always have the machine's starting state as its first component.

Example 2-5. The machine of Fig. 2-30 can be described by the following tape pattern.

1 0 1 0 1 1 0 1 1 0 1 1 1 0 0 1 0 1 1 0 1 0 1 1 0 1 1 0 0 1 1 0 1 1 0 1 1 0 1 0 1 0 0 1 1 1 0 1 1 0 1 0 1 1 0 1

$\quad\quad (q_0, 0, 1, R, q_2)\quad\quad\quad (q_0, 1, 0, R, q_1)\quad\quad\quad (q_1, 1, 1, L, q_0)\quad\quad\quad (q_2, 1, 0, R, q_0)$

Note that the machine's starting state is represented by the first block of the pattern.   □

Now consider the problem of representing the tape patterns of an arbitrary Turing machine for presentation to the universal machine. At first sight this may not seem to be a problem: why not simply present the universal machine with an exact copy of the pattern that appears on the tape of the machine that is to be simulated? But the universal machine is to be a *fixed* machine, having a fixed number of tape symbols to work with. This machine must be able to

	0	1
q_0	$1\ R\ q_2$	$0\ R\ q_1$
q_1		$1\ L\ q_0$
q_2		$0\ R\ q_0$

Figure 2-30

imitate the behavior of any one-tape Turing machine, even one that uses many more tape symbols than the universal machine itself. Thus some means must be adopted for encoding arbitrary tape patterns using a fixed-symbol alphabet.

This encoding will also be accomplished by means of the unary notation. In particular, the pattern of symbols $s_1 s_2 \ldots s_k$ will be represented by stringing together the unary representations of s_1, s_2, ..., s_k with single blank squares between adjacent blocks. Of course, if $s_1 s_2 \ldots s_k$ is the tape pattern of a singly infinite Turing machine, it will be followed by an infinite sequence of blanks. These blanks will *not* be encoded for representation on the tape of the universal machine. Thus the tape pattern 1 3 0 2 1 4 0 0 0 ... will be represented by the encoded pattern 1 1 0 1 1 1 1 0 1 0 1 1 1 0 1 1 0 1 1 1 1 1 1 0 0 0 0 ... on the tape of the universal machine.

General Operation of the Universal Machine

We are now ready to describe, in general terms, how our universal machine carries out its simulations. For convenience, the universal machine will be referred to as the machine **U**, and the machine whose behavior it is simulating will be referred to as the machine **T**.

The tape of the universal machine is divided into three main regions, as indicated in Fig. 2-31. The leftmost square of the tape is permanently inscribed with the special marking symbol A. Following this A is a *buffer region*, which is initially blank. This region is to contain at least $(Q + S + 2)$ squares, where Q and S are the numbers of states and symbols used by the machine **T**. The role of the buffer region is to store a representation of **T**'s current state and scanned symbol. The buffer region is followed by a *machine description region*, which is to contain a permanent copy of the quintuple description of the machine **T**. This region is in turn followed by a *tape description region*, which is used to store the representation of **T**'s current tape pattern.

The universal machine begins its computation with its tape inscribed as follows. The buffer region is blank. The machine description region contains the encoded description of the machine **T** that is to be simulated. This description is preceded by the special marking symbol B and followed by three blank squares. Note that as a result of our convention concerning the ordering of quintuples, the symbol B immediately precedes a block of 1's corresponding to the starting state of **T**. The tape description region contains the encoded description of **T**'s initial tape pattern, with the special marking symbol C

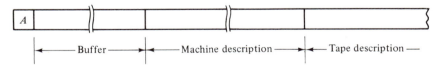

Figure 2-31

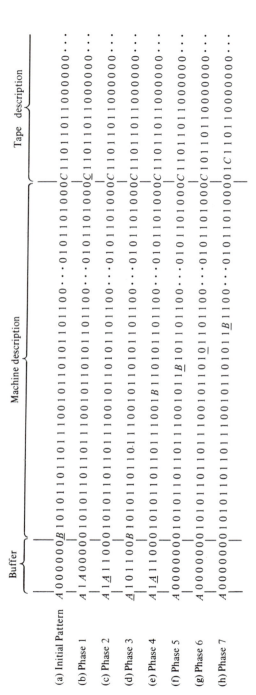

Figure 2-32

immediately preceding the block corresponding to **T**'s initially scanned tape symbol. The reading head of the universal machine initially scans the square containing the marking symbol *B*.

As an example, let us suppose that **T** is the (singly infinite) machine shown in Fig. 2-30, that this machine is presented with the initial tape pattern 1 1 1 0 0 ..., and that the initially scanned symbol is the leftmost 1. In this case, the universal machine's initial tape pattern should have the appearance shown in line (a) of Fig. 2-32. Note especially the use of the symbols *B* and *C* to mark **T**'s initial state and initially scanned tape square.

The universal machine **U** now proceeds to carry out its computation in a series of steps, each of which corresponds to one move in the computation performed by **T**. During each step, **U** uses its description of **T** to update the representation of **T**'s tape pattern and to keep track of **T**'s internal state. If **T** eventually halts, **U** will halt during the corresponding step of its simulation, with the encoded version of **T**'s final tape pattern on its own tape. Each step in the simulation of **T** consists of a number of substeps, or phases, as indicated in the flowchart of Fig. 2-33. The effects of the individual phases are outlined below and illustrated in Fig. 2-32.

Phase 1. The machine **U** copies the block marked by the symbol *B* into the initial portion of the buffer region. It then writes an auxiliary *A* immediately after the new block, erases the symbol *B*, and moves right until it reaches the symbol *C*. Thus if **U**'s initial tape pattern is that shown in line (a) of Fig. 2-32, the completion of Phase 1 yields the pattern shown in line (b).

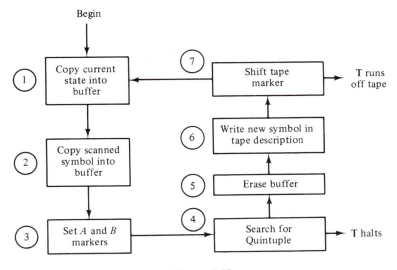

Figure 2-33

Phase 2. The machine **U** next copies the block marked by the symbol C into the portion of the buffer region immediately following the auxiliary A, as illustrated in line (c) of Fig. 2-32.

Phase 3. The machine **U** now erases the auxiliary A from the buffer region and writes a B immediately to the left of the machine description, as shown in line (d) of Fig. 2-32. Note that the buffer region now contains the unary representations of **T**'s current state and scanned symbol, in that order.

Phase 4. The machine **U** next searches through the quintuples of **T**'s description, looking for one whose first two components match the state and symbol representations currently inscribed in the buffer. This search is carried out as follows. The machine begins by comparing, one at a time, the first two blocks to the right of the symbol B with the two blocks in the buffer. If these blocks match, the desired quintuple has been found. If they do not match, **U** moves the B marker to the beginning of the next quintuple and repeats the process. If the end of the machine description is reached without finding a match, the machine **T** must halt during the move currently being simulated, so **U** also halts. In the example of Fig. 2-32, a match is found with the second quintuple. As a result of the manner in which the search process is implemented, the tape is left with A and B marking symbols as shown in line (e) of Fig. 2-32.

Phase 5. Having found the quintuple that specifies **T**'s current move, **U** erases the contents of the buffer and shifts the B marker one block to the right, so that it stands in front of the new-symbol component of the quintuple. (See line (f) in Fig. 2-32.)

Phase 6. The machine **U** next updates its description of **T**'s tape pattern by substituting the new-symbol block following the B marker for the block following the C marker in the tape description. (In general, this will require expanding or contracting the tape description, as discussed in Section 2.1.) Having made this substitution, **U** erases the B marker and moves to the right until it reaches the directional component of the quintuple under consideration, as illustrated in line (g) of Fig. 2-32.

Phase 7. The machine **U** now counts the number of 1's in the directional component and writes a B immediately after it. It then shifts the C marker one block to the left or right, as required. If the shift is to the left, **U** must determine whether **T** runs off its tape; if it does, **U** halts. If the shift is to the right, **U** must determine whether it has reached the end of the encoded part of **T**'s tape pattern; if so, it must add a new block representing the next (blank) square on **T**'s tape. In the example of Fig. 2-32, line (h) shows the tape pattern at the end of this phase.

At this point **U**'s tape is in the same condition it was in prior to Phase 1. The buffer region is blank, the tape description is up to date, the symbol C

marks the currently scanned square, and the symbol B marks the current internal state. The machine is therefore ready to begin Phase 1 of the next step in its simulation by copying the B- and C-marked blocks into the buffer region. The step-by-step simulation process is continued until such time as **T** may either halt or run off its tape, in which case **U** will halt with a representation of **T**'s final condition on its tape.

The reader who understands the examples discussed in Section 2.1 should have little trouble designing a universal machine that operates in the manner described above. Indeed, several of the machines of Section 2.1 can be used as submachines within the universal machine **U**. We conclude that there does exist a one-tape, singly infinite Turing machine that is capable of simulating the computations of any other one-tape, singly infinite Turing machine. Those interested in the details of the construction of the universal machine may find them at the end of this section.

The approach taken here does not represent the only way of designing a universal machine. Nor is our machine **U** the smallest or simplest possible universal machine; the use of certain clever coding techniques can lead to extremely compact universal machines. But the machine **U** does have certain advantages over some of the more elegant universal machines: its operation is relatively easy to understand; it demonstrates that the construction of a universal machine is quite straightforward; and it is easily incorporated into other, more complex Turing machines.

The existence of a universal Turing machine is important for at least two reasons. First, the existence of a universal machine makes it possible to establish certain important properties of Turing machines and the functions they compute. Some of these properties will be discussed in Chapter 3. Second, the fact that one particular machine can imitate the behavior of any other machine implies that there is a limit to the complexity that a Turing machine need have. As long as one is willing to encode tape patterns appropriately, a fixed and relatively simple state structure suffices to carry out any computation that can be performed by any Turing machine. And, as we shall see in Chapter 3, for a large class of computations the necessary encoding can be accomplished very simply.

Detailed Operation of the Universal Machine*

For the sake of completeness, we now provide the details of the construction of the universal machine **U** whose general operation was outlined above.

The first three phases of **U**'s operation are easy to implement. Phase 1 begins with the machine scanning the marking symbol B and requires the machine to copy the block marked by that B into the space immediately to the right of the symbol A in the buffer region. This is precisely the operation that the machine of Fig. 2-9 in Section 2.1 was

---

* Starred material optional.

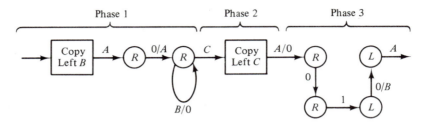

Figure 2-34

designed to perform. We therefore incorporate a copy of that machine in the structure of the universal machine **U**. This copy is denoted by the box labeled "Copy Left *B*" in Fig. 2-34.

The copying submachine ends its operation with the *A* marker under the reading head. The machine **U** must then move to the right, write an auxiliary *A* in the first blank square after the newly formed block, erase the *B* marker, and continue moving to the right until it reaches the *C* marker. These actions are accomplished by the portion of the diagram of Fig. 2-34 marked "Phase 1."

Upon reaching the *C* marker, the machine **U** enters Phase 2, during which it must copy the block marked by the *C* into the space immediately to the right of the auxiliary *A* marker. (See line (c), Fig. 2-32.) This is accomplished by the submachine labeled "Copy Left *C*," which is identical to the first copying submachine except that the symbol *C* plays the role of the symbol *B*. When the second copying has been completed, **U** will be scanning the auxiliary *A* marker. During Phase 3, it must erase that *A*, move to the machine description and write a *B* immediately in front of the description, and then return to the permanent *A* marker. These actions are implemented by the portion of Fig. 2-34 labeled "Phase 3."

The next phase, in which **U** searches for the quintuple corresponding to **T**'s current state and scanned symbol, is somewhat more complicated. To illustrate the method used, we will describe what happens when **U** begins the searching phase with its tape

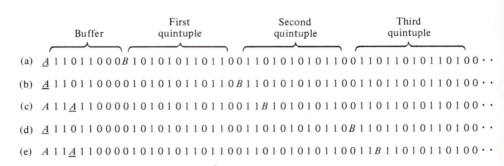

Figure 2-35

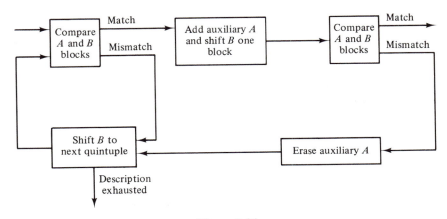

Figure 2-36

inscribed as shown in line (a) of Fig. 2-35.[†] Remember that **U**'s job is to find a quintuple whose first two components match the two blocks that appear in the buffer. The process begins with **U** scanning the *A* marker.

The first thing **U** does is compare the lengths of the blocks following the *A* and *B* markers, using the comparison procedure described in Section 2.1. In the present example, these lengths do not match, so the first quintuple is certainly not the desired one. The machine accordingly moves the *B* marker to the beginning of the next quintuple, as shown in line (b) of Fig. 2-35, and tries again. This time the lengths of the blocks following the *A* and *B* markers do match, although in general several quintuples might have to be tried before this happens.

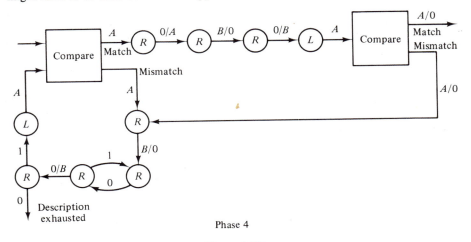

Phase 4

Figure 2-37

[†] This tape pattern is *not* taken from the example of Fig. 2-32.

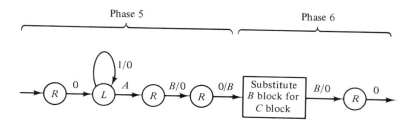

Figure 2-38

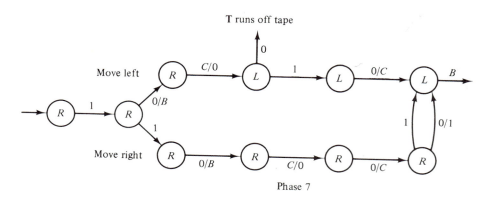

Phase 7

In order to determine whether the desired quintuple has in fact been found, **U** must now compare the second block in the buffer with the second component of the quintuple in question. To do this, it writes an auxiliary A in front of the second block in the buffer, shifts the B marker to the second block in the quintuple, and returns to the auxiliary A, as shown in line (c) of Fig. 2-35. The machine now compares the newly marked blocks in the usual way. If these blocks match, the desired quintuple has been found, and the searching phase is finished. If (as is the case in the present example) these blocks do not match, the machine erases the auxiliary A, shifts the B to the beginning of the next quintuple, and starts all over again. As shown in lines (d) and (e) of Fig. 2-35, the process continues until the proper quintuple has been found.

The whole searching process can be compactly described by the flowchart of Fig. 2-36. Remember that provision must be made for the situation in which the machine **T** halts during the move currently being simulated. This situation is manifested by the absence of any quintuple in the machine description whose first two components match the contents of the buffer. Thus if in the process of shifting the B marker to the "next" quintuple the machine **U** discovers that there is no next quintuple, it should halt.

Figure 2-37 shows a state diagram that implements the flowchart of Fig. 2-36. The boxes labeled "Compare" denote copies of the matching machine described in Fig. 2-12

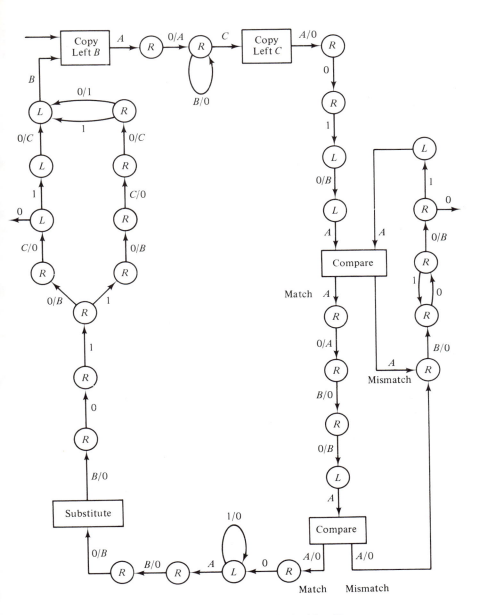

Figure 2-40   The Universal Machine U

of Section 2.1. Remembering that the matching process begins and ends with the machine scanning the pertinent A marker, it is a simple matter to fill in the transitions needed to move the markers around properly. Note that in moving the B marker from one quintuple to the next, the machine detects the end of a quintuple by the appearance of two consecutive blank squares. If they are followed by a third blank square, the end of the machine description has been reached, and U should halt.

Phases 5 and 6 are straightforward. The machine finishes Phase 4 scanning the auxiliary A marker in the buffer region. To complete Phase 5, it must erase the contents of the buffer and shift the B marker one block to the right, so that it stands in front of the block representing T's new tape symbol. These actions are accomplished by the portion of the diagram of Fig. 2-38 labeled "Phase 5." The machine U must now substitute the block following the B for that following the C in the tape description, expanding or contracting the tape description as needed. This is accomplished by a copy of the machine described in Fig. 2-16 of Section 2.1. This copy is denoted by the box marked "Substitute B block for C block" in Fig. 2-38. The substitution process ends with the machine scanning the B marker. To complete Phase 6, the machine erases this marker and moves to the beginning of the next block, which specifies the direction in which T is to move.

The universal machine begins Phase 7 by counting the number of 1's in the directional block and writing a B at the end of that block. If the block contains just one 1, the machine T must shift left. In this case U follows the upper chain of states in the diagram of Fig. 2-39, erasing the C marker, replacing it in front of the immediately preceding block, and returning to the B marker. Note that if the original C is preceded by a blank square, a left shift will cause T to run off its tape. The machine U therefore halts in this situation. If the directional block contains two 1's, T must shift right. In this case U follows the lower chain of states in Fig. 2-39, moving the C marker to the end of the currently marked block. If the square following the new position of the C contains a 1, the machine returns to the B marker. If not, U writes a 1 in that square, thereby creating the encoded representation of a blank square on T's tape, and returns to the B marker.

Having implemented each of the individual phases of the simulation process, we can now provide a complete state diagram for the universal machine U. This diagram, which is a combination of the diagrams of Figs. 2-34, 2-37, 2-38, and 2-39, is shown in Fig. 2-40. The reader is encouraged to check his or her understanding of the operation of the universal machine by tracing the course of a few typical simulation steps through the diagram.

2.4 SUMMARY

A Turing machine is a hypothetical computing device made up of a finite-state control unit coupled to an infinite tape. Each step in a Turing machine computation consists in writing a symbol on the tape, shifting the tape to the left or right, and assuming a new internal state. The particular action to be taken at any step is determined by the current state of the machine and the currently scanned tape symbol. Because of the simple, mechanistic nature of their computations, Turing machines can certainly be viewed as embodying effective algorithms. An understanding of the capabilities of Turing machines may therefore shed some light on the nature of effectively computable functions.

Although the atomic moves available to a Turing machine are very elementary, appropriate sequences of these moves can be used to carry out a wide variety of pattern manipulation operations. These operations include: forming copies of specified blocks or patterns, replacing one block or pattern with another, and comparing specified blocks or patterns. By using these basic operations as subroutines, it is possible to design Turing machines that perform quite sophisticated computations.

There are several ways in which the basic Turing machine model can be modified without affecting its ultimate computing capabilities. Among the generalizations of the model that do not enhance its computing power are the use of tapes having several tracks, the use of more than one reading head, or the use of more than one tape. Among the restrictions that do not decrease the power of the basic model are the use of singly infinite tapes, the limitation to one nonblank tape symbol, or the limitation to two internal states. Thus although one variant of the Turing machine model may prove more efficient or convenient than another in a given application, the fundamental capabilities of Turing machines do not depend on the particular variant adopted.

The technique most commonly used to show that one computational model is as powerful as another is that of simulation. The application of this technique to Turing machines consists in showing that the computations of any machine of one kind can be imitated by an appropriately designed machine of a second, usually more restricted, kind. Two steps are involved: choosing a means of representing the tape patterns of the machine that is to be simulated on the tape of the simulating machine, and designing the simulating machine so that it properly updates these tape pattern representations as the simulation proceeds. In all the cases considered here, the representation schemes are quite simple, and the necessary encoding and decoding of tape patterns can be carried out by Turing machines.

It is also possible to design a single Turing machine that is capable of simulating the behavior of any other Turing machine. Such a machine is called a universal Turing machine. In order to carry out its computations, a universal machine must be provided with two things: a description of the machine it is to simulate, and a description of that machine's initial tape pattern. Since every machine can be represented by a set of ordered quintuples, and since every quintuple can in turn be represented by a series of blocks of 1's, machines are readily described by strings of 0's and 1's. A similar coding scheme suffices for the description of arbitrary tape patterns. Once these representational schemes have been decided upon, the design of the universal machine itself is relatively straightforward. The fact that it is possible to design universal machines means that there is a limit to the physical complexity needed to carry out any of the computations that Turing machines can perform.

2.5 PROBLEMS

Note: In the following problems, the notation 1^n is used to represent a block of n consecutive 1's.

Section 2.1

2.1.1   Convert the Turing machine whose state table is shown below into an equivalent machine in which entry into any given state is always accompanied by motion in the same direction. Give a state diagram for the new machine.

	0	1
q_0	$1\,R\,q_0$	$0\,L\,q_2$
q_1	$0\,L\,q_0$	$1\,R\,q_1$
q_2	$1\,L\,q_1$	

2.1.2   Consider the Turing machine whose state diagram is shown below.

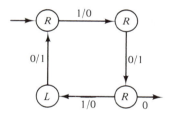

   a) For each of the following initial tape configurations, determine the final tape pattern that the machine will produce and indicate the final head position.

   i) $\ldots \underline{0}\ 1\ 1\ 1\ 0\ 1\ 1\ 1\ 1\ 0 \ldots$

   ii) $\ldots \underline{0}\ 1\ 1\ 1\ 0\ 1\ 1\ 0 \ldots$

   b) What effect will the machine have on an arbitrary initial pattern of the form $\ldots \underline{0}\ 1^m\ 0\ 1^n\ 0 \ldots$, where m and n are positive integers? Explain briefly how the machine works. What is the final position of the reading head?

   c) Show how to modify the given state diagram so the machine will always halt at its starting position.

2.1.3   Consider the Turing machine described by the following state diagram.

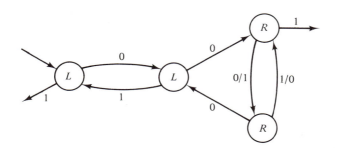

a) Determine the behavior of the machine for each of the following initial tape configurations.

    i)   ... 0 0 0 0 0 0 0̲ 0 0 0 0 0 0 0 ...   (blank tape)

    ii)   ... 0 0 0 0̲ 0 1 0 0 1 0 1 1 0 0 ...

    iii)   ... 0 1 1 0 0 0 1 0̲ 0 0 0 0 0 0 ...

b) Describe as clearly and concisely as you can the initial tape configurations for which the machine will eventually halt.

2.1.4   Give the state diagram for a Turing machine that behaves as follows. When started anywhere on a tape that is blank except for a single 1, the machine eventually halts with that 1 under its reading head. The remainder of the tape is to be blank when the machine halts.

2.1.5   Design a *binary copying machine* that behaves as follows. The machine is to be presented with an initial tape pattern of the form

                ... $A̲$ 1 2 1 2 2 0 ... B 0 0 0 0 0 0 ...

where the symbol A is followed by an arbitrary block of 1's and 2's. The machine is to form a copy of this block immediately to the right of the symbol B, halting with the tape pattern

                ... $A̲$ 1 2 1 2 2 0 ... B 1 2 1 2 2 0 ... .

Assume that the symbol B is originally followed by a sufficient amount of blank space to contain the desired copy. Describe the machine in graphical form, using the tape alphabet $\{0, 1, 2, A, B, X\}$. An eight-state solution is satisfactory, although not minimal.

2.1.6   Give the state diagram for a Turing machine that behaves as follows. When presented with a tape containing an arbitrary string of 1's and 2's (preceded and followed by blanks) and made to scan the first symbol in the string, the machine is to reverse the string. Thus if presented with the tape pattern

$$\ldots 0\ 0\ \underline{1}\ 2\ 1\ 1\ 2\ 1\ 2\ 0\ 0 \ldots$$

the machine should eventually produce the tape pattern

$$\ldots 0\ \underline{0}\ 2\ 1\ 2\ 1\ 1\ 2\ 1\ 0\ 0 \ldots$$

and halt as indicated. The final pattern is to occupy the same region of the tape as the initial pattern. A solution using between six and nine states is reasonable.

2.1.7 A *palindrome* is a string of symbols that reads the same backward as forward. Thus ABA and $ABABBABA$ are palindromes, whereas $ABAB$ and $ABABBA$ are not. Design a Turing machine that determines, and indicates by its final state, whether or not a given string of A's and B's is a palindrome. The machine is to begin its computation scanning the blank square immediately to the left of the given string. The final tape pattern is to be identical to the initial pattern, and the machine is to halt on the initially scanned square. Ten states are sufficient, but not necessarily minimal.

2.1.8 a) Design a Turing machine that, when started on any tape pattern of the form

$$\ldots \underline{0}\ 1^n\ 0\ 1^x\ 0\ 0 \ldots \qquad (\text{where } n > 0,\ x \geq 0)$$

eventually halts with the pattern

$$\ldots \underline{0}\ 1^n\ 0\ 1^{x+n}\ 0\ 0 \ldots$$

on its tape. The new pattern is to start in the same square as the given pattern.

   b) Using the machine of part (a) as a submachine, design a new Turing machine that behaves as follows. When started on any pattern of the form

$$\ldots \underline{0}\ 1^m\ 0\ 1^n\ 0\ 0 \ldots \qquad (\text{where } m > 0,\ n > 0)$$

the machine is eventually to halt with the pattern

$$\ldots \underline{0}\ 1^{mn}\ 0 \ldots$$

on its tape. The location of this final pattern may be chosen to make the design of the machine as simple as possible.

2.1.9 a) Design a Turing machine **M** that behaves as follows. When presented with an initial tape pattern of the form

$$\ldots 0\ \underline{A}\ 1^m\ A\ 1^n\ A\ 0 \ldots$$

M converts that pattern into the pattern

$$\ldots 0 \; \underline{A} \; 1^{\min (m,\, n)} \; A \; 1^{\max (m,\, n)} \; A \; 0 \ldots$$

and halts as shown. The final tape pattern is to appear in the same region of the tape as the initial pattern, and the machine is never to leave that region of the tape. The machine's starting state is to be a right-moving state. Assume that m and n are nonzero.

      b) Using a copy of **M** as a submachine, design a Turing machine **S** that carries out the following sorting operation. When presented with a pattern of the form

$$\ldots 0 \; \underline{A} \; 1^{n_1} \; A \; 1^{n_2} \; A \; 1^{n_3} \; A \ldots A \; 1^{n_k} \; A \; 0 \ldots$$

(where $k \geq 2$), **S** is to rearrange the given blocks of 1's in order of length, with the smallest on the left. The final tape pattern is to have the form

$$\ldots \underline{0} \; 0 \; 1^{m_1} \; 0 \; 1^{m_2} \; 0 \; 1^{m_3} \; 0 \ldots 0 \; 1^{m_k} \; 0 \; 0 \ldots$$

where $m_1 \leq m_2 \leq \ldots \leq m_k$.

2.1.10   a) Design a Turing machine **D** that behaves as follows. When presented with a tape pattern of the form

$$\ldots \underline{0} \; 1^m \; 0 \; 1^n \; 0 \ldots \qquad \text{(where } m > 0,\, n \geq 0)$$

D is eventually to halt with exactly the same tape configuration and is to indicate, by means of its internal state, whether or not m divides n. Explain briefly how the machine works.

      b) Using **D** as a submachine, design a Turing machine **P** that behaves as follows. When presented with a tape pattern of the form

$$\ldots \underline{0} \; 1^k \; 0 \ldots \qquad \text{(where } k \geq 2)$$

P is eventually to halt with exactly the same tape configuration and is to indicate by means of its internal state whether or not k is prime. Explain briefly how the machine works.

Section 2.2

2.2.1   It is desired to design a Turing machine that copies patterns of 1's and 2's in accordance with the following format:

Initial tape pattern: $\ldots \underline{A} \; 1 \; 2 \; 1 \; 1 \; 2 \; 1 \; 2 \; 2 \; 0 \ldots B \; 0 \; 0 \; 0 \; 0 \; 0 \; 0 \; 0 \; 0 \ldots$

Final tape pattern: $\ldots \underline{A} \; 1 \; 2 \; 1 \; 1 \; 2 \; 1 \; 2 \; 2 \; 0 \ldots B \; 1 \; 2 \; 1 \; 1 \; 2 \; 1 \; 2 \; 2 \; 0 \ldots$

      a) First design the machine as a two-track machine. Assume that the given pattern is initially written in the top track and that the bottom track is initially blank. Let the machine use the bottom track to mark each symbol in the given pattern with an X as it is copied. Describe your machine graphically, using the notation of Fig. 2-22(a) for combinations of upper and lower track symbols.

b) Now convert the two-track machine of part (a) into an equivalent one-track machine by assigning single symbols to ordered pairs of upper and lower track symbols. Choose this assignment so that the resulting machine meets the format specified at the beginning of the problem. How many tape symbols does the one-track machine use? What are their roles?

2.2.2 It is desired to design a two-tape Turing machine that behaves as follows. The machine's first tape is initially inscribed with a pattern of the form

$$\ldots \underline{0} \; 1^{n_1} \, 0 \; 1^{n_2} \, 0 \ldots 0 \; 1^{n_k} \, 0 \ldots \qquad \text{(where } k \geq 1\text{)}$$

and the second tape is left blank. The machine is to determine which of the given blocks of 1's is the longest and is to halt with a copy of that block on its second tape. The original pattern is to be left unchanged on the first tape. The machine is to halt scanning the 0 to the right of the given pattern on its first tape and the 0 to the left of the block formed on the second tape.

Design an appropriate machine, using the symbol alphabet $\{0, 1\}$ for each tape. Describe your machine graphically, using the same conventions used for ordinary Turing machines, except that: (1) the symbols scanned and written on the machine's first and second tapes are to be represented by symbol pairs of the form $\frac{s_1}{s_2}$, as in the case of two-track machines; and (2) each state is to be labeled with a pair of direction symbols of the form $\frac{D_1}{D_2}$ to indicate the directions that the machine is to move on its first and second tapes when it enters that state. Each of D_1 and D_2 may be either L, R, or $-$, where the symbol $-$ indicates that the machine does not shift the tape in question. A total of six states is sufficient.

2.2.3 For the purposes of this problem, we define a *big-headed machine* to be a Turing machine whose reading head scans three adjacent squares at the same time. Such a machine may simultaneously change the symbols in all three of its scanned squares, but may only shift its reading head one square at a time. The action to be taken at any given step is determined by the machine's internal state and the symbols appearing in the three currently scanned squares.

a) Show that big-headed machines are no more powerful than ordinary Turing machines.

b) Show that for every ordinary Turing machine there exists an equivalent *one*-state big-headed machine. The initial tape patterns of the two machines are to be identical, the final tape patterns almost identical.

2.2.4 Consider the problem of constructing a singly infinite machine **T** that simulates the behavior of a doubly infinite machine **M**. Assume that the squares of **M**'s tape are numbered as shown below, where square zero is the initially scanned square.

Rather than design **T** as a two-track machine, let us represent the squares of **M**'s tape on **T**'s singly infinite tape in accordance with the following assignment:

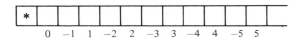

Here ∗ is a special marking symbol used to identify the left-most square of **T**'s tape.

a) Using the representation scheme just described, give a state diagram for the singly infinite counterpart of the machine **M** of Example 2-4.

b) Describe a general procedure for converting an arbitrary doubly infinite machine to its singly infinite counterpart, using the same representation scheme.

2.2.5 A *reset machine* is a singly infinite Turing machine that, at each step in a computation: (i) writes a symbol in the currently scanned square; (ii) *either* moves one square to the right *or* jumps immediately to the left-most square on its tape; and (iii) enters a new state. The particular action to be taken at any given step depends, of course, on the current state and scanned symbol.

a) Can every reset machine be simulated by an ordinary singly infinite machine? Explain.

b) Can every ordinary singly infinite machine be simulated by a reset machine? Explain.

2.2.6 A *quadruple machine* is defined to be a Turing machine that can, at any given step, either write a tape symbol or shift its tape, but not both. Such a machine can be described by a state table in which each entry either contains a new symbol and a new state, or a direction and a new state, or else is blank.

a) Give the state-table description of a quadruple machine that duplicates the behavior of the ordinary machine **M** described in Fig. 2-28(a).

b) Show that the quadruple machine model is equivalent to the standard Turing machine model, in the sense that the behavior of any ordinary machine can be duplicated by an appropriate quadruple machine and vice versa.

2.2.7   The Turing machine **M** uses the tape symbols 0, 1, 2, ..., k. We wish to design a machine **T** that simulates the computations of **M** using only the symbols 0 and 1, as outlined on page 89. Thus each square of **M**'s tape is to be represented by a block of k consecutive squares on **T**'s tape, and **M**'s symbols are to be encoded as follows:

$$0 \to 0\,0\,0\ldots 0, \qquad 1 \to 1\,0\,0\ldots 0, \qquad 2 \to 1\,1\,0\ldots 0, \qquad \ldots, \qquad k \to 1\,1\,1\ldots 1$$

The machine **T** is to begin its computation by scanning the left-most square in the block corresponding to **M**'s starting square.

a) Give an algorithm for converting a tabular description of **M** into a corresponding description of **T**. In particular, explain how each state in **M**, together with the transitions out of that state, is to be replaced by an aggregate of states and transitions in **T**.

b) Assuming that $k = 3$, give a state-diagram description of the portion of **T** corresponding to the following row of **M**'s state table:

	0	1	2	3
q_5	$1\ R\ q_3$	$3\ L\ q_5$	$2\ R\ q_2$	$1\ L\ q_4$

c) If **M** has q states and k tape symbols, how many states might **T** require?

2.2.8   a) Design a machine T_1 that converts each initial pattern of the form

$$\underline{0}\ \underbrace{1\ 1\ 1\ldots 1}_{n}\ 0\ 0\ldots$$

into the final pattern

$$\underline{0}\ 0\ \underbrace{1\ 0\ 0\ 1\ 0\ 0\ 1\ldots 0\ 0\ 1}_{n\ \text{copies of}\ 0\ 0\ 1}\ 0\ 0\ 0\ldots .$$

The machine is to use *only* the symbols 0 and 1. It is to halt on the same square on which it starts and is never to move to the left of that square.

b) Design a machine T_2 that converts each initial pattern of the form

$$\underline{0}\ 0\ \underbrace{1\ 0\ 0\ 1\ 0\ 0\ 1\ldots 0\ 0\ 1}_{m\ \text{copies of}\ 0\ 0\ 1}\ 0\ 0\ 0\ldots$$

into the final pattern

$$\underline{0}\ \underbrace{1\ 1\ 1\ \dots\ 1}_{m}\ 0\ .$$

The machine is to use *only* the symbols 0 and 1. It is to halt on the same square on which it starts and is never to move to the left of that square.

c) A certain machine **M** uses the tape symbols 0, 1, 2, ..., k. Its initial and final tape patterns are all of the form $\underline{0}\ 1^n\ 0\ \dots$. Show that there exists another machine **T** that performs the same computations as **M** but uses only the tape symbols 0 and 1. The initial and final tape patterns of **T** are to be *identical* to those of **M**. (Cf. Problem 2.2.7.)

2.2.9 It is desired to design machines that behave like the doubling machine described in pp. 74–75 but that use (see part a) as few tape symbols as possible, and (see part b) as few internal states as possible.

a) Give the state diagram for a doubling machine that uses only the symbols 0 and 1. (Begin your design from scratch; do not try to convert the machine of Fig. 2-18 into a two-symbol machine.) Five states are sufficient.

b) Give the state table for a doubling machine that has only two internal states and implements the algorithm suggested below. Assume that the machine can write a symbol during the step in which it halts. Entry into a given state need not always be accompanied by motion in the same direction.

Initial tape pattern	...$\underline{0}$	1	1	1	1	0	0	0	0	0...		
	A	$\underline{X}$	2	2	2	A	$\underline{C}$	0	0	0	0	
	A	$\underline{X}$	1	1	1	B	$\underline{C}$	0	0	0	0	
	A	X	$\underline{X}$	2	2	A	A	$\underline{C}$	0	0	0	
	A	X	$\underline{X}$	1	1	B	B	$\underline{C}$	0	0	0	
	A	X	X	$\underline{X}$	2	A	A	A	$\underline{C}$	0	0	
	A	X	X	$\underline{X}$	1	B	B	B	$\underline{C}$	0	0	
	A	X	X	X	$\underline{X}$	A	A	A	A	$\underline{C}$	0	
	A	X	X	X	$\underline{X}$	B	B	B	B	$\underline{C}$	0	
	A	X	X	X	X	X	X	X	X	$\underline{0}$	0	
Final tape pattern	...$\underline{0}$	1	1	1	1	1	1	1	1	1	0	0...

2.2.10 Let **M** be any Turing machine that operates on a doubly infinite tape. Show that there exists another doubly infinite machine $\hat{\mathbf{M}}$ that duplicates each of **M**'s computations in at most half as many steps as **M**, as long as $\hat{\mathbf{M}}$'s initial and final tape patterns are properly encoded. Describe a typical step in $\hat{\mathbf{M}}$'s computation. *Hint:* Let the squares of **M**'s tape be represented on three tracks of $\hat{\mathbf{M}}$'s tape according to the following scheme.

M's tape:

-5	-4	-3	-2	-1	0	1	2	3	4	5

M's tape:

-11	-9	-7	-5	-3	-1	1	3	5	7	9
-10	-8	-6	-4	-2	0	2	4	6	8	10
-9	-7	-5	-3	-1	1	3	5	7	9	11

Section 2.3

2.3.1   It is desired to design a singly infinite Turing machine **V** that behaves as follows. Each initial tape pattern is to consist of a block of $(n + 1)$ 1's, a buffer region, the description of a machine **T**, and the description of **T**'s initial tape pattern, as shown below:

$$\underline{0}\ \underbrace{1\ 1\ 1\ \ldots 1}_{n+1}\ A\ \underbrace{0\ 0\ldots 0}_{\text{Buffer}}\ B\ \underbrace{1\ 0\ 1\ 1\ 0\ 1\ 1\ 0\ 1\ \ldots 0\ 0\ 0}_{\text{Machine description}}\ C\ \underbrace{1\ 0\ 1\ 1\ 0\ 1\ 1\ 0\ \ldots}_{\text{Tape description}}$$

When presented with such a pattern, **V** is to simulate the computation that **T** would perform until **T** either halts or completes n steps, whichever happens first. The machine **V** is then to halt, at which point **T**'s current state and scanned symbol are to be represented in the buffer region of **V**'s tape. Show how to modify the state diagram of the universal machine **U** (Fig. 2-40) so as to obtain the desired machine **V**.

2.3.2   A certain universal Turing machine **W** uses the following format for representing the description and tape pattern of the machine **T** whose behavior it is to simulate:

Machine description:

$$0\ X \leftarrow Q_0 \rightarrow X \leftarrow Q_1 \rightarrow X \quad \ldots \quad X \leftarrow Q_r \rightarrow X$$

Tape description:

$$A \leftarrow S_0 \rightarrow A \leftarrow S_1 \rightarrow A \quad \ldots \quad A \leftarrow S_m \rightarrow A\ 0 \ldots$$

Here the subpattern Q_i corresponds to the ith row of the state table of **T**. It has the form

$$Y \leftarrow t_{i0} \rightarrow Y \leftarrow t_{i1} \rightarrow Y \quad \ldots \quad Y \leftarrow t_{ik} \rightarrow Y$$

where t_{ij} is a unary encoded triple representing the new symbol, direction of motion, and new state that result when **T** is in state q_i and scans symbol s_j. The subpattern S_i is a fixed length pattern of the form $1^j 0^{k-j}$, indicating that the ith square of **T**'s tape contains the symbol s_j. The symbols X, A, and Y are used as punctuation marks.

Discuss the advantages and disadvantages of the representation scheme just outlined as compared with the one used in the text. In particular, what effect does the choice of this new scheme have on the design of the universal machine **W** and upon the "efficiency" of **W**'s operation?

2.3.3 Suppose it is required to design a universal machine for the class of one-tape, *doubly* infinite Turing machines. Describe two or three major ways of arranging the necessary information on the tape of such a universal machine. In each case indicate how the operation of the universal machine would differ from that of the singly infinite universal machine **U**.

2.3.4 Consider the class of Turing machines that can only move their reading heads to the right, not to the left. Explain why this class does not have a universal machine. You may assume that the initial tape patterns of a universal machine would be required to have the form shown below.

Machine description	Tape pattern description	

2.3.5 Consider the class of big-headed Turing machines described in Problem 2.2.3.

a) Show that there exists a one-state big-headed machine $\mathbf{B}_*$ that is capable of simulating the computations of any specified Turing machine **T**.

b) What would have to be done to convert $\mathbf{B}_*$ into a one-state universal machine for the class of big-headed machines?

2.3.6 a) Give a general description of the construction of a Turing machine **Y** which, when given as input the description of an arbitrary machine **M**, simulates the computation that ensues when **M** is started on its own description.

b) What will happen if **Y** is given its own description as input?

Chapter 3

Turing Computability

In this chapter we will investigate the use of Turing machines for evaluating number-theoretic functions. In particular, we will adopt a convention whereby every Turing machine can be thought of as evaluating a number-theoretic function of any desired number of variables. Then upon choosing an indexing scheme for Turing machines, we can view Turing machines as representations of the members of a family of algorithms. This family will turn out to satisfy the five basic properties discussed in Chapter 1. As a result, all the consequences of those properties must apply to Turing machines. We will reexamine some of these consequences in the context of Turing machine computations and establish some related results peculiar to Turing machines.

3.1 NUMBER-THEORETIC COMPUTATIONS AND INDEXING

The purpose of this section is to set the stage for the consideration of Turing machines as members of a family of algorithms. To fulfill this purpose, we must adopt two conventions: one for associating number-theoretic functions with machines, and one for assigning indices to machines. We begin with number-theoretic functions and their evaluation.

Basic Definitions

If Turing machines are to be used to evaluate number-theoretic functions, we must choose a method for representing natural numbers and n-tuples of natural numbers on Turing machine tapes. There are many ways of doing this. But if we seek a representation that is suitable for any Turing machine, regardless of how few tape symbols it uses, the most convenient choice is the unary scheme described in Chapter 2.

Recall that in the unary representation scheme, the natural number n is represented by a block of $n + 1$ consecutive 1's. We shall assume that a Turing machine's nonblank tape symbols are officially denoted 1, 2, 3, etc., so that unary representations can be used on the tape of any machine. For convenience in describing tape patterns, the symbol $\bar{n}$ will be used to denote the unary representation of the natural number n. For example, $\bar{3}$ denotes the tape pattern 1111. As in Chapter 2, the k-tuple $(n_1, \ldots, n_k)$ will be represented by the tape pattern $\bar{n}_1 \, 0 \, \bar{n}_2 \, 0 \cdots 0 \, \bar{n}_k$.

Now suppose that a certain (one-tape) Turing machine is to compute the n-variable number-theoretic function f. We will require that the evaluation of $f(x_1, \ldots, x_n)$ begin with the pattern

$$\cdots \underline{0} \, \bar{x}_1 \, 0 \, \bar{x}_2 \, 0 \cdots 0 \, \bar{x}_n \, 0 \cdots$$

written on the machine's tape. All squares to the left of the block $\bar{x}_1$ are to be blank, as are all squares to the right of $\bar{x}_n$. The machine is to begin by scanning the blank square immediately to the left of the block $\bar{x}_1$, as indicated above. If $f(x_1, \ldots, x_n)$ is defined, the machine is eventually to halt with the pattern

$$\cdots \underline{0} \, \overline{f(x_1, \ldots, x_n)} \, 0 \cdots$$

on its tape, scanned as indicated. The squares to the left and right of this pattern *need not* be blank. If $f(x_1, \ldots, x_n)$ is not defined, the machine should either fail to halt, or else halt in such a way that its tape does not contain a block of 1's immediately to the right of the scanned square.

Note carefully that we have only specified how the machine must behave when presented with the unary representation of n natural numbers. What it does when presented with other initial tape patterns is of no importance. In particular, we do not require the machine to determine whether or not it is being presented with a "legitimate" number-theoretic input.

A number-theoretic function that is evaluated by a Turing machine in the manner just described will be said to be *Turing computable*. As we will see, a great many number-theoretic functions turn out to be Turing computable. For the moment, we settle for a few simple examples.

Example 3-1

a) The one-variable function f_1, where $f_1(x) = x + 2$, is Turing computable. In particular, the Turing machine described in Fig. 3-1(a) evaluates f_1 in accordance with the required conventions. When presented with a tape containing a single block of 1's, the machine simply adds two 1's at the right-hand end of the block, returns to the left end of the block, and halts.

b) The two-variable function f_2, where $f_2(x_1, x_2) = x_1 + x_2$, is Turing computable. In particular, the machine described in Fig. 3-1(b) evaluates this function.

c) The three-variable function f_3, where $f_3(x_1, x_2, x_3) = x_2 + 1$, is Turing computable. It is evaluated by the machine described in Fig. 3-1(c).

d) The two variable function f_4, where

$$f_4(x_1, x_2) = \begin{cases} x_2 - x_1 & \text{if} & x_2 \geq x_1 \\ \uparrow & \text{if} & x_2 < x_1 \end{cases}$$

is Turing computable. It is evaluated by the machine described in Fig. 3-1(d). Note that if $x_2 < x_1$, this machine fails to halt and so yields an undefined function value. □

Since we will often have occasion to work with multitrack tapes, it is appropriate to consider how the foregoing conventions apply to computations on such tapes. In particular, it is appropriate to consider how combinations of symbols from the various tracks are to be denoted by single symbols when we translate from a multitrack representation to a single-track representation. Let us agree that the symbol 0 is to denote the situation in which all the tracks are blank and that the symbol 1 is to denote the situation in which the topmost track contains a 1 and all the other tracks are blank. Under this convention, a computation is initiated on a multitrack tape by writing the unary representations of the arguments in the topmost track and leaving the other tracks blank.

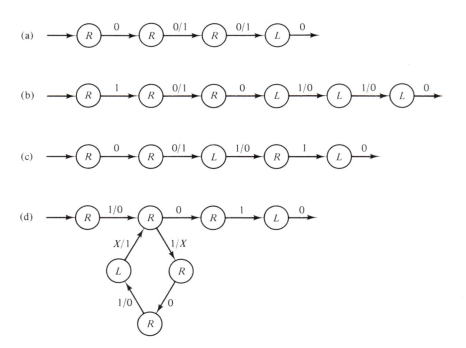

Figure 3-1

Similarly, at the end of the computation, the unary representation of the result is to appear in the topmost track and the corresponding portions of the other tracks are to be blank.

Standard Form

In choosing a format for number-theoretic computations, we deliberately left as much latitude as possible. For example, we did not require the location of the output block to bear any fixed relationship to the location of the argument blocks. Similarly, we placed no restrictions on the portions of the tape that the machine was permitted to visit during a computation, or upon the symbols that it might leave on its tape in addition to the desired output block. This flexibility was retained in order to make sure that as many functions as possible would be included among those considered to be Turing computable.

We now consider a more restrictive, but often more convenient, format for computing number-theoretic functions. In this format we still represent arguments and function values in unary form. However, we impose the following additional requirements.

1.   The machine doing the computation is eventually to halt for a given combination of argument values iff the function in question is defined for that combination of values.

2.   If and when the machine does halt, its tape is to contain only the single block of 1's representing the appropriate function value; all other squares are to be blank.

3.   The block representing the function value is to be aligned so that it begins in the same square as did the block representing the first argument value. Thus the machine is to begin and end its computation scanning the same square.

4.   At no time during the course of its computation is the machine to move to the left of the initially scanned square.

Any machine that operates in accordance with these four constraints will be said to compute its number-theoretic function in *standard form*.

We now set out to show that every Turing-computable function can be computed in standard form by a suitably designed machine. Suppose that M_1 is a Turing machine that evaluates the n-variable number-theoretic function f in accordance with our original specifications for Turing computability. We will first show how to design a new machine M_2 that also computes f, but does so in accordance with constraints 1 through 3 above. Then starting from the machine M_2, we will show how to design still another machine M_3 that also computes f, but does so in accordance with the constraints 1 through 4.

First consider the design of M_2. We think of this machine as operating on a two-track tape. At the beginning of a computation, the unary representations

(a)

0	0	0	0	0	0	0	0	0	0	1	1	1	0	1	1	1	1	0	1	0	0	0	0	0	0	0
0	0	0	0	0	0	0	0	0	0	0	0	0	0	0	0	0	0	0	0	0	0	0	0	0	0	0

(b)

0	0	0	0	0	0	0	0	0	0	1	1	1	0	1	1	1	1	0	1	0	0	0	0	0	0	0
0	0	0	0	0	0	0	0	0	0	*	C	C	C	C	C	C	C	C	C	C	0	0	0	0	0	0

(c)

0	0	0	1	1	1	0	2	2	0	0	0	0	1	1	1	1	1	0	0	1	0	2	0	0	0	0
0	0	C	C	C	C	C	C	C	*	C	C	C	C	C	C	C	C	C	C	C	C	C	C	C	0	0

(d)

0	0	0	1	1	1	0	2	2	0	0	0	0	1	1	1	1	1	0	0	1	0	2	0	0	0	0
0	0	C	C	C	C	C	C	C	*	C	C	A	C	C	C	C	C	C	C	C	C	C	C	C	0	0

(e)

0	0	0	0	0	0	0	0	0	0	0	0	0	1	1	1	1	1	0	0	0	0	0	0	0	0	0
0	0	C	C	C	C	C	C	C	*	C	C	A	C	C	C	C	C	C	C	C	C	C	C	C	0	0

(f)

0	0	0	0	0	0	0	0	0	0	1	1	1	1	1	0	0	0	0	0	0	0	0	0	0	0	0
0	0	C	C	C	C	C	C	C	*	C	C	A	C	C	C	C	C	C	C	C	C	C	C	C	0	0

(g)

0	0	0	0	0	0	0	0	0	0	1	1	1	1	1	0	0	0	0	0	0	0	0	0	0	0	0
0	0	0	0	0	0	0	0	0	0	0	0	0	0	0	0	0	0	0	0	0	0	0	0	0	0	0

Figure 3-2

of the arguments are written in the upper track, as illustrated in line (a) of Fig. 3-2, and the lower track is blank. The first thing that M_2 does is to write the special symbol $*$ in the lower half of the initially scanned square and the special symbol C in the lower half of each square involved in the representation of the arguments, as shown in line (b). Operating on its upper track, M_2 now carries out the computation that M_1 would perform on the given arguments. As it does so, it writes a C in the lower half of each new square that it visits. Thus if M_1's computation terminates, M_2 might arrive at a tape pattern like that shown in line (c) of Fig. 3-2.

The machine M_2 now temporarily marks M_1's final head position, as indicated in line (d) of Fig. 3-2, and checks to see whether that position is followed by a block of 1's in the upper track. If it is not, then the value of the function f is undefined, so M_2 enters a special state in which it moves right indefinitely, never halting. Otherwise, M_2 searches the entire area marked by C's and erases all symbols in the upper track except for the block of 1's immediately following M_1's head position. (See line (e).) Since the unmarked areas of the tape are guaranteed to be blank, this action ensures that the final tape pattern will contain only the block representing the desired function value. M_2 now copies this block into the region just to the right of the symbol ∗ (line f), erases all the symbols in the lower track (line g), and halts.

If we now view M_2 as a one-track machine, it should be clear that M_2 evaluates the function f in accordance with constraints 1 through 3. (Remember that the symbol pairs $\frac{0}{0}$ and $\frac{1}{0}$ are represented by 0 and 1, respectively.) But in the course of its computations, M_2 may very well move to the left of its initially scanned square. To eliminate this possibility, we now make use of the fact that every doubly infinite machine can be simulated by a singly infinite machine. In particular, we will use M_2—still viewed as a one-track machine—as the basis for the design of a new two-track machine M_3 that meets constraint 4 as well as constraints 1 through 3.

The machine M_3 begins its computation with the pertinent argument values written in the upper track of its tape, as illustrated in Fig. 3-3(a). The first thing M_3 does is to write the symbol \$ in the lower half of its initially scanned square, as indicated in Fig. 3-3(b). Note that if the portion of the tape to the left of the initially scanned square is ignored, M_3's tape now contains the correct singly infinite representation of M_2's initial tape pattern. We therefore design M_3 so that it now behaves like the singly infinite version of M_2, thereby ensuring that it never moves to the left of its initially scanned square. Since M_2 ends its computation with nothing but blank tape to the left of its initially scanned square, M_3 will end the corresponding computation with its lower track blank, as shown in Fig. 3-3(c). M_3 then simply erases the symbol \$ and halts, leaving the tape pattern shown in Fig. 3.3(d).

If we view M_3 as a one-track machine, we see that it evaluates the same number-theoretic function as M_2, with the same initial and final format, but without ever moving to the left of its initially scanned square. Thus M_3 evalu-

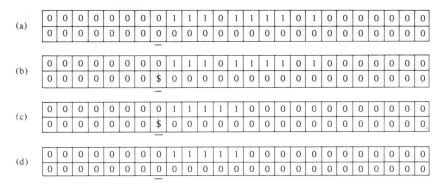

Figure 3-3

ates the given function f in standard form. Since the method outlined above for the design of $\mathbf{M}_2$ and $\mathbf{M}_3$ is a general one, applicable to any Turing-computable function, we have established:

Theorem 3-1. *If f is a Turing-computable function, then there exists some Turing machine that evaluates f in standard form.*

Applications

The fact that every Turing-computable function can be evaluated in standard form often facilitates the job of showing that certain other computations can be carried out by suitably designed Turing machines. We now consider several simple examples.

Lemma 3-1. *If g is an n-variable Turing-computable function, then there exists a Turing machine $\mathbf{T}$ that converts each tape pattern of the form*

$$\cdots \underline{0} \; \bar{x}_1 \; 0 \; \bar{x}_2 \; 0 \cdots 0 \; \bar{x}_n \; 0 \cdots$$

into the pattern

$$\cdots 0 \; \bar{x}_1 \; 0 \; \bar{x}_2 \; 0 \cdots 0 \; \bar{x}_n \; \underline{0} \; \overline{g(x_1, \ldots, x_n)} \; 0 \cdots$$

without ever moving to the left of its initially scanned square.

Argument.[†] The machine $\mathbf{T}$ is designed to behave as follows. When presented with the pattern

$$\cdots \underline{0} \; \bar{x}_1 \; 0 \; \bar{x}_2 \; 0 \cdots 0 \; \bar{x}_n \; 0 \cdots$$

$\mathbf{T}$ first duplicates the entire argument string, so as to produce the pattern:

$$\cdots 0 \; \bar{x}_1 \; 0 \; \bar{x}_2 \; 0 \cdots 0 \; \bar{x}_n \; \underline{0} \; \bar{x}_1 \; 0 \; \bar{x}_2 \; 0 \cdots 0 \; \bar{x}_n \; 0$$

This can be accomplished by repeated applications of the copying procedure of Section 2.1, and does not require the machine to move to the left of its initially scanned square. Starting from the blank square in front of the second set of arguments, $\mathbf{T}$ now evaluates the function g in standard form. Since this evaluation does not carry the machine to the left of the starting point, it is not affected by the presence of the original set of arguments. The final tape pattern will be

$$\cdots 0 \; \bar{x}_1 \; 0 \; \bar{x}_2 \; 0 \cdots 0 \; \bar{x}_n \; \underline{0} \; \overline{g(x_1, \ldots, x_n)} \; 0 \cdots$$

as was required.   □

[†] A rigorous proof would require a detailed explanation of the design and operation of the machine $\mathbf{T}$. Since such an explanation would be very tedious and since it should be clear to the reader that the details can be filled in if required, we will give only the strategy of the proof.

Lemma 3-1 can be extended in a natural way to the evaluation of several functions.

Lemma 3-2. *If $g_1, \ldots, g_m$ are n-variable Turing-computable functions, then there exists a Turing machine* **T** *that converts each tape pattern of the form*

$$\cdots \underline{0} \, \bar{x}_1 \, 0 \cdots 0 \, \bar{x}_n \, 0 \cdots$$

into the pattern

$$\cdots 0 \, \bar{x}_1 \, 0 \cdots 0 \, \bar{x}_n \, \underline{0} \, \overline{g_1(x_1, \ldots, x_n)} \, 0 \cdots 0 \, \overline{g_m(x_1, \ldots, x_n)} \, 0 \cdots$$

without ever moving to the left of its initially scanned square.

Argument. The machine **T** begins its computation by duplicating its argument string and evaluating g_1, exactly as described in Lemma 3-1. This yields the tape pattern:

$$\cdots 0 \, \bar{x}_1 \, 0 \cdots 0 \, \bar{x}_n \, \underline{0} \, \overline{g_1(x_1, \ldots, x_n)} \, 0 \cdots$$

The machine **T** next forms another copy of the argument string just to the right of the block $\overline{g_1(x_1, \ldots, x_n)}$, so as to produce the pattern:

$$\cdots 0 \, \bar{x}_1 \, 0 \cdots 0 \, \bar{x}_n \, 0 \, \overline{g_1(x_1, \ldots, x_n)} \, \underline{0} \, \bar{x}_1 \, 0 \cdots 0 \, \bar{x}_n \, 0 \cdots$$

It now evaluates the function g_2 in standard form, producing the pattern:

$$\cdots 0 \, \bar{x}_1 \, 0 \cdots 0 \, \bar{x}_n \, 0 \, \overline{g_1(x_1, \ldots, x_n)} \, \underline{0} \, \overline{g_2(x_1, \ldots, x_n)} \, 0 \cdots$$

In a similar way, **T** proceeds to evaluate $g_3(x_1, \ldots, x_n)$, $g_4(x_1, \ldots, x_n)$, ..., $g_m(x_1, \ldots, x_n)$. Upon completing the last of these evaluations, it shifts its reading head $m - 1$ blocks to the left and halts. $\qquad\square$

Lemma 3-2 provides the basis for an important result concerning the capabilities of Turing machines.

Theorem 3-2. *Let $g_1, \ldots, g_m$ be n-variable Turing-computable functions, and let h be an m-variable Turing-computable function. Then the composite function $f = h \circ (g_1, \ldots, g_m)$ is also Turing computable.*

Argument. A machine that evaluates f can be designed to operate as follows. Upon being presented with the argument string

$$\cdots \underline{0} \, \bar{x}_1 \, 0 \cdots 0 \, \bar{x}_n \, 0 \cdots$$

the machine first computes $g_1(x_1, \ldots, x_n)$, ..., $g_m(x_1, \ldots, x_n)$ in the manner described in Lemma 3-2. This yields the tape pattern:

$$\cdots 0 \, \bar{x}_1 \, 0 \cdots 0 \, \bar{x}_n \, \underline{0} \, \overline{g_1(x_1, \ldots, x_n)} \, 0 \cdots 0 \, \overline{g_m(x_1, \ldots, x_n)} \, 0 \cdots$$

The machine now computes h in standard form and halts, leaving the pattern

$$\cdots 0 \, \bar{x}_1 \, 0 \cdots 0 \, \bar{x}_n \, \underline{0} \, \overline{h(g_1(x_1, \ldots, x_n), \ldots, g_m(x_1, \ldots, x_n))} \, 0 \cdots$$

on its tape. Thus it has indeed evaluated the function $h \circ (g_1, \ldots, g_m)$ for the arguments $x_1, \ldots, x_n$. □

Thus the set of Turing-computable functions is closed under functional composition. Similar arguments can be used to establish closure under a variety of other rules for defining new functions from old functions. As we will see in Chapters 4 and 5, these closure properties greatly simplify the problem of determining what functions are Turing computable.

Indexing

We are now ready to use the class of Turing machines as the basis for defining a family of algorithms. The first step is to view each Turing machine as computing some n-variable number-theoretic function for each choice of n. This function will be defined in terms of the conventions already established for the evaluation of number-theoretic functions. But for reasons that will become apparent, it is convenient to consider only computations carried out on singly infinite tapes. Since all doubly infinite computations can be simulated on singly infinite tapes, this restriction implies no real loss of generality.

With each Turing machine **T**, and for each positive integer n, we associate the n-variable number-theoretic function $F^{(n)}$ whose values are defined as follows. Let $x_1, \ldots, x_n$ be any choice of natural number arguments. Supply the machine **T** with a singly infinite tape containing the pattern

$$\underline{0} \; \bar{x}_1 \; 0 \cdots 0 \; \bar{x}_n \; 0 \cdots$$

and place the left-most square under the reading head, as indicated. If the machine **T** subsequently halts with its reading head in front of a block of $(m + 1)$ 1's, the value of $F^{(n)}(x_1, \ldots, x_n)$ is defined to be m. If **T** halts in some other configuration, or if it runs off its tape, or if it fails to halt, the value of $F^{(n)}(x_1, \ldots, x_n)$ is undefined. The function $F^{(n)}$ is referred to as the n-variable number-theoretic function *computed* (or evaluated) by **T**.

Several features of this association of functions and machines bear emphasizing. First, for each choice of n, some n-variable number-theoretic function has been assigned to every Turing machine. Second, every function so assigned is necessarily Turing computable. And third, because every Turing-computable function can be evaluated in standard form—and so can be evaluated on a singly infinite tape—every Turing-computable function has been assigned to some machine.

Next we must devise a way of assigning indices to Turing machines and thereby to the functions computed by Turing machines. What we will actually do is assign indices to descriptions of Turing machines, rather than to the machines themselves. As a result, each machine will receive many different indices. But because we are ultimately interested in the association of indices with functions, and because each function will inevitably receive many different indices anyway, such a multiple assignment of indices to machines is not objectionable.

There are many ways in which descriptions of Turing machines might be indexed. One of the most obvious is to arrange the various quintuple descriptions in some convenient order (lexicographic, say) and assign indices to them in sequence. The disadvantage of this approach lies in its implementation. It is going to be necessary to have Turing machines that translate indices into quintuple descriptions and vice versa. And while it is certainly possible to design such machines for the indexing scheme just mentioned, the complexity of their operation renders the whole approach rather unappealing.

A second possible approach makes use of the fact that every quintuple description can be viewed as the binary representation of a natural number. Of course, not every binary representation of a natural number corresponds to a legitimate quintuple description. That is, the patterns of 0's and 1's that make up the binary representations of many numbers will fail to contain the right number of blocks of 1's or will fail to have the proper separations between blocks, and so forth. However, if we relax our format constraints a little, it will be possible to interpret every finite pattern of 0's and 1's as the description of some Turing machine. The key to this interpretation is the universal machine discussed in Section 2.3. With minor modifications in its design, this machine will behave as if every finite pattern of 0's and 1's were the description of some Turing machine.

Consider the ways in which a finite string of 0's and 1's might fail to meet the standard format required of quintuple descriptions. To begin with, the string might have a sequence of three or more consecutive 0's somewhere within it. Referring back to Section 2.3, we find that the universal machine interprets such a sequence of 0's as denoting the end of a machine description and ignores the following symbols. Thus in interpreting an arbitrary string of 0's and 1's as a machine description, we shall ignore all symbols following the first occurrence of three consecutive 0's. Within the portion of the string preceding the first occurrence of three 0's, we interpret the substrings lying between occurrences of two consecutive 0's as representing "quintuples," even though these substrings may not have the standard form for the quintuples of a machine description. Thus the string

$$1\,1\,0\,1\,1\,1\,0\,0\,1\,0\,1\,0\,1\,1\,0\,1\,1\,1\,0\,1\,1\,1\,0\,1\,0\,0\,1\,1\,0\,1\,0\,1\,1\,1\,0\,0\,0\,1\,1\,0\,1\,1\,1\,1\,0\,0\,1$$

"quintuple"   "quintuple"   "quintuple"

contains three "quintuples," as indicated.

There are three ways in which such a "quintuple" can fail to meet the standard format: it can have more than five blocks; it can have fewer than five blocks; or it can have exactly five blocks, with the fourth (directional) block containing more than two 1's.[†] Because of the way in which the universal

---

[†] Note that no matter what lengths the other blocks have, they always denote well-defined states or symbols. If these states or symbols don't happen to be mentioned in any other quintuple, they simply represent states or symbols whose occurrence causes the machine in question to halt.

machine was designed, any extra blocks in a "quintuple" will simply be ignored. A "quintuple" with fewer than five blocks, however, may cause the universal machine to damage or destroy its tape description. We therefore modify the universal machine so that it halts whenever it finds that the next quintuple component it would normally examine is missing. As a result of this modification, quintuples with fewer than five components are treated as if they didn't appear at all. As far as nonstandard direction blocks are concerned, we note that the universal machine will automatically interpret a directional block with more than two 1's as specifying a shift to the right.

Finally, we must consider the case in which two or more quintuples begin with the same first and second components. Such quintuples might specify two different machine moves for the same combination of current state and scanned symbol. Again referring back to the behavior of the universal machine, we find that only the *first* quintuple having a given combination of first two components will ever be interpreted as defining a machine move. Thus no conflict arises from the presence of quintuples with identical first and second blocks.

Thus, with minor modifications, our universal machine treats every finite pattern of 0's and 1's as the description of some Turing machine.[†] Since every natural number has a unique binary representation as a pattern of 0's and 1's, every natural number can be interpreted as representing some Turing machine. We will accordingly consider the number n to be an *index* of the Turing machine **T** iff the binary expansion of n is interpreted by the modified universal machine as a description of **T**. Although somewhat artificial, this scheme has the advantage of making it quite easy to convert back and forth between indices and machine descriptions.

In summary, we have chosen a means of assigning indices to Turing machine descriptions and hence to the functions computed by Turing machines. The adoption of such a scheme enables us to view Turing machines (or their descriptions) as the members of a family of algorithms.

3.2 THE FAMILY OF TURING MACHINES

We now begin to explore the role of Turing machines as implementations of the members of a family of algorithms. This family, whose members are officially descriptions of Turing machines, will be denoted $\mathscr{T}$. Paralleling the notation of Chapter 1, the Turing machine that is assigned the index i will be denoted $\mathbf{T}_i$ and referred to as the ith Turing machine. The n-variable number-theoretic function computed by the machine $\mathbf{T}_i$ will be denoted $F_i^{(n)}$, and i will be referred to as an *index* of this function. In dealing with the family $\mathscr{T}$, it should be remembered that a given function may be computed by many different ma-

---

[†] When it is necessary to make the distinction, we will refer to a description that satisfies the format presented in Section 2.3 as a *standard* Turing machine description.

chines and that a given machine may have many different descriptions. Thus a machine, like a function, may have many different indices.

Our first job is to determine which of the five basic properties discussed in Chapter 1 are satisfied by the family $\mathscr{T}$. Three of the properties can be dealt with quite easily. It is a routine matter to show that the constant, projection, and successor functions are all Turing computable, as is the selection function Λ. Thus the family $\mathscr{T}$ satisfies Property 1 (the Base-Function Property) and Property 4 (the Selection Property). And Theorem 3-2 assures us that the set of Turing-computable functions is closed under functional composition, which means that $\mathscr{T}$ satisfies Property 2 (the Closure Property).

Although Properties 3 and 5 also hold for the family $\mathscr{T}$, this fact is harder to establish. Each of these properties will be discussed in detail in the remainder of this section. Property 3 (the Enumeration Property) follows from the existence of a universal Turing machine but, as we will see, a certain amount of coding and decoding is needed to adapt the behavior of the universal machine to the requirements of Property 3. The fact that Property 3 holds for $\mathscr{T}$ is usually referred to as the "Enumeration Theorem for Turing Machines." Establishing Property 5 (the S-M-N Property) requires a deeper understanding of Turing machine capabilities and will be undertaken last. The fact that Property 5 holds for $\mathscr{T}$ is usually referred to as the "S-M-N Theorem for Turing Machines."

The Enumeration Theorem

We now set out to argue that the family $\mathscr{T}$ satisfies Property 3. To do this, we must show that for each positive integer n, there exists a Turing machine that evaluates the universal function $U^{(n+1)}$, where

$$U^{(n+1)}(i, x_1, \ldots, x_n) = F_i^{(n)}(x_1, \ldots, x_n).$$

We will discuss only the case in which $n = 1$, leaving to the reader the obvious extensions to larger values of n.

Consider, then, the design of a machine $\mathbf{M}_u$ that evaluates the universal function $U^{(2)}$. Since $\mathbf{M}_u$ must determine the result of supplying an arbitrary natural number to an arbitrary Turing machine, it is clear that $\mathbf{M}_u$ must incorporate a universal machine like that described in Section 2.3. But $\mathbf{M}_u$'s initial tape pattern consists of the unary representations of a machine index and an argument value, while the universal machine $\mathbf{U}$ requires specially coded machine and tape descriptions as its inputs. Thus $\mathbf{M}_u$ must begin by converting its unary inputs into the special descriptions needed by the universal machine. It can then proceed to act like the universal machine and simulate the desired computation. When the simulation is over, $\mathbf{M}_u$ converts the resulting tape description back into unary form and halts.

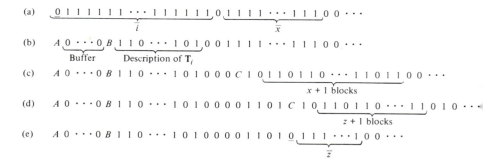

Figure 3-4

Figure 3-4 illustrates the major steps in $\mathbf{M}_u$'s computation. As indicated in line (a) of the figure, the initial tape pattern contains just two blocks of 1's, representing natural numbers i and x. The machine's first job is to replace the unary representation of i by the corresponding description of the machine $\mathbf{T}_i$, together with an appropriate buffer region and A and B markers, as shown in line (b) of the figure. The next step is to replace the unary representation of x by the encoded description of the tape pattern $0\ \bar{x}$, preceded by a C marker. Line (c) of Fig. 3-4 shows the appearance of $\mathbf{M}_u$'s tape after this conversion has been accomplished. The machine $\mathbf{M}_u$ now acts like the universal machine $\mathbf{U}$ of Section 2.3, simulating the computation performed by the machine $\mathbf{T}_i$ for the initial pattern $0\ \bar{x}$. If the simulation terminates, $\mathbf{M}_u$ checks the resulting tape description to determine whether $\mathbf{T}_i$ halts in front of a block of 1's. If it does, as indicated in line (d) of Fig. 3-4, $\mathbf{M}_u$ converts the encoded description of that block back into unary and halts as shown in line (e). Otherwise, $\mathbf{M}_u$ shifts right indefinitely, never halting.

First consider the conversion from the unary representation of i to the description of $\mathbf{T}_i$. In making this conversion, the machine $\mathbf{M}_u$ must leave enough room to the left of the description to serve as a buffer region. Note that the required description of $\mathbf{T}_i$ is just the binary representation of i, and that the binary representation of a number is usually much shorter than the unary representation. Thus the space occupied by the block $\bar{i}$ will in most cases be large enough to contain both the description of $\mathbf{T}_i$ and the necessary buffer region. In particular, whenever $i \geq 10$, the space available for the buffer exceeds the length of the description, and is certainly adequate for recording any state-symbol combination. If $i < 10$, the binary representation of i must have fewer than five blocks of 1's. Since such a pattern describes a machine that halts without making any moves at all, no simulation is required in this case.

The machine $\mathbf{M}_u$ therefore begins its computation by counting the number of 1's in the block representing the index i. If this number is ten or less, $\mathbf{M}_u$

simply halts in front of the block representing x. (Since in this case the machine T_i halts immediately, M_u has produced the correct output.) If the number of 1's in the i-block is greater than ten, M_u erases the last 1 and writes the special symbol A before and after the shortened block. The resulting pattern $A\ 1\ 1 \cdots 1\ A$ will then contain exactly i 1's.

The machine M_u must now transform the block of i 1's into the binary representation of i, placing that representation in the right-hand portion of the region between the A's. The strategy used for making this transformation is based on the usual algorithm for obtaining the binary representation of a natural number. Recall that this algorithm consists in performing successive divisions by two, the first division being performed on the given number, the next on the quotient obtained from the first division, and so on. The successive remainders obtained from this process yield the digits of the desired binary expansion, beginning with the lowest-order digit. The whole process can be implemented by a Turing machine that makes a series of passes across the given block of 1's. During each pass, the machine effects a division by two by marking off alternate 1's. The quotient resulting from this division is represented by the number of 1's left unmarked, while the remainder is indicated by whether or not the last 1 was marked during the pass. Details of the process are left to the reader.

Upon completing the binary expansion, the machine M_u writes a B in front of the binary representation of i and erases the A from the right-hand end. This produces the pattern shown in line (b) of Fig. 3-4, and provides a description of the machine T_i in the form required by the universal machine.

In the second stage of its computation, M_u must convert the unary representation of x into the encoded description of the tape pattern $\underline{0}\ \bar{x}$. To ensure sufficient separation between the machine description and the tape description, M_u shifts the unary representation of x two squares to the right, at which point it writes the symbol C in front of the block. It next erases the second 1 in the block and adds two 1's at the right-hand end, so as to produce the pattern

$$C\ 1\ 0\ \underbrace{1\ 1\ 1 \cdots 1}_{x+1}\ 0\ 0 \cdots .$$

Note that the single 1 following the C is the encoded representation of T's initially scanned blank square. The machine must now expand the block of $(x + 1)$ 1's into a sequence of $x + 1$ blocks of two 1's each. This can be done in a series of passes. During each pass the machine converts one additional 1 into an encoded block and shifts the remaining 1's two squares to the right to make room for the newly created block. This process leads in a routine way to the tape pattern shown in line (c) of Fig. 3-4.

At this point M_u is ready to simulate the behavior of the machine T_i, which it does in the manner of the universal machine described in Section 2.3. If the simulation terminates, $\mathbf{M}_u$ must determine whether $\mathbf{T}_i$'s computation produces a number-theoretic output. That is, $\mathbf{M}_u$ must determine whether the encoded representation of $\mathbf{T}_i$'s final tape pattern has one of the forms:

$$\cdots \underline{C}\; 1\; 0\; \underbrace{1\; 1\; 0\; 1\; 1\; 0\; \cdots\; 0\; 1\; 1\; 0}\; 1\; 0 \cdots$$
$$z + 1 \text{ blocks}$$

or

$$\cdots \underline{C}\; 1\; 0\; \underbrace{1\; 1\; 0\; 1\; 1\; 0\; \cdots\; 0\; 1\; 1}\; 0\; 0\; 0 \cdots .$$
$$z + 1 \text{ blocks}$$

If so, $\mathbf{M}_u$ proceeds to decode $\mathbf{T}_i$'s final tape pattern; if not, it simply shifts right indefinitely, never halting.

To decode $\mathbf{T}_i$'s output value, $\mathbf{M}_u$ must convert either of the preceding tape patterns into the pattern

$$\cdots \underline{0}\; \underbrace{1\; 1\; \cdots\; 1}\; 0.$$
$$(z + 1) \text{ 1's}$$

It begins by erasing the original C, the single 1 that follows it, and (if there is one) the single 1 at the end of the pattern. This converts the original pattern into the pattern:

$$0\; 0\; \underbrace{1\; 1\; 0\; 1\; 1\; 0\; 1\; 1\; \cdots\; 0\; 1\; 1}$$
$$z + 1$$

This sequence of $z + 1$ blocks can now be collapsed into a single block of $(z + 1)$ 1's by a process that is essentially the reverse of that used for the original expansion.

This completes our description of the major steps in the operation of the machine $\mathbf{M}_u$. In light of this description, it should be clear that it is in fact possible to build a Turing machine that evaluates the universal function $U^{(2)}$. The details of the design are left to the reader. (See Problems 3.2.1 through 3.2.3.)

By generalizing the process of encoding tape descriptions, it is also possible to design machines that evaluate the universal functions $U^{(3)}$, $U^{(4)}$, $U^{(5)}$, etc. We can therefore establish:

Theorem 3-3. (*The Enumeration Theorem for Turing Machines*). *For each positive integer n there exists a Turing machine that evaluates the universal function $U^{(n+1)}$, where*

$$U^{(n+1)}(i, x_1, \ldots, x_n) = F_i^{(n)}(x_1, \ldots, x_n).$$

Thus the family $\mathcal{T}$ satisfies Property 3 of Chapter 1.

The S-M-N Theorem

To establish that the S-M-N Property holds for the family $\mathcal{T}$, we must show that for every choice of positive integers m and n there is a Turing-computable function $s_{n,m}$ such that

$$F^{(n)}_{s_{n,m}(i,\,x_1,\,\ldots,\,x_m)}(y_1,\,\ldots,\,y_n) = F^{(m+n)}_i(x_1,\,\ldots,\,x_m,\,y_1,\,\ldots,\,y_n)$$

for all i, $x_1, \ldots, x_m$, and $y_1, \ldots, y_n$. We will consider in detail the special case in which $m = n = 1$. In other words, we will show how to design a machine $\mathbf{M}_s$ to evaluate a function s such that

$$F^{(1)}_{s(i,\,x)}(y) = F^{(2)}_i(x,\,y)$$

for all i, x, and y. Once the operation of this machine is understood, it is a routine matter to design corresponding machines for other values of m and n.

What the machine $\mathbf{M}_s$ must do, when presented with the unary representations of integers i and x, is to compute the *index* of a machine, call it $\mathbf{W}_{i,\,x}$, that yields the same result for the argument y as the machine $\mathbf{T}_i$ would yield for the pair of arguments x and y. Note that for any particular choice of i and x, the function to be computed by the new machine $\mathbf{W}_{i,\,x}$ is completely determined. To evaluate this function, $\mathbf{W}_{i,\,x}$ can proceed as follows. Upon being presented with the unary representation of the number y, $\mathbf{W}_{i,\,x}$ first inserts the unary representation of the (fixed) number x in front of the representation of y. For example, if $x = 4$, $\mathbf{W}_{i,\,x}$ would convert the initial pattern

$$\underline{0}\ 1\ 1\ 1\ 1\ 1\ 1\ 0\ \cdots$$

into the pattern

$$\underline{0}\ 1\ 1\ 1\ 1\ 1\ 0\ 1\ 1\ 1\ 1\ 1\ 1\ 0\ \cdots.$$

Having done this, $\mathbf{W}_{i,\,x}$ acts exactly like the machine $\mathbf{T}_i$, thereby evaluating $F^{(2)}_i(x,\,y)$.

Since the length of the inserted x-block is fixed, the insertion process is easily accomplished. The machine $\mathbf{W}_{i,\,x}$ first shifts the given y-block one square to the right. It then proceeds to write down the members of the x-block one at a time, shifting the y-block one square to the right after writing each 1. When the requisite number of 1's has been written, the machine returns to the left end of the new block, ready to act like the machine $\mathbf{T}_i$. Figure 3-5 shows a state-diagram description for the part of $\mathbf{W}_{i,\,x}$ that carries out the insertion process. Note that the number of states required depends on the value of x. Three states are used for the initial shifting of the y-block, three additional states are used for inserting each 1 in the x-block, and one state is used to reposition the reading head at the end of the insertion process.

Now that we understand the relationship between the machine $\mathbf{W}_{i,\,x}$ and the machine $\mathbf{T}_i$, we can consider the design of the machine $\mathbf{M}_s$. Basically, $\mathbf{M}_s$

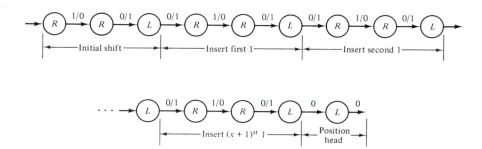

Figure 3-5

will be made to operate as follows. When presented with the unary representations of i and x, M_s first converts the unary representation of i into the quintuple description of the machine T_i. It then uses the given value of x to convert the description of T_i into a description of $W_{i,x}$. Finally, it computes the index of $W_{i,x}$—that is, the unary equivalent of the description of $W_{i,x}$. This index is, of course, the desired value of the function s.

Figure 3-6 illustrates the various stages in the computation performed by M_s. Line (a) shows the form of the initial tape pattern. During phase 1, M_s converts the unary representation of i into binary form, using the technique discussed in connection with the Enumeration Theorem. The resulting quintuple description of the machine T_i is located as shown in line (b) of Fig. 3-6, and delineated by the special marking symbols P and Q. (Note that the portion of the tape between the P and the Q contains just the quintuples of T_i's description—no A or B markers and no buffer region.)

In order to obtain the quintuple description of $W_{i,x}$, the machine M_s must add to the description of T_i a set of quintuples corresponding to the state diagram of Fig. 3-5. If the states involved in these new quintuples were simply numbered 0, 1, 2, 3, etc., some of them might receive numbers already assigned to states in T_i. To avoid such conflicts without having to renumber the states in T_i's description, we shall require that the state numbers used in the new quintuples all be larger than any of the state numbers appearing in T_i's description. In particular, if the largest state number in T_i's description is v, then the states in the added quintuples are to be numbered $v + 1$, $v + 2$, $v + 3$, etc. To this end, Phase 2 of M_s's computation consists in determining the highest numbered state in T_i's description. This is done in the obvious way by making successive comparisons among the current and next state components of the quintuples in the description of T_i. Once M_s has located the largest state block, it forms a copy of this block just to the right of the unary representation of x. It then adds an extra 1 to the new block, so as to obtain the unary representation of $v + 1$, and writes an A in front of the new block, as shown in line (c) of Fig. 3-6.

(a) Initial tape

| 0 | $\overline{i}$ | 0 | $\overline{x}$ | Blank |

(b) Phase 1

| Blank | P | Description of T_i | Q | $\overline{x}$ | Blank |

(c) Phase 2

| Blank | P | Description of T_i | Q | $\overline{x}$ | A | $\overline{v+1}$ | Blank |

(d) Phase 3

| Blank | P | Description of T_i | Q | $\overline{x}$ | 0 | New quintuples | Blank |

(e) Phase 4

| Blank | P | Description of T_i | Q | $\overline{x}$ | 0 | New quintuples | 00 | Description of T_i | Blank |

(f) Phase 5

| Blank | P | Description of T_i | Q | $\overline{x}$ | 0 | Description of $\mathrm{W}_{i,x}$ | 0 | $\overline{s(i,x)}$ | Blank |

Figure 3-6

During Phase 3, the machine $\mathbf{M}_s$ generates the new quintuples that must be added to $\mathbf{T}_i$'s description. In doing this, it uses the length of the unary representation of x to determine how many 1's the machine $\mathbf{W}_{i,\,x}$ must insert in its tape pattern, and consequently how many sets of "1-inserting" quintuples must be generated. The new quintuples are formed in the region of the tape immediately to the right of the x-block, as shown in line (d) of Fig. 3-6. (The block representing $v + 1$ becomes incorporated in the first quintuple.)

To give some idea how Phase 3 is carried out, consider the generation of the quintuples corresponding to the portion of the diagram of Fig. 3-5 that accomplishes the initial shift operation. Numbering the states of the diagram from left to right (starting with $v + 1$) and remembering that certain "self loops" have been omitted from the diagram, we find that the appropriate quintuples are:

$$(v + 1,\, 0,\, 0,\, R,\, v + 1),\ (v + 1,\, 1,\, 0,\, R,\, v + 2),\ (v + 2,\, 1,\, 1,\, R,\, v + 2),$$
$$(v + 2,\, 0,\, 1,\, L,\, v + 3),\ (v + 3,\, 1,\, 1,\, L,\, v + 3)$$

When encoded in unary, these quintuples form the following pattern:

$v + 1\ 0\ 1\ 0\ 1\ 0\ 1\ 1\ 0\ \overline{v + 1}\ 0\ 0\ \overline{v + 1}\ 0\ 1\ 1\ 0\ 1\ 0\ 1\ 1\ 0\ \overline{v + 2}\ 0\ 0\ \overline{v + 2}\ 0\ 1\ 1\ 0\ 1\ 1$
$0\ 1\ 1\ 0\ \overline{v + 2}\ 0\ 0\ \overline{v + 2}\ 0\ 1\ 0\ 1\ 1\ 0\ 1\ 0\ \overline{v + 3}\ 0\ 0\ \overline{v + 3}\ 0\ 1\ 1\ 0\ 1\ 1\ 0\ 1\ 0\ \overline{v + 3}$

Thus $\mathbf{M}_s$ must begin Phase 3 by generating this pattern of 1's and blanks, using the existing block $\overline{v + 1}$ as a starting point.

Figure 3-7 indicates how this is done. Starting with the initial block shown in line (a), the machine first adds on the (fixed) pattern $0\ 1\ 0\ 1\ 0\ 1\ 1\ B$ and returns to the symbol A, as shown in line (b). It then forms a copy of the block marked by the A immediately to the right of the B, as shown in line (c). It next erases the A, changes the B to an A, and adds $0\ B$ to the end of the pattern, as shown in line (d). Another copying (line e) followed by a shift of marking symbols yields the pattern shown in line (f). At this point the first quintuple has been generated, and the first block of the next quintuple has been set up. Thus the machine is now in a position comparable to that of line (a), and can proceed

(a) $\underline{A}\ \overline{v + 1}$

(b) $\underline{A}\ \overline{v + 1}\ 0\ 1\ 0\ 1\ 0\ 1\ 1\ B$

(c) $\underline{A}\ \overline{v + 1}\ 0\ 1\ 0\ 1\ 0\ 1\ 1\ B\ \overline{v + 1}$

(d) $0\ \overline{v + 1}\ 0\ 1\ 0\ 1\ 0\ 1\ 1\ \underline{A}\ \overline{v + 1}\ 0\ B$

(e) $0\ \overline{v + 1}\ 0\ 1\ 0\ 1\ 0\ 1\ 1\ \underline{A}\ \overline{v + 1}\ 0\ B\ \overline{v + 1}$

(f) $0\ \overline{v + 1}\ 0\ 1\ 0\ 1\ 0\ 1\ 1\ 0\ \overline{v + 1}\ 0\ \underline{A}\ \overline{v + 1}$

Figure 3-7

with the generation of the next quintuple in a manner similar to that just outlined. The third, fourth, and fifth quintuples are produced in the same way.

When the quintuples corresponding to the initial shifting operation of Fig. 3-5 have been completed, M_s proceeds to generate the "1-inserting" quintuples. In this phase of the operation it marks the first 1 in the representation of x, generates a sequence of five quintuples that cause the insertion of a single 1 on $W_{i,x}$'s tape, marks another 1 in the x-block, generates another sequence of five quintuples, and so forth. When the x-block is exhausted, M_s adds on the additional quintuples needed to reposition $W_{i,x}$'s reading head. Since the last of these quintuples must cause a transition to T_i's initial state, its fifth block should be a copy of the first block in T_i's description.

During Phase 4, the machine M_s copies T_i's description into the region to the right of the new quintuples, as shown in line (e) of Fig. 3-6. The resulting collection of quintuples constitutes a complete description of the machine $W_{i,x}$. Note that the starting state, $v + 1$, is represented at the beginning of the list of quintuples, as required. During Phase 5, M_s converts the binary description of $W_{i,x}$ into its unary equivalent. (Binary to unary conversion can be accomplished by repeatedly doubling an initial block of length one, adding a single 1 after the kth doubling iff the kth digit in the given binary representation is a 1. See Problem 3.2.5.) The resulting unary representation is formed to the right of the quintuple description of $W_{i,x}$, as shown in line (f) of Fig. 3-6. Since this block represents the index of $W_{i,x}$, and therefore the desired value of $s(i, x)$, the machine M_s now halts.

Although we have only sketched the behavior and construction of the machine M_s, the reader should be able to convince him or herself that the necessary details can be filled in and the machine can be made to operate as advertised. Thus there exists a two-variable Turing-computable function s such that

$$F^{(1)}_{s(i,x)}(y) = F^{(2)}_i(x, y)$$

for all i, x, and y.

The approach used to design the machine M_s is easily extended to deal with functions of arbitrary numbers of variables. In this way we obtain:

Theorem 3-4. (*The S-M-N Theorem for Turing Machines*[†]) *For each choice of positive integers m and n, there exists a total Turing-computable function $s_{n,m}$ such that*

$$F^{(n)}_{s_{n,m}(i, x_1, \ldots, x_m)}(y_1, \ldots, y_n) = F^{(m+n)}_i(x_1, \ldots, x_m, y_1, \ldots, y_n)$$

for all i, $x_1, \ldots, x_m$, $y_1, \ldots, y_n$.

Thus the family $\mathscr{T}$ satisfies Property 5 of Chapter 1.

[†] This theorem is also known as the Iteration Theorem for Turing machines.

Conclusions

When coupled with the fact that the constant, projection, successor, and selection functions are all Turing computable, Theorems 3-2, 3-3, and 3-4 tell us that the family of Turing machines satisfies the five basic properties presented in Chapter 1. As a consequence, the family of Turing machines is an example of a standard family and must share all the features common to such families. Thus, for example, the various domain, totality, and equivalence functions for the family $\mathcal{T}$ cannot be computed by members of $\mathcal{T}$. On the other hand, many index-computing functions are computable by members of $\mathcal{T}$, and $\mathcal{T}$ must satisfy the Recursion Theorem.

The fact that the family $\mathcal{T}$ satisfies Properties 1, 2, and 4 does not depend on the way in which the members of $\mathcal{T}$ are indexed. But the fact that Properties 3 and 5 hold obviously does depend on the choice of an indexing scheme. It is therefore appropriate to ask whether the rather contrived indexing used here is the only one for which these properties hold. If it is, perhaps we should question the significance of these properties and the conclusions that follow from them.

There are actually many indexing schemes for which the Enumeration and S-M-N Theorems hold. From a practical point of view, these schemes include all the "natural" ways of indexing Turing machines or their descriptions, such as the one which results from ordering the legitimate quintuple descriptions lexicographically. The only requirement we need place on a proposed indexing scheme is that there be Turing machines that are capable of converting indices into quintuple descriptions and vice versa. Since this constraint admits a wide range of indexing schemes, we conclude that the Enumeration and S-M-N Theorems are quite fundamental properties of the family $\mathcal{T}$.

3.3 BASIC CAPABILITIES AND LIMITATIONS

In this section we reexamine some of the consequences of the five basic properties, rephrasing the proofs of these consequences in terms of Turing machine computations. In some cases we will find that our knowledge of Turing machines allows us to construct proofs that are more understandable than those given in Chapter 1. And in some cases we will be able to establish additional results that have no direct counterparts in the context of Chapter 1.

Throughout the discussion we will use $\mathbf{T}_i$ to denote the ith Turing machine according to our agreed-upon indexing scheme, and $F_i^{(n)}$ to denote the n-variable number-theoretic function evaluated by $\mathbf{T}_i$. It is important to remember that $\mathbf{T}_i$ denotes a *machine* and $F_i^{(n)}$ a *function*. Thus while we may freely write $F_i^{(1)}(3)$ to denote a function value, the expression $\mathbf{T}_i(3)$ has (for us) no defined meaning. Conversely, while the machine $\mathbf{T}_i$ may be presented with arbitrary tape patterns—including those that do not represent sequences of natural numbers—we cannot speak of applying the function $F_i^{(n)}$ to a tape pattern.

Domain and Halting Functions

Among the functions that cannot be evaluated in any standard family of algorithms are the various domain functions. These include the general domain functions $D^{(2)}$, $D^{(3)}$, $D^{(4)}$, etc., where

$$D^{(n+1)}(i, x_1, \ldots, x_n) = \begin{cases} 1 & \text{if} & F_i^{(n)}(x_1, \ldots, x_n)\downarrow \\ 0 & \text{if} & F_i^{(n)}(x_1, \ldots, x_n)\uparrow \end{cases}$$

as well as the diagonal domain function d, where

$$d(x) = \begin{cases} 1 & \text{if} & F_x^{(1)}(x)\downarrow \\ 0 & \text{if} & F_x^{(1)}(x)\uparrow. \end{cases}$$

The meaning of these domain functions for the family $\mathcal{T}$ follows immediately from the definition of the number-theoretic function evaluated by a given Turing machine. Thus if the machine $\mathbf{T}_i$ eventually halts with the representation of a natural number on its tape after being presented with the tape pattern $\bar{x}$, the value of $D^{(2)}(i, x)$ is 1. If $\mathbf{T}_i$ does not halt when presented with the pattern $\bar{x}$, or if it halts without the representation of a natural number next to its reading head, the value of $D^{(2)}(i, x)$ is 0. Similar remarks apply to the other domain functions for $\mathcal{T}$.

Since $\mathcal{T}$ is a standard family, Theorems 1-3 and 1-4 imply that none of the aforementioned domain functions can be evaluated by members of $\mathcal{T}$. It is instructive to reconsider some of these results in terms of Turing machine behavior. The counterpart of Theorem 1-4, for example, is:

Theorem 3-5. *The function d, where*

$$d(x) = \begin{cases} 1 & \text{if} & F_x^{(1)}(x)\downarrow \\ 0 & \text{if} & F_x^{(1)}(x)\uparrow \end{cases}$$

is not Turing computable.

The proof given in Chapter 1 may be summarized as follows:

General Proof. Assume that the function d is computable. Then the function $\hat{d}$ must also be computable, where

$$\hat{d}(x) = \begin{cases} 0 & \text{if} & d(x) = 0 \\ \uparrow & \text{if} & d(x) \neq 0. \end{cases}$$

Let w be an index for the function $\hat{d}$. Then

$$\hat{d}(w) = \begin{cases} 0 & \text{if} & d(w) = 0 \\ \uparrow & \text{if} & d(w) \neq 0 \end{cases} = \begin{cases} 0 & \text{if} & F_w^{(1)}(w)\uparrow \\ \uparrow & \text{if} & F_w^{(1)}(w)\downarrow \end{cases} = \begin{cases} 0 & \text{if} & \hat{d}(w)\uparrow \\ \uparrow & \text{if} & \hat{d}(w)\downarrow. \end{cases}$$

Thus $\hat{d}(w)$ is defined iff $\hat{d}(w)$ is undefined. Because of this contradiction, we conclude that d cannot in fact be computable. □

This argument is easily rephrased in terms of Turing machine behavior.

Turing Machine Argument. Assume that the function d is Turing computable and that **T** is a Turing machine that computes it. Modify **T** so as to form a new machine **T̂** that behaves as follows. **T̂** begins its computations by acting like **T**. If **T** produces an output of 0, **T̂** does likewise and halts. But if **T** produces an output of 1, **T̂** enters a state in which it keeps moving to the right, never halting. (Thus **T̂** evaluates the function $\hat{d}$.) Now suppose that w is an index of **T̂**, and consider what happens if **T̂** is presented with the tape pattern $\bar{w}$. By design, **T̂** will ultimately halt and produce a natural number as its output iff **T** would produce the natural number 0 when presented with the tape pattern $\bar{w}$. But by assumption, **T** will produce a 0 in this situation iff the machine **T**$_w$—that is, the machine **T̂**—fails to produce a natural number when presented with $\bar{w}$. Thus **T̂** produces a natural number when presented with $\bar{w}$ iff it does not produce a natural number when presented with $\bar{w}$. From this contradiction we conclude that the function d cannot in fact be Turing computable. □

The reader may rephrase the proof of Theorem 1-3 in a similar way.

Closely related to the domain functions are a number of functions that simply indicate whether a specified Turing machine halts, without reference to its final tape pattern. Consider, for example, a function h that indicates whether the machine **T**$_x$ eventually halts after being presented with the unary representation of x. We define h so that

$$h(x) = \begin{cases} 1 & \text{if} & \mathbf{T}_x \text{ halts for the initial pattern } \bar{x} \\ 0 & \text{if} & \mathbf{T}_x \text{ does not halt for the initial pattern } \bar{x}. \end{cases}$$

This function will be called the *diagonal halting function* for Turing machines.

The diagonal halting function has no counterpart in arbitrary families of algorithms, and was not covered by the discussion of Chapter 1. Like the diagonal domain function, however, it is not Turing computable. The proof of this fact is almost identical to that of Theorem 3-5.

Theorem 3-6. *The function h, where*

$$h(x) = \begin{cases} 1 & \text{if} & \mathbf{T}_x \text{ halts for the initial pattern } \bar{x} \\ 0 & \text{if} & \mathbf{T}_x \text{ does not halt for the initial pattern } \bar{x} \end{cases}$$

is not Turing computable.

Argument. Assume that h is computed by some machine $\mathbf{T}$. Modify $\mathbf{T}$ so as to obtain a new machine $\hat{\mathbf{T}}$, exactly as described in the proof of Theorem 3-5. Again let w be an index for $\hat{\mathbf{T}}$, and consider what happens when $\hat{\mathbf{T}}$ is presented with the pattern $\bar{w}$. By construction, $\hat{\mathbf{T}}$ will eventually halt iff $\mathbf{T}$ produces a 0. But by assumption, $\mathbf{T}$ produces a 0 when presented with the pattern $\bar{w}$ iff the machine $\mathbf{T}_w$—that is, the machine $\hat{\mathbf{T}}$—fails to halt when presented with $\bar{w}$. Thus $\hat{\mathbf{T}}$ halts when presented with $\bar{w}$ iff it fails to halt when presented with $\bar{w}$. From this contradiction we conclude that h cannot be Turing computable.   □

By analogy with the domain functions, we may also introduce a series of "general" halting functions. For each positive integer n, we define the $(n + 1)$-variable function $H^{(n+1)}$ so that

$$H^{(n+1)}(i, x_1, \ldots, x_n) = \begin{cases} 1 & \text{if} \quad \mathbf{T}_i \text{ halts for the pattern } \bar{x}_1 0 \cdots 0 \bar{x}_n \\ 0 & \text{if} \quad \mathbf{T}_i \text{ does not halt for the pattern } \bar{x}_1 0 \cdots 0 \bar{x}_n. \end{cases}$$

The function $H^{(n+1)}$ will be referred to as the *n-input halting function* for Turing machines. Arguments similar to that of Theorem 3-6 can be used to show that none of the functions $H^{(2)}$, $H^{(3)}$, ... is Turing computable. The details are left to the reader.

As a final example, we consider a halting function that is of special interest in the study of Turing machines. This is the *blank tape halting function*, which is denoted h_0 and defined as follows:

$$h_0(x) = \begin{cases} 1 & \text{if } \mathbf{T}_x \text{ halts after being started on blank tape} \\ 0 & \text{if } \mathbf{T}_x \text{ does not halt after being started on blank tape} \end{cases}$$

As might be expected, this function is not Turing computable either. But because of the blank tape input requirement, the noncomputability of h_0 is not readily established by applying some machine to its own index and thereby obtaining a contradiction. Instead we will make use of a reduction argument, showing that the computability of h_0 would imply that of h.

We begin by introducing a set of auxiliary machines that allow us to relate the functions h_0 and h. For each natural number x, we define a Turing machine $\mathbf{W}_x$ that behaves as follows. When started on blank tape, $\mathbf{W}_x$ first writes down a block of $(x + 1)$ 1's, then returns to the beginning of the block and proceeds to behave exactly like the machine $\mathbf{T}_x$. (Such a machine $\mathbf{W}_x$ is easily formed from $\mathbf{T}_x$ by the addition of approximately x states.) Now observe that $\mathbf{W}_x$ eventually halts when started on blank tape iff $\mathbf{T}_x$ eventually halts when presented with the pattern $\bar{x}$. Thus if we had a means of determining whether an arbitrary machine halts when started on blank tape, we could apply it to the machine $\mathbf{W}_x$ to determine whether the machine $\mathbf{T}_x$ halts when presented with $\bar{x}$.

Theorem 3-7. *The function h_0, where*

$$h_0(x) = \begin{cases} 1 & \text{if } \mathbf{T}_x \text{ halts when started on blank tape} \\ 0 & \text{otherwise} \end{cases}$$

is not Turing computable.

Argument. Assume that h_0 is computed by some Turing machine **T**. Design a new machine $\hat{\mathbf{T}}$ that operates as follows. When presented with the tape pattern $\bar{x}$, $\hat{\mathbf{T}}$ first constructs the quintuple description of the machine $\mathbf{W}_x$ defined above. It does this in much the same way that the machine $\mathbf{M}_s$ discussed in Section 3.2 constructs a description of the machine $\mathbf{W}_{i,\,x}$. Specifically, it first obtains the description of $\mathbf{T}_x$, then finds the highest numbered state in $\mathbf{T}_x$, then appends the quintuples needed to make $\mathbf{W}_x$ write down $\bar{x}$. The machine $\hat{\mathbf{T}}$ next converts the quintuple description of $\mathbf{W}_x$ into unary form, and erases the rest of its tape, leaving just the index of $\mathbf{W}_x$. Finally, $\hat{\mathbf{T}}$ behaves exactly as **T** would when presented with this index, leaving **T**'s final tape pattern as its own final tape pattern. Now, by definition of **T**, $\hat{\mathbf{T}}$ will produce an output of 1 iff $\mathbf{W}_x$ halts when started on blank tape. But $\mathbf{W}_x$ halts when started on blank tape iff $\mathbf{T}_x$ halts when presented with $\bar{x}$. Thus $\hat{\mathbf{T}}$ produces an output of 1 for the input $\bar{x}$ if $\mathbf{T}_x$ halts when presented with $\bar{x}$, and an output of 0 otherwise. In other words, $\hat{\mathbf{T}}$ evaluates the diagonal halting function h. Since h is known not to be Turing computable, it follows that h_0 cannot be Turing computable either.   □

We see, then, that the arguments used in Chapter 1 to establish the non-computability of the various domain functions can be applied directly to the case of Turing machines. And when rephrased in terms of Turing-machine behavior, these arguments can be used to establish the noncomputability of various functions that are peculiar to the Turing machine context.

Index Computation

The S-M-N Property, together with some of the other basic properties, ensures the computability of indices for a variety of interesting functions. We now review some of these index-computing capabilities as they pertain to Turing machines.

Consider first the result that establishes the computability of indices for constant functions. This result may be rephrased as:

Theorem 3-8. *There exists a total, one-variable Turing-computable function k such that $F_{k(x)}^{(1)} = C_x^{(1)}$ for all x.*

Before discussing the Turing machine aspects of this result, we repeat the proof given in Chapter 1.

General Proof. Let k be the function $s \circ (C_w^{(1)}, P_1^{(1)})$, where s is the S-M-N function for $m = n = 1$, and w is an index for $P_1^{(2)}$. This function k is obviously computable. Moreover,

$$F_{k(x)}^{(1)}(y) = F_{s(w,\ x)}^{(1)}(y) = F_w^{(2)}(x,\ y) = P_1^{(2)}(x,\ y) = x = C_x^{(1)}(y)$$

for all x and y.   □

Although this proof is valid for any standard family, including $\mathcal{T}$, our detailed knowledge of Turing machines permits us to present a somewhat more understandable argument that the result holds for the family $\mathcal{T}$. In particular, it permits us to describe the operation of a machine that evaluates the function k.

Turing-Machine Argument.   Let $\mathbf{T}$ be a Turing machine designed to behave as follows. When presented with the initial pattern $\bar{x}$, $\mathbf{T}$ proceeds to generate the description of a machine $\mathbf{M}_x$ which, when presented with the initial pattern $\bar{y}$, erases that pattern and writes in its place the pattern $\bar{x}$. After generating the description of $\mathbf{M}_x$, the machine $\mathbf{T}$ converts that description into the corresponding index and gives the index as its output. (The reader should convince him or herself that it is possible to construct a machine $\mathbf{T}$ that behaves in this manner.) Now note that when presented with the initial pattern $\bar{x}$, the machine $\mathbf{T}$ produces an index of a machine that computes the one-variable constant function $C_x^{(1)}$. Thus $\mathbf{T}$ evaluates the desired function k.   □

The next theorem provides another simple example of a function that computes indices. In this case we provide only a proof in Turing machine terms.

Theorem 3-9.   *There exists a total, one-variable Turing-computable function g such that*

$$F_{g(x)}^{(1)}(y) = F_x^{(1)}(y + 1)$$

for all x and y.

Turing-Machine Argument.   Let $\mathbf{T}$ be a Turing machine designed to behave as follows. When presented with the initial pattern $\bar{x}$, $\mathbf{T}$ proceeds to generate the description of a machine $\mathbf{M}_x$ that operates as follows. When presented with the initial tape pattern $\bar{y}$, $\mathbf{M}_x$ appends a single 1 to that pattern, then acts exactly like the machine $\mathbf{T}_x$. After generating this description, the machine $\mathbf{T}$ converts the description into an index and halts. Thus when given $\bar{x}$, $\mathbf{T}$ produces an index of a machine that computes the function $F_x \circ (S)$; that is, $\mathbf{T}$ evaluates the desired function g.   □

The advantage of formal proofs of the existence of index-computing functions lies in their precision and conciseness. But formal proofs can be unintuitive and therefore unsatisfying. Although Turing-machine arguments are often cumbersome and imprecise, they usually indicate how the functions in question can actually be evaluated, and thus are sometimes more illuminating than formal proofs. Moreover, arguments based on the construction of Turing machines are not limited to number-theoretic computations. This last point is worth illustrating again.

Suppose that **T** is some fixed Turing machine. Then for each natural number x we may define a new machine $\mathbf{M}_{\mathbf{T}, x}$ that behaves as follows. When started on completely blank tape, $\mathbf{M}_{\mathbf{T}, x}$ first writes the pattern $\bar{x}$ on its tape. It then returns to the beginning of that pattern and proceeds to act exactly like the machine **T**. Thus $\mathbf{M}_{\mathbf{T}, x}$ halts when started on blank tape iff **T** halts when presented with the pattern $\bar{x}$. Moreover, whenever $\mathbf{M}_{\mathbf{T}, x}$ does halt after being started on blank tape, its final tape pattern will be identical to that produced by **T** after being presented with $\bar{x}$.

Our next theorem guarantees the Turing computability of an index for the machine $\mathbf{M}_{\mathbf{T}, x}$.

Theorem 3-10. *Let **T** be any given Turing machine. Then there exists a total, one-variable Turing-computable function f such that, for all x, $f(x)$ is an index for the machine $\mathbf{M}_{\mathbf{T}, x}$.*

Argument. A machine **W** that evaluates f can be designed to behave as follows. Upon being presented with the pattern $\bar{x}$, **W** first generates the quintuple description of a machine that, starting on blank tape, writes the pattern $\bar{x}$. It then appends the (fixed) quintuple description of the machine **T**. Thus **W**'s tape now contains a complete description of the machine $\mathbf{M}_{\mathbf{T}, x}$. (The state numbering problem is handled as in the case of the S-M-N Theorem, except that here the value of v is fixed and need not be computed.) To complete its computation, **W** simply converts the description of $\mathbf{M}_{\mathbf{T}, x}$ into the corresponding index. $\square$

By suitably choosing the machine **T**, Theorem 3-10 can be made to yield a variety of interesting results. Here are three simple examples.

Corollary 3-10.1. *There exists a total Turing-computable function f_1 such that when the machine $\mathbf{T}_{f_1(x)}$ is started on blank tape, it writes a block of $2x$ 1's and halts.*

Proof. Let **T** be a machine that, when presented with a block of 1's, first doubles the length of that block, then erases the last two 1's and halts. Applying the theorem to this machine **T** yields the desired result. $\square$

Corollary 3-10.2. *There exists a total Turing-computable function f_2 such that when the machine $\mathbf{T}_{f_2(x)}$ is started on blank tape, it writes down the binary expansion of x and halts.*

Proof. Let **T** be a machine that performs unary to binary conversion, and apply the theorem. $\square$

Corollary 3-10.3. *There exists a total Turing-computable function f_3 such that when the machine $\mathbf{T}_{f_3(x)}$ is started on blank tape, it writes down a standard description of $\mathbf{T}_x$ and halts. (Recall that a standard machine description is one in which every quintuple is represented by a sequence of five unary-encoded blocks, as described in Section 2.3.)*

Proof. Let $\mathbf{T}$ be a machine that first converts the unary representation of an index x into its binary equivalent, and then "cleans up" the resulting binary description by deleting redundant or deficient quintuples and superfluous quintuple components. Again apply the theorem.   □

Totality and Equivalence Functions

The S-M-N Property also implies the noncomputability of certain functions, including various totality and equivalence functions. Consider first the (one-argument) totality function t, where

$$t(x) = \begin{cases} 1 & \text{if} & F_x^{(1)} \text{ is total} \\ 0 & \text{if} & F_x^{(1)} \text{ is not total.} \end{cases}$$

In Chapter 1, we showed that this function is not computable by reduction from the diagonal domain function. In other words, we showed that if the totality function were algorithmic, the diagonal domain function would also have to be algorithmic. Let us now rephrase this reduction argument in terms of Turing machines.

The basic strategy is simple. With every Turing machine $\mathbf{T}_x$ we associate another machine $\mathbf{W}_x$ such that:

1. $\mathbf{W}_x$ computes a total one-variable number-theoretic function iff $\mathbf{T}_x$ eventually produces a natural number after being presented with the pattern $\bar{x}$.

2. There exists a Turing-computable function that yields an index for $\mathbf{W}_x$ when applied to the number x.

Once such an association is established, any assumed procedure for computing the totality function would provide a procedure for computing the diagonal domain function. Specifically, to evaluate the domain function for input x, we would only have to compute an index for the machine $\mathbf{W}_x$ and apply the assumed procedure to that index.

Taking the approach used in Chapter 1, we define $\mathbf{W}_x$ to be a machine that computes the constant function $C_0^{(1)}$ if $F_x^{(1)}(x)\downarrow$ and the empty function $\Phi^{(1)}$ if $F_x^{(1)}(x)\uparrow$. The next step is to establish the computability of an index for $\mathbf{W}_x$.

Lemma 3-3. *There exists a total, one-variable Turing-computable function q such that $\mathbf{T}_{q(x)}$ evaluates the function $C_0^{(1)}$ if $F_x^{(1)}(x)\downarrow$ and the function $\Phi^{(1)}$ if $F_x^{(1)}(x)\uparrow$.*

Argument. A machine $\mathbf{Q}$ that evaluates the function q can be made to operate as follows. When presented with the pattern $\bar{x}$, $\mathbf{Q}$ generates the description of a machine $\mathbf{W}$ that behaves as follows.

> When presented with the representation of any natural number, $\mathbf{W}$ erases that representation and writes the pattern $\bar{x}\ 0\ \bar{x}$ instead. It then acts like the machine $\mathbf{M}_u$ described in the proof of the Enumeration Theorem; that is, it converts the $\bar{x}$ blocks into encoded descriptions of $\mathbf{T}_x$ and the tape pattern $\bar{x}$ and simulates the computation that $\mathbf{T}_x$ performs for the initial pattern $\bar{x}$. If the result of this computation is a natural number, $\mathbf{W}$ halts with $\bar{0}$ as its output. If the result is not a natural number, or if $\mathbf{T}_x$ fails to halt, $\mathbf{W}$ also fails to halt.

Thus $\mathbf{W}$ serves as the machine $\mathbf{W}_x$ defined above. Once $\mathbf{Q}$ has generated the description of this machine, it converts that description into the corresponding index and halts. □

Lemma 3-3 provides the basis for:

Theorem 3-11. *The totality function*

$$t(x) = \begin{cases} 1 & \text{if} & F_x^{(1)} \text{ is total} \\ 0 & \text{if} & F_x^{(1)} \text{ is not total} \end{cases}$$

is not Turing computable.

Proof. Assume the contrary. Then the function $t \circ (q)$, where q is the index-computing function of Lemma 3-3, must also be Turing computable. But

$$t(q(x)) = \begin{cases} 1 & \text{if} & F_{q(x)}^{(1)} \text{ is total} \\ 0 & \text{if} & F_{q(x)}^{(1)} \text{ is not total.} \end{cases}$$

Since $F_{q(x)}^{(1)}$ is total iff $F_x^{(1)}(x)$ is defined, $t \circ (q)$ is just the diagonal domain function, which is known *not* to be Turing computable. □

Instead of trying to determine whether a Turing machine produces a natural-number output for every natural-number input, we might simply try to determine whether it *halts* for every natural-number input. This possibility leads us to define a new function t_*, such that

$$t_*(x) = \begin{cases} 1 & \text{if} & \mathbf{T}_x \text{ halts for every initial pattern of the form } \bar{n} \\ 0 & \text{otherwise.} \end{cases}$$

As might be expected, t_* is not Turing computable either. Indeed, Lemma 3-3 suffices to reduce the computability of d to that of t_*.

Theorem 3-12. *The function t_* is not Turing computable.*

Proof. Assume the contrary. Then the function $t_* \circ (q)$ must be computable, where again q is the index-computing function of Lemma 3-3. By definition of

t_*, we have

$$t_*(q(x)) = \begin{cases} 1 & \text{if} \quad \mathbf{T}_{q(x)} \text{ halts for every pattern of the form } \bar{n} \\ 0 & \text{otherwise.} \end{cases}$$

But by definition of q, $\mathbf{T}_{q(x)}$ halts for every pattern $\bar{n}$ iff $F_x^{(1)}(x)$ is defined. Thus $t_* \circ (q)$ is in fact the diagonal domain function d, which is known not to be Turing computable.   □

We now turn briefly to a discussion of equivalence functions. Recall that in Chapter 1 we defined the equivalence function e as follows:

$$e(x, y) = \begin{cases} 1 & \text{if} \quad F_x^{(1)} = F_y^{(1)} \\ 0 & \text{if} \quad F_x^{(1)} \neq F_y^{(1)} \end{cases}$$

Since $\mathscr{T}$ is a standard family, it follows from the results of Chapter 1 that its equivalence function is not Turing computable. This fact can be established by reduction from the diagonal domain function or by reduction from the totality function t. In either case, formulation of the appropriate constructive arguments is left to the reader.

Instead of trying to determine whether two machines compute the same number-theoretic function, we might ask whether they yield the same final tape pattern whenever they are presented with the same natural-number representation. Thus we might define a new equivalence function e_*, where

$$e_*(x, y) = \begin{cases} 1 & \text{if, for each initial pattern } \bar{n}, \text{ either} \\ & \mathbf{T}_x \text{ and } \mathbf{T}_y \text{ both fail to halt, or else they} \\ & \text{both halt with the same final tape pattern} \\ 0 & \text{otherwise.} \end{cases}$$

Alternatively, we might define a blank tape equivalence function e_0, where

$$e_0(x, y) = \begin{cases} 1 & \text{if, when started on blank tape, } \mathbf{T}_x \\ & \text{and } \mathbf{T}_y \text{ both fail to halt, or else both halt} \\ & \text{with the same final tape pattern} \\ 0 & \text{otherwise.} \end{cases}$$

Not surprisingly, neither of these new equivalence functions turns out to be Turing computable. This fact does not follow directly from our general results about abstract families, since neither e_* nor e_0 deals exclusively with number-theoretic computations. However, the reader should have little trouble finding appropriate reduction arguments that relate the evaluation of e_* and e_0 to that of other functions already known not to be Turing computable.

The Recursion Theorem

We conclude our discussion of the capabilities and limitations of the family of Turing machines by considering the Recursion Theorem. In terms of Turing machines, this theorem says that for every total, Turing-computable, one-variable function f, and for each positive integer n, there exists a natural number m such that the machines $\mathbf{T}_m$ and $\mathbf{T}_{f(m)}$ compute the same n-variable number-theoretic function.

This result may be justified informally for the case $n = 1$ by paraphrasing an argument used in Chapter 1. Note that it must be possible to design a Turing machine $\mathbf{R}$ that behaves as follows. Upon being presented with the initial tape pattern $\bar{u}$, $\mathbf{R}$ does two things:

1.   Constructs the description of a Turing machine $\mathbf{Q}_u$ that, when presented with any input pattern, first simulates the computation of the machine $\mathbf{T}_u$ for the input $\bar{u}$ and, if that computation terminates with a numerical value, proceeds to act like the machine $\mathbf{T}_{F_u(u)}$ for the given input pattern.

2.   Converts the description of the machine $\mathbf{Q}_u$ into an index i and evaluates $f(i)$.

Let v be an index of the machine $\mathbf{R}$.

Now consider the machine $\mathbf{Q}_v$—that is, the machine whose description $\mathbf{R}$ constructs when given the input $\bar{v}$. Suppose that m is the index corresponding to this description of $\mathbf{Q}_v$; suppose, in other words, that $\mathbf{Q}_v$ is the machine $\mathbf{T}_m$. Note that when $\mathbf{Q}_v$ is supplied with the initial pattern $\bar{x}$, it will produce the same result as the machine $\mathbf{T}_{F_v(v)}$ would produce when supplied with $\bar{x}$. But since F_v is the one-variable function evaluated by the machine $\mathbf{R}$ and since the application of $\mathbf{R}$ to the input $\bar{v}$ yields the value $f(m)$, it follows that $\mathbf{T}_{F_v(v)}$ is the machine $\mathbf{T}_{f(m)}$. Thus $\mathbf{T}_m$ (that is, the machine $\mathbf{Q}_v$) evaluates the same one-variable function as the machine $\mathbf{T}_{f(m)}$.

The formal proof of the Recursion Theorem rests on an auxiliary result which we rephrase here as:

Lemma 3-4. *For each positive integer n there exists a total, one-variable Turing-computable function ρ such that:*

$$F^{(n)}_{\rho(u)} = \begin{cases} F^{(n)}_{F_u(u)} & \text{if} \quad F_u(u)\downarrow \\ \Phi^{(n)} & \text{if} \quad F_u(u)\uparrow \end{cases}$$

In other words, if $F_u(u)$ is defined, $\mathbf{T}_{\rho(u)}$ computes the same n-variable function as $\mathbf{T}_{F_u(u)}$; while if $F_u(u)$ is undefined, $\mathbf{T}_{\rho(u)}$ computes the empty function of n variables.

An abstract proof of this lemma has already been given in Chapter 1. In Turing machine terms, the argument amounts to showing that it is possible to design a machine that, given $\bar{u}$, generates a description of the machine $\mathbf{Q}_u$ and converts that description into the corresponding index. The details are left to the reader.

Lemma 3-4 provides the foundation for:

Theorem 3-13. (*The Recursion Theorem for Turing Machines*) *Let* f *be any total, Turing-computable, one-variable number-theoretic function and let* n *be any positive integer. Then there exists a natural number* m *such that* $\mathbf{T}_m$ *and* $\mathbf{T}_{f(m)}$ *compute the same* n-*variable number-theoretic function.*

Proof. Since f is total and Turing computable, so is the composite function $f \circ (\rho)$, where ρ is the index-computing function described in Lemma 3-4. Let v be any index for $f \circ (\rho)$ and choose $m = \rho(v)$. Since $f \circ (\rho)$ is total, $F_v(v)$ is defined. Therefore

$$F_m^{(n)}(x_1, \ldots, x_n) = F_{\rho(v)}^{(n)}(x_1, \ldots, x_n) = F_{F_v(v)}^{(n)}(x_1, \ldots, x_n)$$

$$= F_{f(\rho(v))}^{(n)}(x_1, \ldots, x_n) = F_{f(m)}^{(n)}(x_1, \ldots, x_n)$$

for all $x_1, \ldots, x_n$. Thus $\mathbf{T}_m$ and $\mathbf{T}_{f(m)}$ do in fact compute the same n-variable number-theoretic function. $\square$

By modifying Lemma 3-4 slightly, we can establish other versions of the Recursion Theorem that are peculiar to Turing machines. For this purpose we must introduce some new terminology. Let A be any designated finite alphabet of symbols. Let $\mathbf{M}_1$ and $\mathbf{M}_2$ be any two Turing machines, and suppose that these machines are provided with identical initial tape patterns. Then $\mathbf{M}_1$ and $\mathbf{M}_2$ will be said to *behave alike with respect to the alphabet* A if

1. they both fail to halt; or
2. they both halt with identical tape patterns consisting solely of symbols from A; or
3. they both halt with tape patterns (not necessarily identical) that include symbols not belonging to A.

In other words, the two machines behave alike with respect to A iff, whenever either machine halts with a pattern of symbols from A, the other machine also halts with the same pattern on its tape.

Lemma 3-5. *Let* A *be any finite alphabet of symbols. Then there exists a total, one-variable, Turing-computable function* γ *such that:*

1. *If* $F_u(u)$ *is defined, the machines* $\mathbf{T}_{\gamma(u)}$ *and* $\mathbf{T}_{F_u(u)}$ *behave alike with respect to* A *after being started on blank tape.*
2. *If* $F_u(u)$ *is undefined, the machine* $\mathbf{T}_{\gamma(u)}$ *fails to halt after being started on blank tape.*

Argument. A machine Γ that evaluates the function γ can be designed to operate as follows. Given the initial pattern $\bar{u}$, Γ generates the description of a machine $\mathbf{W}$ that behaves as follows.

When started on blank tape, $\mathbf{W}$ first writes down the (fixed) pattern $\bar{u}\ 0\ \bar{u}$. It then proceeds to simulate the action of the machine $\mathbf{T}_u$ for the initial tape pattern $\bar{u}$. If $\mathbf{T}_u$ fails to halt in this situation, or if it halts but does not produce a natural number, $\mathbf{W}$ fails to halt. If $\mathbf{T}_u$ halts and produces the natural number k, $\mathbf{W}$ next simulates the action of the machine $\mathbf{T}_k$ on initially blank tape. If $\mathbf{T}_k$ fails to halt, $\mathbf{W}$ must of course do likewise. If $\mathbf{T}_k$ does halt, $\mathbf{W}$ then determines whether $\mathbf{T}_k$'s final tape pattern consists solely of symbols from A. If so, $\mathbf{W}$ decodes the representation of that pattern into the actual pattern and halts; if not, it simply prints a symbol that is not in A and halts.[†]

Once Γ has generated the description of $\mathbf{W}$, it converts that description into the corresponding index and halts. To see that Γ in fact evaluates γ, we note that if $F_u(u)$ is defined, the machines $\mathbf{W}$ and $\mathbf{T}_{F_u(u)}$ behave alike with respect to A after being started on blank tape; while if $F_u(u)$ is undefined, $\mathbf{W}$ fails to halt. □

As Lemma 3-4 provided the basis for Theorem 3-13, so Lemma 3-5 yields:

Theorem 3-14. (*The Blank Tape Recursion Theorem*) *Let A be any finite alphabet of symbols, and let f be any total, one-variable, Turing-computable function. Then there exists a natural number m such that the machines $\mathbf{T}_m$ and $\mathbf{T}_{f(m)}$ behave alike with respect to A after being started on blank tape.*

Proof. Since f is total and Turing computable, so is the function $f \circ (\gamma)$, where γ is the index-computing function of Lemma 3-5. Let v be any index of $f \circ (\gamma)$ and choose $m = \gamma(v)$. Since $f \circ (\gamma)$ is total, $F_v(v)$ is defined. Thus by definition of γ, the machines $\mathbf{T}_{\gamma(v)}$ and $\mathbf{T}_{F_v(v)}$ behave alike with respect to A after starting on blank tape. Now, by definition of m, $\mathbf{T}_{\gamma(v)}$ is the machine $\mathbf{T}_m$. And since $F_v(v) = f(\gamma(v)) = f(m)$, $\mathbf{T}_{F_v(v)}$ is the machine $\mathbf{T}_{f(m)}$. Thus the machines $\mathbf{T}_m$ and $\mathbf{T}_{f(m)}$ do in fact behave alike with respect to A after starting on blank tape. □

Used in conjunction with Theorem 3-10, the Blank Tape Recursion Theorem yields a variety of unexpected results. We content ourselves here with a single example. Recall that Corollary 3-10.3 ensures the existence of a total

---

[†] It is at this point that we need the constraint associated with the choice of a particular alphabet A. Since the number of tape symbols used by $\mathbf{W}$ must be fixed by the value u, it is not in general possible to design $\mathbf{W}$ to decode representations of arbitrarily large alphabets.

Turing-computable function f_3 such that when the machine $\mathbf{T}_{f_{3(x)}}$ is started on blank tape, it writes a standard quintuple description of the machine $\mathbf{T}_x$ and halts. Applying the Blank Tape Recursion Theorem to this function f_3, we have:

Corollary 3-14.1. *There exists a Turing machine that, when started on blank tape, eventually writes a standard quintuple description of itself and halts.*

Proof. Let A be the alphabet $\{0, 1\}$. Then according to Theorem 3-14, there exists a natural number m such that $\mathbf{T}_m$ and $\mathbf{T}_{f_{3(m)}}$ behave alike with respect to A after being started on blank tape. But according to Corollary 3-10.3, when $\mathbf{T}_{f_{3(m)}}$ is started on blank tape it writes a description of $\mathbf{T}_m$ and halts. Since this description consists entirely of 0's and 1's, it is in fact a pattern of symbols from the alphabet A. Thus after being started on blank tape the machine $\mathbf{T}_m$ must eventually halt with the same final tape pattern, i.e., a description of $\mathbf{T}_m$. $\square$

We have seen in this section that the results of Chapter 1 are readily translated into Turing-machine terms. Thus the class of Turing machines provides a convenient representation of a standard family of algorithms. From our point of view, the most important feature of this representation is the fact that the functions defined by Turing machines are all effectively computable. In the next two chapters we turn to the problem of finding a mathematical characterization of these functions.

3.4 SUMMARY

In order to use Turing machines to evaluate number-theoretic functions, it is necessary to choose a convention for representing natural numbers. We adopt the unary notation, in which the number n is represented by a block of $(n + 1)$ 1's. A number-theoretic function is then said to be Turing computable if there exists a Turing machine that converts the unary representations of the arguments of the function into the unary representation of the value of the function. Every Turing-computable function can be evaluated in a standard form in which: the final tape pattern contains only the representation of the function value; that representation appears in the position originally occupied by the first argument block; and the machine never moves to the left of its starting point. This fact simplifies the job of showing that certain functions are Turing computable. In particular, it makes it easy to show that the class of Turing-computable functions is closed under functional composition.

The conventions used for number-theoretic computations provide a natural way of viewing every Turing machine as evaluating some n-variable function for each n. When it comes to indexing Turing machines, several schemes are possible. The one adopted here is based on the fact that, with minor modifications, the universal machine of Chapter 2 will interpret any pattern of 0's and 1's as the description of some machine. The index i is therefore assigned to the machine that is described by the binary expansion of i. The principal advantage of this scheme is the ease with which indices can be converted into machine descriptions and vice versa. The assignment of functions and indices to Turing machines makes it possible to treat Turing machines as representations of the members of a family of algorithms. This family is denoted $\mathcal{T}$.

Since the constant, projection, successor, and selection functions are all Turing computable, and since the Turing-computable functions are closed under composition, the family $\mathcal{T}$ satisfies Properties 1, 2, and 4 of the five basic properties discussed in Chapter 1. Turing machines that evaluate the universal functions for $\mathcal{T}$ can be obtained by combining the universal machine of Chapter 2 with components that encode indices and arguments into machine and tape descriptions and that decode tape descriptions into numerical representations. Machines that evaluate the S-M-N functions for $\mathcal{T}$ can be obtained by combining index encoding and decoding components with machines that modify descriptions of other machines. Thus the family $\mathcal{T}$ also satisfies Properties 3 and 5, and so forms a standard family of algorithms. As a result, all the consequences of the five basic properties must hold in $\mathcal{T}$.

Among these consequences are the facts that the domain, totality, and equivalence functions for the family $\mathcal{T}$ are not Turing computable. While the proofs given in Chapter 1 suffice to establish these results, the arguments can also be phrased in Turing-machine terms. These Turing-machine arguments can be extended to yield results that have no direct counterpart for arbitrary families of algorithms. Among the latter results are the facts that the general, diagonal, and blank tape halting functions for Turing machines are not Turing computable. As in the case of the general results of Chapter 1, the two basic techniques used to show that functions are not Turing computable are those of diagonalization and reduction.

Another consequence of the basic properties is the Recursion Theorem, which must also hold for the family $\mathcal{T}$. According to this theorem, if f is a total Turing-computable function, and if f is thought of as operating on indices so as to map Turing machines into Turing machines, then for each n there must exist a machine that evaluates the same n-variable function as its image under f. There are also variations of this theorem that are peculiar to Turing machines. Perhaps the most interesting of these variations is the Blank Tape Recursion Theorem. Roughly speaking, this theorem says that there must exist a machine that, upon being started on blank tape, yields the same final outcome as its image under f. One of the implications of the Blank Tape Recursion Theorem is that there must exist a Turing machine that writes down its own description after being started on blank tape.

3.5 PROBLEMS

Section 3.1

3.1.1 Consider the following alternative conventions for number-theoretic computations. A machine that evaluates the n-variable function f is to begin its computation with the pattern $\cdots 0\ \bar{x}_1\ 0\ \bar{x}_2\ 0 \cdots 0\ \bar{x}_n\ 0 \cdots$ written on an otherwise blank tape. If $f(x_1, \ldots, x_n)$ is undefined, the machine is not to halt. If $f(x_1, \ldots, x_n)$ is defined, the machine is to halt with a final tape pattern containing a *total* of $f(x_1, \ldots, x_n)$ 1's, which may be distributed in *any way* throughout the pattern. Thus, for example, the final tape pattern

$$\cdots 0\ 1\ 2\ 0\ 1\ 1\ 0\ 2\ 3\ 1\ 0\ 1\ 1\ 0\ 2\ 3\ 0\ 0 \cdots$$

represents the value 6. Show that the set of number-theoretic functions that are computable according to these conventions is identical to the set of functions that are computable according to the conventions used in the text.

3.1.2 For each of the functions described below, design a Turing machine that evaluates that function in standard form. You may use the tape symbols 0, 1, and X.

a) $f(x) = \left\lfloor \dfrac{x}{2} \right\rfloor$[†]   (6 states are sufficient)

b) $f(x, y) = y$ (6 states are sufficient)

c) $f(x) = \begin{cases} x/3 & \text{if 3 divides } x \\ \uparrow & \text{otherwise} \end{cases}$ (8 states are sufficient)

3.1.3 In this problem we drop the requirement that Turing machines be presented with tapes that are blank except for their argument blocks. We say that a Turing machine computes the n-variable number-theoretic function f in *pseudo-standard form* if, upon being presented with the pattern $0\ \bar{x}_1\ 0\ \bar{x}_2\ 0 \cdots 0\ \bar{x}_n\ 0$, which may be preceded and followed by arbitrary strings of symbols, the machine meets the following constraints.

1. The machine halts iff $f(x_1, \ldots, x_n)\downarrow$.

2. If $f(x_1, \ldots, x_n) = z$, the machine halts in front of a block of $(z + 1)$ 1's, thus: $\cdots 0\ \bar{z}\ 0 \cdots$. This block may be preceded and followed by arbitrary strings of symbols. The block $\bar{z}$ begins in the same square as the argument block $\bar{x}_1$, so that the machine begins and ends its computation by scanning the same square.

Show that every Turing-computable function can be computed in pseudo-standard form.

[†] The expression "$\lfloor a \rfloor$" denotes the integer part of the quantity a.

3.1.4 A certain Turing machine **M** computes the two-variable function f, where $f(x, y) = \max (x, y)$. Assume that **M** evaluates f in standard form and that it begins its computations by moving right without changing state. It is not known what number-theoretic functions **M** computes for other numbers of arguments. Using **M** as a submachine, design a Turing machine **T** whose associated n-variable number-theoretic function $F^{(n)}$ is specified by the equation

$$F^{(n)}(x_1, \ldots, x_n) = \max (x_1, \ldots, x_n).$$

The machine **T** is to perform all its computations in standard form.

3.1.5 Let g be any total, one-variable Turing-computable function. Show that the function f where

$$f(x) = \sum_{i=0}^{x} g(i)$$

is also Turing computable.

3.1.6 For each two-variable number-theoretic function f, we define the one-variable function $\hat{f}$ as follows. If there exists a natural number z such that $f(x, z) = 0$, then $\hat{f}(x)$ is the smallest such z; if there is no natural number z such that $f(x, z) = 0$, then $\hat{f}(x)$ is undefined. Show that whenever f is a total, two-variable Turing-computable function, the associated function $\hat{f}$ must also be Turing computable. What if f is not total?

3.1.7 a) Suppose that the total one-variable number-theoretic function f can be evaluated in standard form by an n-state Turing machine **M** that uses only the tape symbols 0 and 1. Show that for each natural number k, there exists a $(2n + k + 2)$-state machine that (i) uses only the symbols 0 and 1; and (ii) after being started on blank tape, eventually produces a block of $(f(f(k)) + 1)$ 1's and halts. (You may assume that **M**'s starting state is a right-moving state.)

b) Assume that the exponentiation function f, where $f(x) = 2^x$, can be evaluated in standard form by a 10-state machine that uses only the tape symbols 0 and 1. Show that there exists a 36-state machine that, after starting on blank tape, writes a block of $(2^{65,536} + 1)$ 1's and halts.

Section 3.2

3.2.1   Design a Turing machine M_a that effects unary to binary conversion in accordance with the following format.

Initial tape pattern: $A\ 1\ 1\ 1\ \cdots\ \cdots\ \cdots\ \cdots\ 1\ 1\ A$

$\underbrace{}_{i\ 1\text{'s}}$

Final tape pattern: $A\ 0\ 0\ 0\ \cdots\ \cdot\ 0\ 1\ 1\ 0\ 1\ 0\ \cdots\ \cdot\ 1\ A$

$\underbrace{}_{\text{binary representation of }i}$

Use the strategy suggested in the text. Specifically, let M_a carry out its computation by making a series of passes from right to left across its tape, adding one new digit to the binary representation of i after each pass. The machine is not to leave the region of the tape bounded by the A's.

3.2.2   In this problem we show that it is possible to construct Turing machines that convert back and forth between unary blocks and the encoded representations used by the universal machine of Section 2.3.

a) Design a machine M_b that encodes argument blocks in accordance with the following format.

Initial tape pattern: $\underline{0}\ 1\ 1\ 1\ 1\ \cdots\ 1\ 0\ 0\ 0\ \cdots\ 0\ 0\ 0\ 0\ \cdots$

$\underbrace{}_{(x+1)\ 1\text{'s}}$

Final tape pattern: $\underline{C}\ 1\ 0\ 1\ 1\ 0\ 1\ 1\ 0\ 1\ 1\ 0\ \cdots\ 1\ 1\ 0\ 0\ \cdots$

$\underbrace{}_{x+1\text{ blocks}}$

The machine is to begin and end its computation on the same square, and is never to move to the left of that square.

b) Design a machine M_c that decodes the representation of an arbitrary block in accordance with the following format.

Initial tape pattern: $\underline{C}\ 1\ 0\ 1\ 1\ 0\ 1\ 1\ 0\ 1\ 1\ 0\ \cdots\ 1\ 1\ 0\ 0\ \cdots$

$\underbrace{}_{z+1\text{ blocks}}$

Final tape pattern: $\underline{0}\ 1\ 1\ 1\ \cdots\ 1\ 0\ 0\ 0\ \cdots\ \cdot\ 0\ 0\ \cdots$

$\underbrace{}_{(z+1)\ 1\text{'s}}$

The machine is to begin and end its computation on the same square, and is never to move to the left of that square.

3.2.3   Give a state diagram for the machine $\mathbf{M}_u$ that evaluates the universal function $U^{(2)}$ in the manner described in the text. You may use as submachines the universal machine $\mathbf{U}$ of Section 2.3 and the machines $\mathbf{M}_a$, $\mathbf{M}_b$, and $\mathbf{M}_c$ of Problems 3.2.1 and 3.2.2.

3.2.4   It is desired to modify the machine $\mathbf{M}_u$ discussed in the text (and in Problem 3.2.3) so as to obtain a new machine $\mathbf{M}_*$ with the following property: For each $n \geq 1$, the $(n + 1)$-variable number-theoretic function evaluated by $\mathbf{M}_*$ is to be the universal function $U^{(n+1)}$. Explain what phases of $\mathbf{M}_u$'s operation must be changed, and in what ways. You need not give a state diagram for $\mathbf{M}_*$.

3.2.5   Design a Turing machine $\mathbf{M}$ that effects binary to unary conversion in accordance with the following format:

Initial tape pattern: $A\ \underline{1\ 0\ 1\ 1\ 0\ \cdots\ 1\ 0}\ A$

binary representation of n

Final tape pattern: $A\ 1\ 0\ 1\ 1\ 0\ \cdots\ 1\ 0\ A\ 0\ \underline{1\ 1\ 1\ \cdots\ 1\ 0\ 0}$

$(n + 1)$ 1's

You may use the doubling machine of Fig. 2-18 as a submachine. Assume that $n > 0$.

3.2.6   For each natural number n, let $\mathbf{K}_n$ denote a Turing machine that behaves as follows. When started on blank tape, $\mathbf{K}_n$ moves toward the right, writing down a block of $n + 1$ consecutive 1's, and halts at the end of the block. Thus if $\mathbf{K}_7$ is started on the tape

$$\cdots 0\ \underline{0}\ 0\ 0\ 0\ 0\ 0\ 0\ 0\ 0\ 0 \cdots$$

it will halt with the pattern

$$\cdots 0\ 1\ 1\ 1\ 1\ 1\ 1\ 1\ 1\ \underline{0}\ 0 \cdots .$$

Give the state diagram for a machine $\mathbf{T}$ that, when presented with the pattern $\bar{x}$, generates the quintuple description of the machine $\mathbf{K}_x$, in accordance with the following format.

Initial tape pattern:

$$\underline{0}\ \underbrace{1\ 1\ 1\ \cdots\ 1}\ 0\ 0\ 0\ 0\ 0\ 0\ 0\ 0\ 0\ 0 \cdots 0\ 0\ 0\ 0\ 0\ 0\ 0 \cdots 0\ 0\ 0\ 0 \cdots$$

$(x + 1)$ 1's

Final tape pattern:

$$0\ 0\ 0\ 0\ \cdots\ 0\ \underline{A}\ 0\ \underbrace{1\ 0\ 1\ 0\ 1\ 1\ 0\ 1\ 1\ \cdots\ 0\ 1\ 1\ 0\ 1\ 1\ 1\ \cdots\ 1\ 0\ 0\ 0} \cdots$$

description of $\mathbf{K}_x$

The states of $\mathbf{K}_x$ are to be numbered 0, 1, 2, ... You may use two kinds of submachines within $\mathbf{T}$:

i) The copying machine of Fig. 2-8, to be denoted

ii) Any submachine of the form

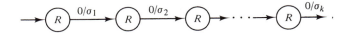

to be denoted

3.2.7 For each choice of natural numbers k and q, let $\mathbf{M}_{k,\,q}$ denote a Turing machine that evaluates the function $f(x) = x + k$ in standard form and whose states are numbered q, $q + 1$, $q + 2$, etc., with q as starting state. Give the state diagram of a Turing machine $\mathbf{T}$ that, when presented with the inputs $\bar{k}$ and $\bar{q}$, generates a quintuple description of the machine $\mathbf{M}_{k,\,q}$ in accordance with the following format.

Initial tape pattern:

$$0\;\underline{0}\;\underbrace{1\;1\;1\;\cdots\;1}\;0\;\underbrace{1\;1\;\cdots\;1}\;0\;0\;0\;\cdots$$
$$\quad\quad(k+1)\text{ 1's}\quad(q+1)\text{ 1's}$$

Final tape pattern:

$$0\;0\;0\;0\;0\;\cdots\;0\;\underline{0}\;\underbrace{1\;1\;\cdots\;1\;0\;1\;0\;1\;0\;1\;1\;\cdots\;1\;1\;1}\;0\;0\;0\;\cdots$$
$$\text{description of }\mathbf{M}_{k,\,q}$$

You may use the copying and writing submachines specified in Problem 3.2.6. Assume that $k > 0$.

3.2.8 Let $\mathbf{M}_s$ denote the Turing machine that evaluates the index-computing function $s_{1,\,1}$ as described in the text. Show that $\mathbf{M}_s$ can be modified to form a new machine $\mathbf{M}_*$ whose $(m + 1)$-variable function serves as the index-computing function $s_{n,\,m}$ for all $n \geq 1$. In particular, explain what phases of $\mathbf{M}_s$'s operation must be changed, and in what ways.

Section 3.3

3.3.1   Use Turing-machine arguments similar to those given for Theorems 3-5 and 3-6 to show that neither of the following functions is Turing computable.

a) The general domain function $D^{(n+1)}$, where

$$D^{(n+1)}(i, x_1, \ldots, x_n) = \begin{cases} 1 & \text{if} & F_i^{(n)}(x_1, \ldots, x_n)\downarrow \\ 0 & \text{if} & F_i^{(n)}(x_1, \ldots, x_n)\uparrow. \end{cases}$$

b) The general halting function $H^{(n+1)}$, where

$$H^{(n+1)}(i, x_1, \ldots, x_n) = \begin{cases} 1 & \text{if } T_i \text{ halts for the initial} \\ & \text{pattern } \underline{0}\ \bar{x}_1\ 0 \cdots 0\ \bar{x}_n\ 0 \cdots \\ 0 & \text{if } T_i \text{ does not halt for the initial} \\ & \text{pattern } \underline{0}\ \bar{x}_1 0 \cdots 0\ \bar{x}_n 0 \cdots \end{cases}$$

3.3.2   With each Turing machine **T** we may associate the special halting function h_T, where

$$h_T(x) = \begin{cases} 1 & \text{if } \mathbf{T} \text{ eventually halts after being started} \\ & \text{on the initial tape pattern } \cdots \underline{0}\ \bar{x}\ 0 \cdots \\ 0 & \text{otherwise.} \end{cases}$$

a) Show that the function h_T is Turing computable when **T** is the machine described by the following state diagram.

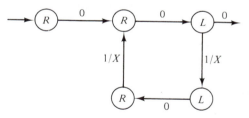

b) Show that there exists a Turing machine **T** for which the function h_T is not Turing computable. You need not give a state diagram for this machine, but you should describe what the machine does.

3.3.3   There are many functions whose computability would imply that of the diagonal domain function. Consider the function η defined as follows. For each natural number x, consider all those machines T_i, where $i \leq x$, that eventually halt after being started on the initial pattern $\underline{0}\ \bar{i}\ 0 \cdots$. Now let $\eta(x)$ be the sum of the numbers of steps executed by these machines before halting. Show that η is not Turing computable.

3.3.4 In this problem we consider an extension of the Enumeration and S-M-N Theorems to functions of zero variables, where the "arguments" of a zero-variable function are assumed to be represented by blank tape.

a) Show that the universal function $U^{(1)}$ is Turing computable, where $U^{(1)}(i)$ is the value produced when the machine $\mathbf{T}_i$ is started on a singly infinite blank tape.

b) Show that the S-M-N functions $s_{0,m}$ are Turing computable, where $s_{0,m}(i, x_1, \ldots, x_m)$ is the index of a machine which, when started on a singly infinite blank tape, eventually produces the value $F_i^{(m)}(x_1, \ldots, x_m)$.

3.3.5 In each of the following parts, establish the existence of a total Turing-computable function g that satisfies the specified condition. A general description of the behavior of a machine that evaluates g will be sufficient.

a) $F_{g(n)}^{(n)} = P_n^{(n)}$

b) $F_{g(x, y)}^{(1)} = F_x^{(1)} \circ (F_y^{(1)})$

c) $\mathbf{T}_{g(x)}$ halts for all finite initial patterns of 0's and 1's iff $F_x(x)\!\downarrow$.

3.3.6 In each of the following parts, determine whether there exists a total Turing-computable function g that satisfies the specified condition. Justify your answers.

a) $F_{g(x)}^{(1)}$ is the successor function iff $F_x(x)\!\downarrow$

b) $F_{g(x)}(y) = \begin{cases} 0 & \text{if} \quad F_x(y)\!\uparrow \\ \uparrow & \text{otherwise} \end{cases}$

c) The function g is one-to-one, and for each x the machine $\mathbf{T}_{g(x)}$ halts when started on blank tape.

d) A one-variable Turing-computable function f is total iff $f = F_{g(x)}^{(1)}$ for some x.

3.3.7 For each of the functions described below, use a reduction argument of the specified type to show that the function in question is not Turing computable. Briefly outline the operation of any auxiliary machines involved in your arguments.

a) $f(x) = \begin{cases} 1 & \text{if } \mathbf{T}_x \text{ eventually halts with its tape blank after} \\ & \text{having been started on a doubly infinite blank tape} \\ 0 & \text{otherwise,} \end{cases}$

by reduction from the blank-tape halting function h_0.

b) $f(x) = \begin{cases} 1 & \text{if } F_x^{(1)} \text{ is monotone increasing (that is, if} \\ & F_x^{(1)}(n+1) > F_x^{(1)}(n)) \\ 0 & \text{otherwise,} \end{cases}$

by reduction from the diagonal halting function h.

c) $f(x) = \begin{cases} 1 & \text{if } \mathbf{T}_x \text{ writes the symbol 1 at some step in} \\ & \text{its computation after being started on} \\ & \text{a doubly infinite blank tape} \\ 0 & \text{otherwise,} \end{cases}$

by reduction from the blank-tape halting function h_0.

d) $f(x) = \begin{cases} 1 & \text{if } \mathbf{T}_x \text{ eventually runs off its tape after} \\ & \text{being started on the first square of} \\ & \text{a singly infinite blank tape} \\ 0 & \text{otherwise,} \end{cases}$

by reduction from the diagonal domain function d.

e) $f(x) = \begin{cases} 1 & \text{if } \mathbf{T}_x \text{ ever makes three consecutive right shifts} \\ & \text{after starting on a doubly infinite blank tape} \\ 0 & \text{otherwise,} \end{cases}$

by reduction from the blank-tape halting function h_0.

3.3.8 Show that none of the functions described below are Turing computable. Briefly outline the operation of any auxiliary machines used in your arguments.

a) $f(x) = \begin{cases} 1 & \text{if there is some initial tape pattern on the} \\ & \text{alphabet of } \mathbf{T}_x \text{ for which } \mathbf{T}_x \text{ eventually halts} \\ 0 & \text{otherwise} \end{cases}$

b) $f(x, n) = \begin{cases} 1 & \text{if } \mathbf{T}_x \text{ eventually halts for all initial tape} \\ & \text{patterns of the form } \underline{0}\ \bar{m}\ 0 \cdots, \text{ where } 0 \le m \le n \\ 0 & \text{otherwise} \end{cases}$

c) $f(x) = \begin{cases} 1 & \text{if } \mathbf{T}_x \text{ halts in an odd number of steps after} \\ & \text{starting on a doubly infinite blank tape} \\ 0 & \text{otherwise} \end{cases}$

d) $f(x, y) = \begin{cases} 1 & \text{if, after starting on doubly infinite blank} \\ & \text{tapes, either } \mathbf{T}_y \text{ fails to halt or else } \mathbf{T}_x \\ & \text{halts in fewer steps than } \mathbf{T}_y \\ 0 & \text{otherwise} \end{cases}$

3.3.9 This problem is concerned with the class of Turing machines that use only the symbols 0 and 1, and that eventually halt after being started on a doubly infinite blank tape. Such machines will be called *two-symbol halting machines*. Note that because the number of n-state, two-symbol halting machines is finite, the number of 1's that such a machine can leave on its tape when it halts is bounded. The *Busy Beaver function* W is defined to be that function whose value for argument n is the maximum number of 1's that any n-state, two-symbol halting machine leaves on its tape after being started on blank tape. In this context, we assume that a machine can write a symbol during the step in which it halts.

    a) Show that the function W is monotone increasing—in other words, that $W(n + 1) > W(n)$.

    b) Use the result of Problem 3.1.7(a) to show that $W(n) > 2n$ for large n. *Hint:* Consider the function f such that $f(x) = 2x$.

    c) Let f be any monotone-increasing, Turing-computable function such that $f(n) > 2n$ for large n. Use the results of Problem 2.2.8(c) and 3.1.7(a) to show that $W(n) > f(n)$ for large n.

    d) Show that W is not Turing computable.

3.3.10 Again consider the class of two-symbol halting machines introduced in Problem 3.3.9. The function M is defined so that $M(n)$ is the maximum number of moves that any n-state, two-symbol halting machine makes after being started on doubly infinite blank tape. In this problem we show that M is not Turing computable, and that its noncomputability is equivalent to that of the Busy Beaver function W.

    a) Outline an argument showing that the Turing computability of M would imply that of W, and thereby establish that M is not Turing computable.

    b) Suppose that $\mathbf{T}$ is an n-state, two-symbol machine that, after starting on blank tape, makes a total of m moves and halts. Describe the construction of a two-symbol machine $\hat{\mathbf{T}}$ that, after starting on blank tape, eventually halts with exactly m 1's on its tape. Explain why the machine $\mathbf{T}$ needs at most $an + b$ states, where a and b are constants.

    c) Outline an argument showing that the Turing computability of W would imply that of M.

3.3.11 In each of the following parts, establish the existence of a natural number m (or natural numbers m_1 and m_2) satisfying the stated condition.

    a) $F_m^{(1)}(y) = F_y^{(1)}(m)$ for all y.

    b) After being started on blank tape, $\mathbf{T}_m$ halts with $(m^2 + 5)$ 1's on its tape.   *(continued)*

c) After being started on blank tape, $\mathbf{T}_m$ halts with the decimal representation of m on its tape.

d) When $\mathbf{T}_{m_1}$ is started on blank tape, it eventually halts with a description of $\mathbf{T}_{m_2}$ on its tape. And when $\mathbf{T}_{m_2}$ is started on blank tape, it eventually halts with a description of $\mathbf{T}_{m_1}$ on its tape.

3.3.12   In this problem let $\mathbf{G}_u$ denote a Turing machine that, upon being started on blank tape, first evaluates $F_u(u)$ and, if that computation terminates, imitates the behavior of the machine $\mathbf{T}_{F_u(u)}$ on initially blank tape. Using this nomenclature, describe the major steps in the operation of a Turing machine $\mathbf{M}$ that, after starting on blank tape, writes its own description and halts. Verify that $\mathbf{M}$ does in fact produce its own description.

3.3.13   a) Prove the following generalization of the Recursion Theorem for Turing machines.

> *Theorem*   Let A be any fixed, finite set of tape symbols, and let f be any total one-variable Turing-computable function. Then there must exist a natural number m such that the machines $\mathbf{T}_m$ and $\mathbf{T}_{f(m)}$ behave alike with respect to the alphabet A whenever they are started on identical tape patterns consisting solely of symbols from A.

b) Show that for each total, one-variable Turing-computable function f there exists a natural number m such that $F_m^{(n)}$ and $F_{f(m)}^{(n)}$ are identical for all $n > 0$.

c) Show that there exists a natural number m such that $\mathbf{T}_m$ halts when presented with the initial tape pattern $\bar{x}$ iff $x = m$. Why isn't the basic Recursion Theorem (Theorem 3-13) adequate for this proof?

Chapter 4

Primitive Recursive Functions

The Turing machine model provides a way of establishing the effective computability of a broad class of functions. However, because our discussion of Turing machines has concentrated on the mechanical aspects of computation, we do not yet have a mathematical characterization of the Turing-computable functions. It is now time to address the problem of providing mathematical definitions for various classes of functions. We begin in this chapter by introducing the class of primitive recursive functions. This class includes virtually all the effectively computable functions of everyday interest but not all the Turing-computable functions. In Chapter 5 we will define a larger class of functions, called general recursive functions, and show that this class is identical to the class of Turing-computable functions.

4.1 INTRODUCTION

Formal definitions of classes of functions are almost always inductive in nature. In making such a definition, we specify certain *base functions* that are to belong to the class in question and certain *composition rules* by means of which the other members of the class are to be built up from the base functions. Of special interest is the case in which the base functions are effectively computable and in which all the composition rules preserve effective computability, for then all the functions in the class must be effectively computable.

If a class of functions is to be useful for practical computation, it should certainly include the constant, projection, and successor functions and it should certainly be closed under functional composition. Thus one of the simplest classes of functions we might consider is that generated by choosing the constant, projection, and successor functions as base functions and choosing ordinary functional composition as the only composition rule. Unfortunately, the resulting class of functions is rather restricted and uninteresting, as the reader may readily discover. In order to obtain a larger and more useful class of functions, it is necessary either to include additional base functions or to permit additional rules of composition. We shall take the latter approach.

Primitive Recursion

Recursive definitions provide a variety of useful composition rules. Suppose, for example, that we wish to define a one-variable number-theoretic function f. This might be done by specifying two things: the value of $f(0)$, and a rule for obtaining the value of $f(y + 1)$ from the values of y and $f(y)$. To be more precise, we might specify a natural number α and a total two-variable number-theoretic function h and stipulate that:

$$f(0) = \alpha \tag{4-1a}$$

$$f(y + 1) = h(y, f(y)) \qquad \text{for all } y \tag{4-1b}$$

For instance, if $\alpha = 1$ and $h(a, b) = (a + 1)b$, the function f will be the familiar factorial function.

The definitional scheme of Eqs. (4-1) is easily extended to functions of two or more variables. Suppose that g and h are total number-theoretic functions of n and $n + 2$ variables, respectively. Then g and h can be used to define a new $(n + 1)$-variable function f as follows:

$$f(x_1, \ldots, x_n, 0) = g(x_1, \ldots, x_n) \tag{4-2a}$$

$$f(x_1, \ldots, x_n, y + 1) = h(x_1, \ldots, x_n, y, f(x_1, \ldots, x_n, y)) \tag{4-2b}$$

This new function f is said to be defined from the given functions g and h by *primitive recursion*. The variable y is referred to as the *recursion variable* of the definition and f is said to be obtained by (primitive) recursion *on* y. The variables $x_1, \ldots, x_n$ are referred to as the *parameters* of the (primitive) recursion. Evidently Eqs. (4-1) represent a degenerate form of primitive recursion in which there are no parameters and g is reduced to a function of no variables— i.e., a constant.

Note that definition by primitive recursion can be viewed as a composition rule for obtaining a new function f from given functions g and h. This composition rule will play an important role in our study of effectively computable functions.

Example 4-1. The familiar two-variable addition function can be defined by primitive recursion from the one-variable function g and the three-variable function h, where

$$g(a) = a$$

and

$$h(a, b, c) = c + 1.$$

To see this, let f be the function defined so that

$$f(x, 0) = g(x)$$
$$f(x, y + 1) = h(x, y, f(x, y)).$$

Then

$$f(x, 0) = x$$
$$f(x, y + 1) = f(x, y) + 1$$

and an easy induction argument establishes the fact that $f(x, y) = x + y$. ☐

It must be emphasized that, in order to constitute a definition by primitive recursion, the specification of a function f must take precisely the form exhibited in Eqs. (4-2). In particular, the last argument of the function h in Eq. (4-2b) must be $f(x_1, \ldots, x_n, y)$. Thus a specification of the form

$$f(x, 0) = g(x)$$
$$f(x, y + 1) = h(x, y, f(x, \max (y - 3, 0)))$$

is not a definition by primitive recursion. Neither is

$$f(x, 0) = g(x)$$
$$f(x, y + 1) = h(x, y, f(y, x)).$$

Indeed, it is not immediately clear whether this last pair of equations actually specifies a well-defined function f.

The preceding remark prompts us to ask whether primitive recursion always yields well-defined functions. That is, given the total functions g and h, does there necessarily exist a unique function f satisfying Eqs. (4-2)? As intuition suggests, the answer is yes, but to establish this fact rigorously requires a careful set-theoretic argument. The general strategy is to form the intersection of all the $(n + 2)$-ary relations that satisfy Eqs. (4-2) and to prove three things about the resulting relation: that it also satisfies Eqs. (4-2); that it is a total function; and that it is the only total function satisfying Eqs. (4-2). Since the details are not particularly instructive, we omit them and present the conclusion as:

Theorem 4-1. *Let g and h be total number-theoretic functions of n and $n + 2$ variables, respectively. Then there exists a unique total $(n + 1)$-variable function f such that*

$$f(x_1, \ldots, x_n, 0) = g(x_1, \ldots, x_n)$$

and

$$f(x_1, \ldots, x_n, y + 1) = h(x_1, \ldots, x_n, y, f(x_1, \ldots, x_n, y)).$$

This theorem provides the formal justification for viewing primitive recursion as a composition rule for obtaining a new function from two given functions.

The form of Eqs. (4-2) suggests a method of evaluating any function that is defined by primitive recursion. To determine $f(x_1, \ldots, x_n, y)$, first obtain $f(x_1, \ldots, x_n, 0)$ by evaluating $g(x_1, \ldots, x_n)$, then obtain $f(x_1, \ldots, x_n, 1)$ by evaluating $h(x_1, \ldots, x_n, 0, f(x_1, \ldots, x_n, 0))$, then obtain $f(x_1, \ldots, x_n, 2)$ by evaluating $h(x_1, \ldots, x_n, 1, f(x_1, \ldots, x_n, 1))$, and so on until the desired value $f(x_1, \ldots, x_n, y)$ is reached. The existence of this procedure implies that the function f must be effectively computable whenever the functions g and h are.

Rephrasing this last observation in terms of Turing computability, we get:

Theorem 4-2. *Suppose that the $(n + 1)$-variable function f is defined by primitive recursion from the total n-variable Turing-computable function g and the total $(n + 2)$-variable Turing-computable function h. Then f is also Turing computable.*

Argument. A Turing machine **T** that evaluates the function f can be designed to operate as follows. Given the initial pattern $0 \, \bar{x}_1 \, 0 \cdots 0 \, \bar{x}_n \, 0 \, \bar{y} \, 0$, the machine first erases a single 1 from the right-hand end of the block representing the argument y. It then forms a copy of the entire argument string immediately to the right of the original string and erases the last 1 in the new y-block. Thus the tape now contains the pattern

$$0 \, \bar{x}_1 \, 0 \cdots 0 \, \bar{x}_n \, 0 \, \overline{y - 1} \, 0 \, \bar{x}_1 \, 0 \cdots 0 \, \bar{x}_n \, 0 \, \overline{y - 2} \, 0.$$

The machine next forms a copy of the string $\bar{x}_1 \, 0 \cdots 0 \, \bar{x}_n \, 0 \, \overline{y - 2}$ and again erases the last 1 in the new y-block. It continues to form successive copies in this way, each time decreasing the length of the y-block by one, until finally the y-block has been completely removed. At this point the tape contains the pattern

$$0 \, \bar{x}_1 \, 0 \cdots 0 \, \bar{x}_n \, 0 \, \overline{y - 1} \, 0 \, \bar{x}_1 \, 0 \cdots 0 \, \bar{x}_n \, 0 \, \overline{y - 2} \, 0 \cdots$$
$$0 \, \bar{x}_1 \, 0 \cdots 0 \, \bar{x}_n \, 0 \, \bar{0} \, 0 \, \bar{x}_1 \, 0 \cdots 0 \, \bar{x}_n \, 0.$$

The machine **T** now uses the last (right-most) copy of the x's as the arguments for a standard-form evaluation of $g(x_1, \ldots, x_n)$, thereby obtaining the value of $f(x_1, \ldots, x_n, 0)$ and producing the tape pattern

$$0 \, \bar{x}_1 \, 0 \cdots 0 \, \bar{x}_n \, 0 \, \overline{y - 1} \, 0 \, \bar{x}_1 \, 0 \cdots 0 \, \bar{x}_n \, 0 \, \overline{y - 2} \, 0 \cdots$$
$$0 \, \bar{x}_1 \, 0 \cdots 0 \, \bar{x}_n \, 0 \, \bar{0} \, 0 \, \overline{f(x_1, \ldots, x_n, 0)} \, 0.$$

It next uses the last remaining copy of the x's, together with the blocks $\bar{0}$ and $\overline{f(x_1, \ldots, x_n, 0)}$, as the arguments for a standard-form evaluation of $h(x_1, \ldots, x_n, 0, f(x_1, \ldots, x_n, 0))$, thereby obtaining the value of $f(x_1, \ldots, x_n, 1)$ and producing the tape pattern

$$0 \, \bar{x}_1 \, 0 \cdots 0 \, \bar{x}_n \, 0 \, \overline{y - 1} \, 0 \, \bar{x}_1 \, 0 \cdots 0 \, \bar{x}_n \, 0 \, \overline{y - 2} \, 0 \cdots$$
$$0 \, \bar{x}_1 \, 0 \cdots 0 \, \bar{x}_n \, 0 \, \bar{1} \, 0 \, \overline{f(x_1, \cdots, x_n, 1)} \, 0.$$

The machine next uses the pattern $\bar{x}_1 \, 0 \cdots 0 \, \bar{x}_n \, 0 \, \overline{1} \, 0 \, \overline{f(x_1, \ldots, x_n, 1)}$ to evaluate $h(x_1, \ldots, x_n, 1, f(x_1, \ldots, x_n, 1)) = f(x_1, \ldots, x_n, 2)$, then the pattern $\bar{x}_1 \, 0 \cdots 0 \, \bar{x}_n \, 0 \, \overline{2} \, 0 \, \overline{f(x_1, \ldots, x_n, 2)}$ to evaluate $h(x_1, \ldots, x_n, 2, f(x_1, \ldots, x_n, 2)) = f(x_1, \ldots, x_n, 3)$, and so on. The machine continues to work its way to the left in this manner until the tape pattern has been reduced to

$$\bar{x}_1 \, 0 \cdots 0 \, \bar{x}_n \, 0 \, \overline{y-1} \, 0 \, \overline{f(x_1, \ldots, x_n, y-1)},$$

at which point it evaluates $h(x_1, \ldots, x_n, y-1, f(x_1, \ldots, x_n, y-1)) = f(x_1, \ldots, x_n, y)$ and halts. $\qquad\qquad\qquad\qquad\qquad\qquad\qquad\Box$

Thus the process of defining new functions by primitive recursion is one that preserves both effective computability and Turing computability.

Primitive Recursive Functions

We now define the *primitive recursive functions* to be those functions that can be obtained from the constant, projection, and successor functions by means of functional composition and primitive recursion. To make sure that this definition is understood, we spell it out in detail.

1. The constant, projection, and successor functions are all primitive recursive functions.

2. If $g_1, \ldots, g_m$ are n-variable primitive recursive functions, and if h is an m-variable primitive recursive function, then the composite function $h \circ (g_1, \ldots, g_m)$ is also a primitive recursive function.

3. If g and h are n- and $(n+2)$-variable primitive recursive functions, then the $(n+1)$-variable function f defined from g and h by primitive recursion is also a primitive recursive function.

No functions other than those that can be defined by means of a finite number of applications of rules 1 through 3 are primitive recursive functions.

To illustrate the use of this definition, we will show that the familiar two-variable addition and multiplication functions are primitive recursive. First consider the functions g_1 and h_1, where

$$g_1(a) = a$$

and

$$h_1(a, b, c) = c + 1.$$

These functions are primitive recursive since g_1 is a projection function and h_1 is the composition of a projection function and the successor function. Specifically,

$$g_1 = P_1^{(1)}$$

and

$$h_1 = S \circ (P_3^{(3)}).$$

Now recall from Example 4-1 that the addition function—call it f_1—can be obtained from g_1 and h_1 by primitive recursion:

$$f_1(x, 0) = g_1(x)$$

$$f_1(x, y + 1) = h_1(x, y, f_1(x, y))$$

Thus addition is a primitive recursive function.

Next consider the functions g_2 and h_2, where

$$g_2(a) = 0$$

and

$$h_2(a, b, c) = a + c.$$

These functions are primitive recursive because g_2 is a constant function and h_2 is the composition of two projection functions and the addition function, which we have just shown to be primitive recursive. Specifically,

$$g_2 = C_0^{(1)}$$

and

$$h_2 = f_1 \circ (P_1^{(3)}, P_3^{(3)}).$$

The two-variable multiplication function—call it f_2—can now be obtained by primitive recursion from g_2 and h_2, as the reader may easily verify:

$$f_2(x, 0) = g_2(x)$$

$$f_2(x, y + 1) = h_2(x, y, f_2(x, y))$$

Thus multiplication is also primitive recursive.

Before giving further examples of specific primitive recursive functions, we consider some simple but important facts about primitive recursive functions in general. Note that if f is a primitive recursive function, it must be possible to form a sequence of functions, with f as the last member, such that each function in the sequence is either a constant, projection, or successor function, or else is obtained from preceding members of the sequence by functional composition or primitive recursion. Such a sequence of functions is called a *formal* (*primitive recursive*) *derivation* of the function f. Formal derivations are usually annotated to show how each of their members is obtained.

Example 4-2.

a) The following is a formal derivation of the addition function f_1 discussed above:

$$P_1^{(1)}, \qquad P_3^{(3)}, \qquad S, \qquad h_1, \qquad f_1$$

Note that the first two members of this list are projection functions, the third is the successor function, the fourth is obtained from the second and third by composition, and the last is obtained from the first and fourth by primitive recursion. Thus an appropriate annotated version of the derivation might be:

$$P_1^{(1)} \quad \text{projection function}$$

$$P_3^{(3)} \quad \text{projection function}$$

$$S \quad \text{successor function}$$

$$h_1 \quad \text{where } h_1 = S \circ (P_3^{(3)})$$

$$f_1 \quad \text{where } \begin{cases} f_1(x, 0) = P_1^{(1)}(x) \\ f_1(x, y+1) = h_1(x, y, f_1(x, y)) \end{cases}$$

b) The following is a formal derivation of the multiplication function f_2, the annotation being left to the reader:

$$P_1^{(1)}, \quad P_3^{(3)}, \quad S, \quad h_1, \quad f_1, \quad C_0^{(1)}, \quad P_1^{(3)}, \quad h_2, \quad f_2 \qquad \square$$

In practice, formal derivations are too detailed to be very useful for describing primitive recursive functions. The notion of a formal derivation is an important one nevertheless. For one thing, it makes more precise the definition of a primitive recursive function, since a function f is primitive recursive iff there exists a formal primitive recursive derivation of f. Moreover, the notion of a formal derivation often plays a key role in establishing general properties of primitive recursive functions.

Theorem 4-3. *Every primitive recursive function f is total.*

Proof. The proof is accomplished by induction on the length of the shortest derivation of f.

Basis. If f has a derivation of length one, it must be a constant function, a projection function, or the successor function. In each case, f is certainly total.

Induction step. Assume that all primitive recursive functions having derivations of length n or less are total and let f require a derivation of length $n+1$. If f is obtained from the primitive recursive functions $g_1, \ldots, g_m$, and h by functional composition, then each of these latter functions must have derivations of length n or less. The induction hypothesis then ensures that $g_1, \ldots, g_m$ and h are total, from which it follows that f is total. Similarly, if f is obtained from the primitive recursive functions g and h by primitive recursion, the induction hypothesis implies that g and h are total, whence Theorem 4-1 guarantees that f is total. $\qquad \square$

Theorem 4-3 has several obvious consequences. For instance, we see from Theorem 1-2 that the class of primitive recursive functions cannot contain a universal function, no matter what indexing scheme we might choose. Perhaps more to the point in the present context is the conclusion that not all Turing-computable functions are primitive recursive. (In fact, not all total Turing-computable functions are primitive recursive, but this is not obvious.) On the other hand, we can rather easily establish:

Theorem 4-4. *Every primitive recursive function f is Turing computable.*

Proof. As in the case of Theorem 4-3, the proof is accomplished by induction on the length of the shortest derivation of f.

Basis. If f has a derivation of length one, it must be a constant, projection, or successor function, in which case it is certainly Turing computable.

Induction step. Assume that all primitive recursive functions having derivations of length n or less are Turing computable, and let f require a derivation of length $n + 1$. Then either f is obtained from functions $g_1, \ldots, g_m$, and h by functional composition, or else it is obtained from functions g and h by primitive recursion. In either case, the induction hypothesis implies that the functions from which f is obtained are Turing computable. In the case of composition Theorem 3-2 then ensures that f is Turing computable, while in the case of primitive recursion Theorem 4-2 ensures that f is Turing computable. $\square$

Useful Primitive Recursive Functions

In order to illustrate the richness of the class of primitive recursive functions, as well as to provide some tools that will be needed later on, we now present a number of useful primitive recursive functions. In describing these functions, we will find it convenient to adopt the notation a^+ as an abbreviation for $a + 1$. Note that once the nature of a primitive recursive definition is understood, it is possible to be relatively informal in presenting the derivation of a primitive recursive function. This point can be illustrated by the addition function f_1.

Recall that to obtain a formal derivation of f_1, we first defined the auxiliary functions $g_1 = P_1^{(1)}$ and $h_1 = S \circ (P_3^{(3)})$ and established their primitive recursiveness, then defined f_1 by primitive recursion from g_1 and h_1. Substituting the values of g_1 and h_1, and making use of the superscript $^+$ notation, we may put this primitive recursion in the form:

$$f_1(x, 0) = x$$
$$f_1(x, y^+) = (f_1(x, y))^+ \tag{4-3}$$

Since the definitions of g_1 and h_1 are implicit in Eqs. (4-3), and since the primitive recursiveness of g_1 and h_1 is obvious from their definitions, Eqs. (4-3) provide an adequate informal derivation of f_1. Reverting to the familiar infix notation, we can recast Eqs. (4-3) in the form:

$$x + 0 = x$$
$$x + y^+ = (x + y)^+$$

In practice, it is this last pair of equations that would be used to justify the contention that addition is primitive recursive.

We now present informal derivations for a variety of primitive recursive functions. (The addition and multiplication functions are included for the sake of completeness.) In each case the reader should be sure he understands how to convert the given derivation into a formal one. Note that once a function has been shown to be primitive recursive, it can be used in the derivations of other functions.

Addition. The addition function is defined by primitive recursion using the equations:

$$x + 0 = x$$
$$x + y^+ = (x + y)^+ \tag{4-4}$$

Multiplication. Multiplication, which is really nothing more than repeated addition, can be defined by primitive recursion in terms of addition. The appropriate equations are:

$$x \cdot 0 = 0$$
$$x \cdot y^+ = x + x \cdot y \tag{4-5}$$

This informal derivation should be compared with the formal one given earlier.

Exponentiation. Just as multiplication can be viewed as repeated addition, so exponentiation can be viewed as repeated multiplication. In particular, exponentiation can be obtained from multiplication by primitive recursion, as follows:

$$x^0 = 1$$
$$x^{y^+} = x(x^y) \tag{4-6}$$

The factorial function. The factorial function is also a form of repeated multiplication, and it too can be obtained from multiplication by means of primitive recursion.

$$0! = 1$$
$$(y^+)! = y^+ \cdot y! \tag{4-7}$$

Note that, despite the appearance of y^+ on the right-hand side of the second equation, the value of $(y^+)!$ can indeed be expressed as a primitive recursive function of y and $y!$.

The predecessor function. Primitive recursive functions are not limited to the familiar "algebraic" functions. One of the simplest and most useful primitive recursive functions is a one-variable function that acts as a partial inverse for the successor function. Called the *predecessor function*, and denoted by the prefix symbol Pd, it is defined as follows:

$$\text{Pd}(y) = \begin{cases} 0 & \text{if} & y = 0 \\ y - 1 & \text{if} & y > 0 \end{cases}$$

To show that the predecessor function is primitive recursive, we note that it can be derived very simply by primitive recursion from the constant 0 and the projection function $P_1^{(2)}$:

$$\text{Pd}(0) = 0$$
$$\text{Pd}(y^+) = y \tag{4-8}$$

Proper subtraction. We next consider a function that serves as a partial inverse for addition. This function, called *proper subtraction* and represented by the infix symbol $\dot{-}$, is defined as follows.

$$x \dot{-} y = \begin{cases} x - y & \text{if} & x \geq y \\ 0 & \text{if} & x < y \end{cases}$$

In other words, $x \dot{-} y$ is identical to $x - y$ except that $x \dot{-} y$ assumes the value 0 whenever $x - y$ is negative. (The expression $x \dot{-} y$ is sometimes read "x monus y.") To show that proper subtraction is a primitive recursive function, we note that it can be obtained by primitive recursion as follows.

$$x \dot{-} 0 = x$$
$$x \dot{-} y^+ = \text{Pd}(x \dot{-} y) \tag{4-9}$$

Absolute difference. The two-variable function whose value is the magnitude of the difference between its argument values is called the *absolute difference function*. Its value for the arguments x and y will usually be denoted in the obvious way, namely, $|x - y|$. This function can be obtained by functional composition from addition and proper subtraction, since

$$|x - y| = (x \dot{-} y) + (y \dot{-} x). \tag{4-10}$$

It is therefore primitive recursive.

The sign functions. It is often convenient to be able to determine whether or not a given variable has the value zero. For this purpose we introduce two one-variable functions, denoted sg and $\overline{sg}$, defined as follows.

$$sg(y) = \begin{cases} 1 & \text{if} & y > 0 \\ 0 & \text{if} & y = 0 \end{cases} \qquad \overline{sg}(y) = \begin{cases} 0 & \text{if} & y > 0 \\ 1 & \text{if} & y = 0 \end{cases}$$

These functions sg and $\overline{sg}$ will be referred to as the *sign function* and the *cosign function*, respectively. Both are primitive recursive, for both can be defined very simply by primitive recursion.

$$sg(0) = 0$$
$$sg(y^+) = 1 \qquad\qquad (4\text{-}11)$$

$$\overline{sg}(0) = 1$$
$$\overline{sg}(y^+) = 0 \qquad\qquad (4\text{-}12)$$

Comparison functions. The sign and cosign functions can be used to construct a variety of functions that serve to compare the values of two variables. The most common of these comparison functions are the *less than function*, denoted ls; the *greater than function*, denoted gr; and the *equality function*, denoted eqs. These three functions are defined as follows. $(0,2)$

$$ls(x, y) = \begin{cases} 1 & \text{if} & x < y \\ 0 & \text{if} & x \geq y \end{cases}$$

$$gr(x, y) = \begin{cases} 1 & \text{if} & x > y \\ 0 & \text{if} & x \leq y \end{cases}$$

$$eqs(x, y) = \begin{cases} 1 & \text{if} & x = y \\ 0 & \text{if} & x \neq y \end{cases}$$

All three functions can be obtained from the sign functions by functional composition:

$$ls(x, y) = sg(y \dot- x) \qquad\qquad (4\text{-}13)$$
$$gr(x, y) = sg(x \dot- y) \qquad\qquad (4\text{-}14)$$
$$eqs(x, y) = \overline{sg}(|x - y|) \qquad\qquad (4\text{-}15)$$

Hence all three are primitive recursive.

The utility of the comparison functions is perhaps best illustrated by an example.

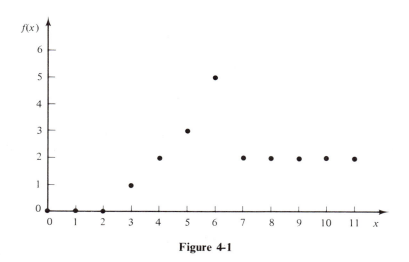

Figure 4-1

Example 4-3. Consider the one-variable function f described in Fig. 4-1, where $f(x)$ is assumed to have the value 2 for $x > 11$. Making use of the fact that for any given value of x at most one of the quantities $\text{ls}(x, 6)$, $\text{eqs}(x, 6)$, and $\text{gr}(x, 6)$ can be nonzero, we write

$$f(x) = (x \div 2) \cdot \text{ls}(x, 6) + 5 \cdot \text{eqs}(x, 6) + 2 \cdot \text{gr}(x, 6).$$

To see that this is a valid description of f, note first that if $x < 6$, the equation reduces to $f(x) = x \div 2$, which does correctly describe the behavior of $f(x)$ in the range $0 \le x < 6$. Similarly, if $x = 6$ the equation reduces to $f(x) = 5$, while if $x > 6$ it reduces to $f(x) = 2$, in each case giving a correct description of the values of $f(x)$. Thus f can be obtained by composition from the primitive recursive functions ls, eqs, gr, $C_2^{(1)}$, $C_6^{(1)}$, $C_5^{(1)}$, multiplication, addition, and proper subtraction. Consequently the function f is also primitive recursive. $\square$

Example 4-3 suggests that a wide variety of functions can be shown to be primitive recursive by breaking them up into "pieces." Of special interest are the functions that are zero for all but a finite number of combinations of argument values, for these functions are all primitive recursive.

Example 4-4. Consider the two-variable function g, where:

$$g(x, y) = \begin{cases} 3 & \text{if} \quad x = 1 \text{ and } y = 7 \\ 5 & \text{if} \quad x = 2 \text{ and } y = 0 \\ 1 & \text{if} \quad x = 4 \text{ and } y = 4 \\ 0 & \text{otherwise} \end{cases}$$

Noting that a term of the form $a \cdot \text{eqs}(x, b) \cdot \text{eqs}(y, c)$ will have the value a if $x = b$ and $y = c$ and the value 0 otherwise, we can write

$$g(x, y) = 3 \cdot \text{eqs}\ (x, 1) \cdot \text{eqs}\ (y, 7) + 5 \cdot \text{eqs}\ (x, 2) \cdot \text{eqs}\ (y, 0)$$
$$+ \text{eqs}\ (x, 4) \cdot \text{eqs}\ (y, 4).$$

Thus g is a primitive recursive function.   □

Since the technique used in Example 4-4 can obviously be extended to apply to a function of any number of variables, we have

Theorem 4-5. *Every number-theoretic function whose value is zero for all but a finite number of combinations of argument values is primitive recursive.*

Bounded Sums and Products

It is often desired to form the sum or product of the values of a given function for a succession of argument values. As a simple example, let g be a given one-variable total function and let f be the new one-variable function defined so that

$$f(y) = \sum_{i=0}^{y} g(i). \tag{4-16}$$

This function f is referred to as a *bounded sum* of the function g. Note that f can be obtained from g by primitive recursion, for:

$$f(0) = g(0)$$
$$f(y^+) = f(y) + g(y^+)$$

Thus f is primitive recursive whenever g is.

Example 4-5. Let g be the projection function $P_1^{(1)}$, and define f as in Eq. (4-16). Then f is the primitive recursive function whose value is given by

$$f(y) = 0 + 1 + 2 + \cdots + y = \frac{y(y + 1)}{2}.$$ □

The bounded summation introduced in Eq. (4-16) can be generalized in two ways. The first is to allow for one or more extra variables, or parameters, in the summation. Thus if g is a total $(n + 1)$-variable function, we may define the new $(n + 1)$-variable function f so that

$$f(x_1, \ldots, x_n, y) = \sum_{i=0}^{y} g(x_1, \ldots, x_n, i). \tag{4-17}$$

Note that Eq. (4-16) represents a special case of Eq. (4-17) in which $n = 0$.

Theorem 4-6. *Let g be a primitive recursive function of* $n + 1$ *variables. Then the function f defined from g as specified in Eq. (4-17) must also be primitive recursive.*

Proof. The function f can be obtained from g by primitive recursion:

$$f(x_1, \ldots, x_n, 0) = g(x_1, \ldots, x_n, 0)$$
$$f(x_1, \ldots, x_n, y^+) = f(x_1, \ldots, x_n, y) + g(x_1, \ldots, x_n, y^+)$$

Thus f is necessarily primitive recursive.   ☐

A second way of generalizing bounded summation is to replace the limits of summation by "computed values." These computed limits may be functions of any or all of the arguments of the function being defined. Thus if g, w_1, and w_2 are functions of $n + 1$, n, and n variables, respectively, we may define the new n-variable function f so that

$$f(x_1, \ldots, x_n) = \sum_{i = w_1(x_1, \ldots, x_n)}^{w_2(x_1, \ldots, x_n)} g(x_1, \ldots, x_n, i). \qquad (4\text{-}18)$$

In making this definition we adopt the convention that $f(x_1, \ldots, x_n) = 0$ whenever $w_2(x_1, \ldots, x_n) < w_1(x_1, \ldots, x_n)$.

It is not surprising to find that f is primitive recursive whenever g, w_1, and w_2 are.

Corollary 4-6.1. *Let g, w_1, and w_2 be primitive recursive functions of $n + 1$, n, and n variables, respectively. Then the function f defined from g, w_1, and w_2 as specified in Eq. (4-18) must also be primitive recursive.*

Proof. To save space, we provide a proof for the case $n = 1$; extension to the general case is trivial. We begin by noting that the value of the desired function f can be expressed in the form

$$f(x) = \overline{sg}(w_1(x) \doteq w_2(x))(g(x, w_1(x)) + g(x, w_1(x) + 1) + \cdots$$
$$+ g(x, w_1(x) + (w_2(x) \doteq w_1(x))))$$
$$= \overline{sg}(w_1(x) \doteq w_2(x)) \sum_{i=0}^{w_2(x) \doteq w_1(x)} g'(x, i)$$

where $g'(x, i) = g(x, w_1(x) + i)$. Since g' is obviously primitive recursive, Theorem 4-6 ensures that the function f', where

$$f'(x, y) = \sum_{i=0}^{y} g'(x, i)$$

is primitive recursive. Then since

$$f(x) = \overline{sg}(w_1(x) \doteq w_2(x))f'(x, w_2(x) \doteq w_1(x))$$

it follows that f must also be primitive recursive.   ☐

In summary, we see that the definitional schemes represented by Eqs. (4-16), (4-17), and (4-18) all yield primitive recursive functions whenever the functions to which they are applied are primitive recursive. These schemes will all be referred to as definitions by *bounded summation*.

Similar results are obtained when multiplication is used as the repeated operation, rather than addition. For instance, if g is a given one-variable total function, we can define a new function f so that

$$f(y) = \prod_{i=0}^{y} g(i).$$

As in the case of bounded summation, this scheme can be generalized to include the use of parameters. Thus if g is a total function of $n + 1$ variables, we may define a new $(n + 1)$-variable function f so that

$$f(x_1, \ldots, x_n, y) = \prod_{i=0}^{y} g(x_1, \ldots, x_n, i). \tag{4-19}$$

The reader may easily establish the following counterpart of Theorem 4-6.

Theorem 4-7. *Let g be a primitive recursive function of $n + 1$ variables. Then the function f defined from g as specified in Eq. (4-19) is also primitive recursive.*

The use of repeated products can also be generalized by allowing computed limits. In this case it is convenient to require that the resulting function assume the value 1 whenever the lower limit of the product exceeds the upper limit. Thus we write

$$f(x_1, \ldots, x_n) = \prod_{i=w_1(x_1, \ldots, x_n)}^{w_2(x_1, \ldots, x_n)} g(x_1, \ldots, x_n, i) \tag{4-20}$$

to indicate that the value of $f(x_1, \ldots, x_n)$ is to be the product of the values $g(x_1, \ldots, x_n, w_1(x_1, \ldots, x_n))$ through $g(x_1, \ldots, x_n, w_2(x_1, \ldots, x_n))$ when $w_1(x_1, \ldots, x_n) \leq w_2(x_1, \ldots, x_n)$, and 1 otherwise. Minor modifications in the proof of Corollary 4-6.1 then yield:

Corollary 4-7.1. *Let g, w_1, and w_2 be primitive recursive functions of $n + 1$, n, and n variables, respectively. Then the function f defined from g, w_1, and w_2 as specified by Eq. (4-20) must also be primitive recursive.*

The schemes represented by Eqs. (4-19) and (4-20) will be referred to as definitions by *bounded products*.

Bounded Minimalization

It is often necessary to determine the smallest value of a variable for which a specified condition holds. For example, given an $(n + 1)$-variable function g and a particular choice of values of the variables $x_1, \ldots, x_n$, we might wish to know the smallest value of the variable z for which $g(x_1, \ldots, x_n, z) = 0$. This value is commonly represented by the expression

$$\mu z[g(x_1, \ldots, x_n, z) = 0]$$

which is read, "the least z such that $g(x_1, \ldots, x_n, z)$ equals zero." Since this value will in general depend on the choice of $x_1, \ldots, x_n$, we can think of defining a new n-variable function f such that

$$f(x_1, \ldots, x_n) = \mu z[g(x_1, \ldots, x_n, z) = 0].$$

For instance, if g is the two-variable function such that $g(x, z) = x \dotminus (2z + 1)$, then f will be the one-variable function whose value for the argument x is the integer part of $x/2$.

Unfortunately, the function f defined in this way need not be primitive recursive, even though g is. In fact, f need not even be total, since for some choices of $x_1, \ldots, x_n$ there may be no value of z for which $g(x_1, \ldots, x_n, z) = 0$. However, it is possible to modify the process of selecting a "least value" so that the function f is primitive recursive whenever g is.

Suppose that g is any total function and that $x_1, \ldots, x_n$, and y are any natural numbers. We use the expression

$$\overset{y}{\mu}z[g(x_1, \ldots, x_n, z) = 0]$$

to denote the smallest value of z in the range $0 \leq z \leq y$ for which $g(x_1, \ldots, x_n, z) = 0$ whenever such a value of z exists, and $y + 1$ whenever no such value of z exists. Since the value of

$$\overset{y}{\mu}z[g(x_1, \ldots, x_n, z) = 0]$$

depends on $x_1, \ldots, x_n$, and y, we are led to define the $(n + 1)$-variable function f so that

$$f(x_1, \ldots, x_n, y) = \overset{y}{\mu}z[g(x_1, \ldots, x_n, z) = 0]. \tag{4-21}$$

In so doing, we think of $\overset{y}{\mu}z$ as representing an operator that converts the given total function g into the new function f. We call this operator the *bounded minimalization operator*, and say that f is defined from g by *bounded minimalization*. The value y is called the (upper) bound, or *limit*, of the minimalization.

It is clear from the definition of bounded minimalization that the function f appearing in Eq. (4-21) is necessarily total. We now propose to show that f must in fact be primitive recursive whenever g is. We begin with an intermediate result.

Lemma 4-1. *Let g be any primitive recursive function of $n + 1$ variables. Define the $(n + 1)$-variable function h so that:*

$$h(x_1, \ldots, x_n, k) = \begin{cases} 1 & \text{if } g(x_1, \ldots, x_n, z) \text{ differs from 0 for each} \\ & \text{value of } z \text{ in the range } 0 \leq z \leq k \\ 0 & \text{otherwise} \end{cases}$$

Then h is also primitive recursive.

Proof. Note that $h(x_1, \ldots, x_n, k)$ is to have the value 1 iff each of the terms sg $(g(x_1, \ldots, x_n, 0))$ through sg $(g(x_1, \ldots, x_n, k))$ has the value 1. We may therefore write

$$h(x_1, \ldots, x_n, k) = \prod_{i=0}^{k} \text{sg} \ (g(x_1, \ldots, x_n, i)).$$

Since g and sg are primitive recursive, it follows from Theorem 4-7 that h must also be primitive recursive. $\square$

Lemma 4-1 leads directly to our desired result.

Theorem 4-8. *Let g be any primitive recursive function of $n + 1$ variables, and let f be defined from g by bounded minimalization, so that*

$$f(x_1, \ldots, x_n, y) = \overset{y}{\mu z}[g(x_1, \ldots, x_n, z) = 0].$$

Then f is also primitive recursive.

Proof. The value of $f(x_1 \ldots, x_n, y)$ is either the smallest z less than or equal to y for which $g(x_1, \ldots, x_n, z) = 0$ or, if there is no such z, $y + 1$. In either case, the value of $f(x_1, \ldots, x_n, y)$ is just the number of values of k in the range $0 \leq k \leq y$ such that $g(x_1, \ldots, x_n, z)$ differs from 0 for all z less than or equal to k. Thus we can write

$$f(x_1, \ldots, x_n, y) = \sum_{k=0}^{y} h(x_1, \ldots, x_n, k)$$

where h is the function defined in Lemma 4-1. Since h is primitive recursive whenever g is, f must also be primitive recursive. $\square$

Bounded minimalization can be generalized in at least two ways. First, the upper limit of the minimalization can be replaced by a computed value, just as in the case of bounded sums or bounded products. As might be expected, the resulting function f is still primitive recursive whenever g is.

Corollary 4-8.1. *Let g and w be primitive recursive functions of $n + 1$ and n variables, respectively. Define the n-variable function f so that*

$$f(x_1, \ldots, x_n) = \overset{w(x_1, \ldots, x_n)}{\mu z} [g(x_1, \ldots, x_n, z) = 0].$$

Then f is also primitive recursive.

The proof is straightforward and is left to the reader.

A second way of generalizing bounded minimalization is to replace the condition $g(x_1, \ldots, x_n, z) = 0$ by a condition of the form $g_1(x_1, \ldots, x_n, z) = g_2(x_1, \ldots, x_n, z)$. If g_1 and g_2 are specified functions, the expression

$$\overset{y}{\mu z}[g_1(x_1, \ldots, x_n, z) = g_2(x_1, \ldots, x_n, z)]$$

denotes the smallest z in the range $0 \le z \le y$ for which $g_1(x_1, \ldots, x_n, z)$ and $g_2(x_1, \ldots, x_n, z)$ are equal, or $y + 1$ if there is no such value of z. Since $g_1(x_1, \ldots, x_n, z) = g_2(x_1, \ldots, x_n, z)$ iff $|g_1(x_1, \ldots, x_n, z) - g_2(x_1, \ldots, x_n, z)| = 0$, we have:

Corollary 4-8.2. *Let g_1, g_2, and w be primitive recursive functions of $n + 1, n + 1$, and n variables, respectively, and define the n-variable function f so that*

$$f(x_1, \ldots, x_n) = \overset{w(x_1, \ldots, x_n)}{\mu z} \ [g_1(x_1, \ldots, x_n, z) = g_2(x_1, \ldots, x_n, z)].$$

Then f is also primitive recursive.

The Division Functions

Bounded minimalization in its various forms provides a powerful technique for obtaining new primitive recursive functions from old ones. We shall illustrate this fact by deriving a number of primitive recursive functions associated with division and divisibility. These functions will play an important role in later sections.

We begin with a pair of closely related two-variable functions: the *quotient function*, denoted "quo," and the *remainder function*, denoted "rem." The quotient function is defined so that quo(x, y) is the quotient obtained upon dividing y by x, as long as $x \ne 0$; if $x = 0$, quo(x, y) is defined to be zero. Similarly, the remainder function is defined so that rem(x, y) is the remainder obtained upon dividing y by x, as long as $x \ne 0$; if $x = 0$, rem(x, y) is defined to be y. Thus quo(3, 7) = 2, rem(3, 7) = 1, quo(5, 23) = 4, rem(5, 23) = 3, quo(4, 0) = 0, rem(4, 0) = 0, etc.

To establish the primitive recursiveness of the quotient function, we note that the desired value of quo(x, y) is just the value of z that makes xz come as close as possible to y without exceeding y. More to the point, quo(x, y) is the smallest value of z such that $x(z + 1)$ is strictly greater than y. Since this value of z will never exceed y, we are led to try

$$\overset{y}{\mu z}[\mathrm{gr}(x(z + 1), y) = 1]$$

as an expression for the value of quo(x, y). A little thought shows that this expression is correct except when $x = 0$, in which case it yields the value $y + 1$ rather than the desired value 0. The remedy is to multiply by sg(x), thereby forcing the value of the expression to be 0 whenever $x = 0$. Thus we have

$$\mathrm{quo}(x, y) = \mathrm{sg}(x)\left(\overset{y}{\mu z}[\mathrm{gr}(x(z + 1), y) = 1] \right) \tag{4-22}$$

from which it follows that quo is primitive recursive.

Before going on, it may be worthwhile to explain in some detail exactly how Eq. (4-22) establishes the primitive recursiveness of the quotient function.

We first define the auxiliary functions g_1, g_2, and w so that:

$$g_1(x, y, z) = gr(x(z + 1), y)$$
$$g_2(x, y, z) = 1$$
$$w(x, y) = y$$

The functions g_2 and w are obviously primitive recursive. And since g_1 is formed by composition from functions known to be primitive recursive, it too must be primitive recursive. Corollary 4-8.2 therefore ensures that the function f is primitive recursive, where

$$f(x, y) = {}^{w(x, y)}\mu z[g_1(x, y, z) = g_2(x, y, z)] = \overset{y}{\mu}z[gr(x(z + 1), y) = 1].$$

(In this application of Corollary 4-8.2, $n = 2$, $x_1 = x$, and $x_2 = y$.) Finally, we note that the function quo is obtained from f, sg $\circ$ $(P_1^{(2)})$, and multiplication by means of functional composition, since

$$quo(x, y) = sg(x)f(x, y).$$

The function quo is therefore primitive recursive.

    Of course, we will not usually resort to such detailed derivations of primitive recursive functions, the informal approach of Eq. (4-22) being quite satisfactory in practice. It is nevertheless important to understand the logic underlying our informal applications of bounded minimalization, and the reader should make sure that he is able to supply the missing details.

    Once the primitive recursiveness of the quotient function has been established, that of the remainder function follows easily, since

$$rem(x, y) = y \div x \ quo(x, y). \tag{4-23}$$

Thus the remainder function is obtained by composition from various functions known to be primitive recursive.

    We now introduce four functions associated with the divisibility relation. The first is the *divides function*, denoted "div" and defined so that

$$div(x, y) = \begin{cases} 1 & \text{if } x > 0, y > 0, \text{ and } x \text{ divides } y \\ 0 & \text{otherwise.} \end{cases}$$

Referring to the definition of the remainder function, we see that $div(x, y)$ is to have the value 1 iff $y > 0$ and $rem(x, y) = 0$. We can therefore write

$$div(x, y) = \overline{sg} \ (rem \ (x, y)) \cdot sg \ (y) \tag{4-24}$$

from which we conclude that div is primitive recursive.

The second function is the *number of divisors function*, denoted "ndiv" and defined so that ndiv(x) equals the number of divisors of x, with 1 and x being included in the count. Thus ndiv(3) = 2, ndiv(12) = 6, etc. Evidently

$$\text{ndiv}(x) = \sum_{i=0}^{x} \text{div}(i, x) \tag{4-25}$$

so this function is also primitive recursive.

The third function is the *prime function*. This is a one-variable function denoted "pr" and defined as follows.

$$\text{pr}(x) = \begin{cases} 1 & \text{if } x \text{ is a prime} \\ 0 & \text{otherwise.} \end{cases}$$

Since a number is prime iff it has exactly two divisors, we can write

$$\text{pr}(x) = \text{eqs}(\text{ndiv}(x), 2) \tag{4-26}$$

and pr is also primitive recursive.

The fourth function requires some notation. We shall often use p_i to denote the ith prime number, starting with 2 as the "0th" prime. Thus $p_0 = 2$, $p_1 = 3$, $p_2 = 5$, $p_3 = 7$, $p_4 = 11$, and so on. (The subscript i is sometimes referred to as the *index* of the prime p_i.) We now define the one-variable function "pn" so that $\text{pn}(x) = p_x$. This function will be called the *n*th *prime function*.

To show that the function pn is primitive recursive, we use the fact that p_{y+1} never exceeds $w(p_y) = p_y! + 1$ in value.[†] Thus $\text{pn}(y^+)$ is the smallest number less than or equal to $w(\text{pn}(y))$ that is prime and greater than $\text{pn}(y)$. We can therefore define pn by primitive recursion as follows.

$$\text{pn}(0) = 2$$
$$\text{pn}(y^+) = \overset{w(\text{pn}(y))}{\mu z}[\text{pr}(z) \cdot \text{gr}(z, \text{pn}(y)) = 1] \tag{4-27}$$

To make it clear that Eqs. (4-27) represent a valid primitive recursion, note that we have expressed $\text{pn}(y^+)$ in the form $h(y, \text{pn}(y))$, where

$$h(a, b) = \overset{w(b)}{\mu z}[\text{pr}(z) \cdot \text{gr}(z, b) = 1].$$

Corollary 4-8.2 ensures that h is primitive recursive, whence it follows that pn is primitive recursive.

It should be evident from the preceding examples that many useful functions turn out to be primitive recursive. In order to establish the primitive recursiveness of more complex functions, we must examine some general information representing schemes.

[†] Note that $p_y! + 1$ is not divisible by any of the primes $p_0, p_1, \ldots, p_y$. Therefore $p_y! + 1$ is either prime or else divisible by a prime greater than p_y. In either case $p_y! + 1 \geq p_{y+1}$.

4.2 CANTOR AND GÖDEL NUMBERINGS

There are many situations in which it is desirable to be able to represent ordered n-tuples of natural numbers by single natural numbers. In this section we introduce two basic methods for encoding n-tuples into natural numbers and show how these encoding schemes can be used to simplify the derivations of many primitive recursive functions. The same methods will be used in Chapter 5 to establish the computability of a broad class of functions.

Cantor Numbering

We begin with the problem of representing ordered pairs of natural numbers by single natural numbers. Since we wish to be able not only to encode ordered pairs into natural numbers but also to decode natural numbers into ordered pairs, what we seek is a one-to-one mapping from N^2 to N. Because some of our applications will require that each natural number be assigned to some ordered pair, we will insist that the mapping be a bijection from N^2 to N. And, of course, we would like to choose a mapping for which the encoding and decoding processes are as simple as possible; certainly we would like the encoding and decoding to be accomplished by primitive recursive functions.

One of the simplest bijections from N^2 to N is that shown in Fig. 4-2, where the components of an ordered pair are represented by row and column coordinates and the natural number assigned to a given pair is listed at the appropriate point within the array. Note that the natural numbers are assigned in sequence along the various diagonals of the array, running from upper right to lower left along each diagonal, and working outward diagonal by diagonal from the upper left corner of the array. This particular mapping will be denoted by π.

Such a diagonal-by-diagonal method of assigning natural numbers to ordered pairs will be called a *Cantor-numbering scheme*, after the manner in which Cantor showed that the sets N^2 and N have the same cardinality. The number $\pi(x, y)$ will be called the *Cantor number* of the ordered pair (x, y). Thus the Cantor number of $(1, 2)$ is 7, the Cantor number of $(4, 3)$ is 32, and so on.

Although Fig. 4-2 provides a convenient tabular representation of the function π, there are many occasions when an algebraic description is needed. It is not difficult to show that the values of π are given by the equation

$$\pi(x, y) = \frac{(x + y)(x + y + 1)}{2} + x = \frac{1}{2}(x^2 + 2xy + y^2 + 3x + y). \quad (4\text{-}28)$$

Thus π is in fact a primitive recursive function. At this point Eq. (4-28) may be taken to be the formal definition of π and Fig. 4-2 relegated to the position of an informal description of what π "looks like." To be rigorous, of course, we should formally establish that the function defined by Eq. (4-28) is a bijection from N^2 to N. This can be done by a series of inductive arguments, but the process is not instructive and we shall not undertake it.

x \ y	0	1	2	3	4	5	6	7	8
0	0	1	3	6	10	15	21	28	36
1	2	4	7	11	16	22	29	37	
2	5	8	12	17	23	30	38		
3	9	13	18	24	31				
4	14	19	25	32					
5	20	26	33						
6	27	34							
7	35								

Figure 4-2

We turn next to the problem of decoding a given Cantor number $\pi(x, y)$ into the ordered pair (x, y). Let us define the functions σ_1 and σ_2 so that $\sigma_1(z)$ is the first component of the ordered pair whose Cantor number is z and $\sigma_2(z)$ is the second component of the ordered pair whose Cantor number is z. For instance, referring to Fig. 4-2, we see that $\sigma_1(7) = 1$, $\sigma_2(7) = 2$, $\sigma_1(30) = 2$, $\sigma_2(30) = 5$, and so on. The functions σ_1 and σ_2 will be called the *decoding functions* for the Cantor numbering scheme of Fig. 4-2. The basic relationship among the functions σ_1, σ_2, and π is expressed by:

$$\pi(\sigma_1(z), \sigma_2(z)) = z \qquad (4\text{-}29)$$

In fact, since π is known to be a bijection, Eq. (4-29) can be taken as the formal definition of σ_1 and σ_2.[†]

Although σ_1 and σ_2 do not have simple algebraic descriptions, they are necessarily primitive recursive. To establish this fact, we first observe that $x \leq \pi(x, y)$ and $y \leq \pi(x, y)$ for every choice of x and y. Therefore $\sigma_1(z)$ must be the least value (in fact the only value) of x in the range $0 \leq x \leq z$ for which there exists a value of y such that $\pi(x, y) = z$. Next note that when such a value

---

[†] A mapping π from N^2 to N for which there exist functions σ_1 and σ_2 satisfying Eq. (4-29) is called a *pairing function*. Our Cantor-numbering function π is only one of several pairing functions that appear in the literature.

of y does exist it must be unique and it too must lie in the range $0 \le y \le z$. This means that for each choice of x and z the quantity

$$\sum_{i=0}^{z} \text{eqs}\,(\pi(x, i), z)$$

will assume the value 1 if there exists a y such that $\pi(x, y) = z$ and the value 0 otherwise. We may therefore write:

$$\sigma_1(z) = \mu x \left[\sum_{i=0}^{z} \text{eqs}(\pi(x, i), z) = 1 \right] \qquad (4\text{-}30)$$

Since Eq. (4-30) provides a definition of σ_1 from the primitive recursive functions π and eqs using bounded summation and bounded minimalization, it follows that σ_1 is primitive recursive. A completely analogous argument shows that

$$\sigma_2(z) = \mu y \left[\sum_{i=0}^{z} \text{eqs}(\pi(i, y), z) = 1 \right] \qquad (4\text{-}31)$$

and hence that σ_2 is also primitive recursive.

Before discussing applications of the Cantor-numbering function π, let us see how similar numbering schemes can be devised for ordered triples, ordered quadruples, etc. First consider the problem of numbering ordered triples—i.e., the problem of establishing a simple bijection from N^3 to N. Since the ordered triple (w, x, y) can always be represented by the ordered pair $((w, x), y)$, such a bijection is readily constructed from the function π. To avoid confusion, let us rename π with the symbol π^2 to indicate that the arguments of this function are ordered pairs. Then the new function π^3, where

$$\pi^3(w, x, y) = \pi^2(\pi^2(w, x), y) \qquad (4\text{-}32)$$

is easily seen to be a bijection from N^3 to N. Making the obvious generalization, we will refer to $\pi^3(w, x, y)$ as the *Cantor number* of the triple (w, x, y). Thus the Cantor number of the triple $(0, 3, 1)$ is $\pi^2(\pi^2(0, 3), 1) = \pi^2(6, 1) = 34$.

Decoding functions for π^3 are readily obtained from the decoding functions for π^2. First we rename the functions σ_1 and σ_2 with the symbols σ_1^2 and σ_2^2 to indicate that they are the decoding functions for π^2. Then we define σ_1^3, σ_2^3, and σ_3^3 to be the decoding functions for π^3. That is, we define $\sigma_1^3(z)$, $\sigma_2^3(z)$, and $\sigma_3^3(z)$ to be respectively the first, second, and third components of the triple whose Cantor number is z. Thus $\sigma_1^3(34) = 0$, $\sigma_2^3(34) = 3$, and $\sigma_3^3(34) = 1$, for example. It follows immediately from the definition of π^3 that

$$\sigma_1^3(z) = \sigma_1^2(\sigma_1^2(z))$$
$$\sigma_2^3(z) = \sigma_2^2(\sigma_1^2(z)) \qquad (4\text{-}33)$$
$$\sigma_3^3(z) = \sigma_2^2(z).$$

Therefore σ_1^3, σ_2^3, and σ_3^3, as well as π^3, are all primitive recursive functions.

The preceding technique is easily extended to ordered quadruples, quintuples, etc. In general, we use π^n to denote the Cantor-numbering function for ordered n-tuples, and $\sigma_1^n, \ldots, \sigma_n^n$ to denote the corresponding decoding functions. The Cantor-numbering functions are defined inductively so that

$$\pi^{n+1}(x_1, \ldots, x_n, x_{n+1}) = \pi^2(\pi^n(x_1, \ldots, x_n), x_{n+1}). \qquad (4\text{-}34)$$

The decoding functions are therefore given by:

$$\sigma_1^{n+1}(z) = \sigma_1^n(\sigma_1^2(z))$$

$$\sigma_2^{n+1}(z) = \sigma_2^n(\sigma_1^2(z))$$

$$\vdots \qquad\qquad (4\text{-}35)$$

$$\sigma_n^{n+1}(z) = \sigma_n^n(\sigma_1^2(z))$$

$$\sigma_{n+1}^{n+1}(z) = \sigma_2^2(z)$$

Thus for each choice of n, the functions π^n and $\sigma_1^n, \ldots, \sigma_n^n$ are all primitive recursive.

Simultaneous Recursion

It is often convenient to be able to define two or more functions together. Consider, for example, the simultaneous definition of the one-variable functions γ_1 and γ_2 by means of the equations:

$$\gamma_1(0) = 0$$

$$\gamma_2(0) = 1$$

$$\gamma_1(y^+) = \gamma_2(y) \dot- \gamma_1(y) \qquad (4\text{-}36)$$

$$\gamma_2(y^+) = \gamma_1(y) + \gamma_2(y) + y$$

Note carefully that these equations are not just two primitive recursions written together, for the value $\gamma_1(y^+)$ is expressed as a function of both $\gamma_1(y)$ and $\gamma_2(y)$, and the value $\gamma_2(y^+)$ is expressed as a function of both $\gamma_1(y)$ and $\gamma_2(y)$. Thus γ_1 and γ_2 are really being defined "simultaneously."

The basic idea introduced in Eqs. (4-36) is readily generalized to functions of more than one variable. We say that the $(n + 1)$-variable functions f_1 and f_2 are defined by *simultaneous recursion* whenever f_1 and f_2 are specified by equations of the form

$$f_1(x_1, \ldots, x_n, 0) = g_1(x_1, \ldots, x_n)$$

$$f_2(x_1, \ldots, x_n, 0) = g_2(x_1, \ldots, x_n)$$

$$f_1(x_1, \ldots, x_n, y^+) = h_1(x_1, \ldots, x_n, y, f_1(x_1, \ldots, x_n, y), f_2(x_1, \ldots, x_n, y))$$

$$f_2(x_1, \ldots, x_n, y^+) = h_2(x_1, \ldots, x_n, y, f_1(x_1, \ldots, x_n, y), f_2(x_1, \ldots, x_n, y))$$

$$(4\text{-}37)$$

where g_1 and g_2 are given n-variable functions and h_1 and h_2 are given $(n + 3)$-variable functions. It is intuitively clear, and can be rigorously shown, that such a recursive specification defines a unique pair of functions f_1 and f_2. Moreover, these new functions will be effectively computable as long as the given functions g_1, g_2, h_1, and h_2 are.

Example 4-6. In the case of the definition of Eqs. (4-36), where $n = 0$, the functions g_1 and g_2 reduce to the constants 0 and 1, respectively. The functions h_1 and h_2 are the three-variable functions whose values are:

$$h_1(a, b, c) = c \,\dot-\, b$$

$$h_2(a, b, c) = b + c + a$$

The values of γ_1 and γ_2 can be obtained by repeated application of the defining equations. Thus

$$
\begin{aligned}
\gamma_1(1) &= \gamma_2(0) \,\dot-\, \gamma_1(0) & &= 1 \,\dot-\, 0 & &= 1 \\
\gamma_2(1) &= \gamma_1(0) + \gamma_2(0) + 0 & &= 0 + 1 + 0 & &= 1 \\
\gamma_1(2) &= \gamma_2(1) \,\dot-\, \gamma_1(1) & &= 1 \,\dot-\, 1 & &= 0 \\
\gamma_2(2) &= \gamma_1(1) + \gamma_2(1) + 1 & &= 1 + 1 + 1 & &= 3
\end{aligned}
$$

and so on. The following table shows the values obtained in this manner for $y \le 8$.

y	0	1	2	3	4	5	6	7	8	...
$\gamma_1(y)$	0	1	0	3	2	9	8	23	22	...
$\gamma_2(y)$	1	1	3	5	11	17	31	45	75	...

□

In general, a definition by simultaneous recursion does not provide an obvious way of obtaining separate descriptions of the two functions being defined. For instance, even though each of the functions γ_1 and γ_2 of Example 4-6 can be represented by an independent algebraic expression, the defining equations do not tell us what these expressions are. On the other hand, it is easy to show that as long as the functions g_1, g_2, h_1, and h_2 are primitive recursive, each of the functions f_1 and f_2 defined by simultaneous recursion from g_1, g_2, h_1, and h_2 must also be primitive recursive. To see how this is done, let us return to Eqs. (4-36) and the functions γ_1 and γ_2 that they define.

Note that in effect Eqs. (4-36) provide a way of generating the ordered pairs $(\gamma_1(0), \gamma_2(0))$, $(\gamma_1(1), \gamma_2(1))$, $(\gamma_1(2), \gamma_2(2))$, etc. In particular, the first two equations specify the pair $(\gamma_1(0), \gamma_2(0))$ and the last two equations specify the pair $(\gamma_1(y^+), \gamma_2(y^+))$ in terms of the pair $(\gamma_1(y), \gamma_2(y))$ and the value of y. If we represent each ordered pair by its Cantor number, it should be possible to express the Cantor number of $(\gamma_1(y^+), \gamma_2(y^+))$ as a function of the Cantor number of $(\gamma_1(y), \gamma_2(y))$ and the value of y. In this way we might expect to obtain the sequence of Cantor numbers $\pi^2(\gamma_1(0), \gamma_2(0))$, $\pi^2(\gamma_1(1), \gamma_2(1))$, $\pi^2(\gamma_1(2), \gamma_2(2))$, ... by a simple primitive recursion.

To this end, we define the auxiliary function θ so that

$$\theta(y) = \pi^2(\gamma_1(y), \gamma_2(y)). \tag{4-38}$$

Thus $\theta(0) = \pi^2(0, 1) = 1$, $\theta(1) = \pi^2(1, 1) = 4$, $\theta(2) = \pi^2(0, 3) = 6$, and so on. The function θ provides in effect a simultaneous representation of the two functions γ_1 and γ_2. Moreover, the values of γ_1 and γ_2 can be "extracted" from those of θ by means of the decoding functions σ_1^2 and σ_2^2. Specifically:

$$\gamma_1(y) = \sigma_1^2(\theta(y))$$
$$\gamma_2(y) = \sigma_2^2(\theta(y))$$

Thus the primitive recursiveness of γ_1 and γ_2 will be assured once we establish the primitive recursiveness of θ.

As already noted, the value of $\theta(y^+)$ can be determined from the values of $\theta(y)$ and y. Beginning with Eq. (4-38) and making use of the defining equations for γ_1 and γ_2, we have

$$\theta(y^+) = \pi^2(\gamma_1(y^+), \gamma_2(y^+))$$
$$= \pi^2(\gamma_2(y) \doteq \gamma_1(y), \gamma_1(y) + \gamma_2(y) + y)$$
$$= \pi^2(\sigma_2^2(\theta(y)) \doteq \sigma_1^2(\theta(y)), \sigma_1^2(\theta(y)) + \sigma_2^2(\theta(y)) + y).$$

Thus θ can be defined by primitive recursion in the form

$$\theta(0) = 1$$
$$\theta(y^+) = h(y, \theta(y))$$

where $h(a, b) = \pi^2(\sigma_2^2(b) \doteq \sigma_1^2(b), \sigma_1^2(b) + \sigma_2^2(b) + a)$. Since h is primitive recursive, θ must also be primitive recursive, which in turn implies that γ_1 and γ_2 are primitive recursive.

This technique is readily generalized to apply to any two functions defined by simultaneous recursion.

Theorem 4-9. Let g_1 and g_2 be n-variable primitive recursive functions, and let h_1 and h_2 be $(n + 3)$-variable primitive recursive functions. Then the functions f_1 and f_2 defined by simultaneous recursion from g_1, g_2, h_1, and h_2 as specified in Eqs. (4-37) are both primitive recursive.

Proof. Let θ denote the auxiliary $(n + 1)$-variable function whose values are given by the equation

$$\theta(x_1, \ldots, x_n, y) = \pi^2(f_1(x_1, \ldots, x_n, y), f_2(x_1, \ldots, x_n, y)).$$

This function θ can be defined by primitive recursion, since

$$\theta(x_1, \ldots, x_n, 0)$$
$$= \pi^2(g_1(x_1, \ldots, x_n), g_2(x_1, \ldots, x_n))$$
$$\theta(x_1, \ldots, x_n, y^+)$$
$$= \pi^2(h_1(x_1, \ldots, x_n, y, \sigma_1^2(\theta(x_1, \ldots, x_n, y)), \sigma_2^2(\theta(x_1, \ldots, x_n, y))),$$
$$h_2(x_1, \ldots, x_n, y, \sigma_1^2(\theta(x_1, \ldots, x_n, y)), \sigma_2^2(\theta(x_1, \ldots, x_n, y)))).$$

Equivalently, we may write

$$\theta(x_1, \ldots, x_n, 0) = g'(x_1, \ldots, x_n)$$
$$\theta(x_1, \ldots, x_n, y^+) = h'(x_1, \ldots, x_n, y, \theta(x_1, \ldots, x_n, y))$$

for appropriately defined primitive recursive functions g' and h'. Thus the function θ is primitive recursive. And since

$$f_1(x_1, \ldots, x_n, y) = \sigma_1^2(\theta(x_1, \ldots, x_n, y))$$
$$f_2(x_1, \ldots, x_n, y) = \sigma_2^2(\theta(x_1, \ldots, x_n, y))$$

it follows that f_1 and f_2 are also primitive recursive.   □

Example 4-7. Consider the one-variable function f that assumes the value 1 for the argument 0, the value 2 for the next two argument values, the value 3 for the next three argument values, and so on, in general assuming the value n for n consecutive argument values. The first few values of this function are shown in the table below.

y	0	1	2	3	4	5	6	7	8	9	10	11	12	13	14	15
$f(y)$	1	2	2	3	3	3	4	4	4	4	5	5	5	5	5	6

There are several ways of showing that f is primitive recursive. One is to define it simultaneously with the auxiliary function k described below. This function serves to count the number of consecutive argument values for which f must assume a given value.

y	0	1	2	3	4	5	6	7	8	9	10	11	12	13	14	15
$f(y)$	1	2	2	3	3	3	4	4	4	4	5	5	5	5	5	6
$k(y)$	1	1	2	1	2	3	1	2	3	4	1	2	3	4	5	1

Note that $f(y^+)$ equals $f(y)$ unless $f(y) = k(y)$, in which case $f(y^+)$ equals $f(y) + 1$. And $k(y^+) = k(y) + 1$ as long as $k(y) < f(y)$; when $k(y) = f(y)$, $k(y^+)$ assumes the value 1. Thus we can write:

$$f(0) = k(0) = 1$$

$$f(y^+) = f(y) + \text{eqs}(f(y), k(y)) \qquad\qquad = h_1(y, f(y), k(y))$$

$$k(y^+) = (k(y) + 1) \cdot \text{ls}(k(y), f(y)) + \text{eqs}(k(y), f(y)) = h_2(y, f(y), k(y))$$

where h_1 and h_2 are suitably defined primitive recursive functions. Since these equations constitute a definition of f and k by simultaneous recursion, we conclude that both f and k are primitive recursive. □

The use of simultaneous recursion is readily extended to the definition of more than two functions at a time. In general, we say that the functions $f_1, \ldots, f_m$ are defined by *simultaneous recursion* from the functions $g_1, \ldots, g_m$ and $h_1, \ldots, h_m$ if:

$$f_1(x_1, \ldots, x_n, 0) = g_1(x_1, \ldots, x_n)$$

$$\vdots \qquad\qquad \vdots$$

$$f_m(x_1, \ldots, x_n, 0) = g_m(x_1, \ldots, x_n) \qquad\qquad (4\text{-}39)$$

$$f_1(x_1, \ldots, x_n, y^+) = h_1(x_1, \ldots, x_n, y, f_1(x_1, \ldots, x_n, y), \ldots, f_m(x_1, \ldots, x_n, y))$$

$$\vdots \qquad\qquad \vdots$$

$$f_m(x_1, \ldots, x_n, y^+) = h_m(x_1, \ldots, x_n, y, f_1(x_1 \ldots, x_n, y), \ldots, f_m(x_1, \ldots, x_n, y))$$

By extending the Cantor-numbering trick of Theorem 4-9 to ordered m-tuples of the form $(f_1(x_1, \ldots, x_n, y), \ldots, f_m(x_1, \ldots, x_n, y))$ we get:

Theorem 4-10. *Let $g_1, \ldots, g_m$ be n-variable primitive recursive functions and let $h_1, \ldots, h_m$ be $(n + m + 1)$-variable primitive recursive functions. If the $(n + 1)$-variable functions $f_1, \ldots, f_m$ are defined from $g_1, \ldots, g_m$ and $h_1, \ldots, h_m$ by simultaneous recursion, as specified in Eqs. (4-39), then $f_1, \ldots, f_m$ are all primitive recursive.*

The proof is straightfoward and is left to the reader.

Gödel Numbering

For each natural number n, the Cantor-numbering function π^n provides a simple means of representing all the ordered n-tuples of natural numbers. Or, since an n-tuple can be viewed as a sequence of n elements, we can say that π^n provides a means of representing all sequences of natural numbers of length n. But it is often necessary to have a numbering scheme that provides representations for all finite sequences of natural numbers, not just those of a single fixed length. The Cantor-numbering method can be adapted to this purpose, but the necessary encoding and decoding become rather cumbersome and it is usually more convenient to take a completely different approach.

One of the simplest and most common methods for representing arbitrary sequences of natural numbers makes use of prime decompositions. In this scheme the sequence $x_0, x_1, \ldots, x_n$ is represented by the natural number

$$p_0^{x_0} p_1^{x_1} \cdots p_n^{x_n}$$

where as usual p_i denotes the ith prime. Thus the sequence 3, 1, 2 is represented by the natural number $2^3 3^1 5^2 = 600$, the sequence 1, 1, 0, 2 is represented by $2^1 3^1 5^0 7^2 = 294$, and so on. This scheme, of which there are many variations, is called a *Gödel numbering* in honor of its originator, Kurt Gödel. The number $p_0^{x_0} p_1^{x_1} \cdots p_n^{x_n}$ is called the *Gödel number* of the sequence $x_0, x_1, \ldots, x_n$ and is commonly denoted $\langle x_0, x_1, \ldots, x_n \rangle$.

Unfortunately, many different sequences can have the same Gödel number. For example, the sequences (1, 2, 3), (1, 2, 3, 0), (1, 2, 3, 0, 0), (1, 2, 3, 0, 0, 0), etc., all have the same Gödel number, namely, $2^1 3^2 5^3 = 2250$. But since every nonzero natural number has a unique prime decomposition, the only way that two sequences can have the same Gödel number is for them to agree except for one or more terminal 0's. In most of our applications the context will specify the length of the sequence being represented by a given Gödel number, thereby precluding any possible ambiguity. For instance, if we know that 2250 is the Gödel number of a sequence of length five, that sequence can only be (1, 2, 3, 0, 0).

If n is fixed the value $\langle x_0, x_1, \ldots, x_n \rangle$ is obviously a primitive recursive function of $x_0, \ldots, x_n$. For instance, if $n = 3$,

$$\langle x_0, x_1, x_2, x_3 \rangle = 2^{x_0} 3^{x_1} 5^{x_2} 7^{x_3}.$$

If n is not fixed, we can still view the computation of a Gödel number as a primitive recursive process as long as the members of the sequence $x_0, \ldots, x_n$ can be generated by a primitive recursive function. In particular, suppose that f is a one-variable primitive recursive function and that x_i is the value of $f(i)$, for $i = 0, 1, \ldots$. Then the Gödel number of the sequence formed by the first $n + 1$ values $x_0, \ldots, x_n$ is given by

$$\langle x_0, \ldots, x_n \rangle = \prod_{i=0}^{n} \mathrm{pn}(i)^{f(i)}.$$

Thus the value of $\langle x_0, \ldots, x_n \rangle$ is a primitive recursive function of n. This fact will be used in most of our applications of Gödel numbering.

Now consider the problem of decoding Gödel numbers; i.e., the problem of obtaining the values of $x_0, \ldots, x_n$ from the single natural number $z = \langle x_0, \ldots, x_n \rangle$. For this purpose we introduce a two-variable function E, defined so that for all $z > 0$, $E(i, z)$ is the exponent of p_i in the prime decomposition of z. Thus $E(0, 40) = 3$, since the prime decomposition of 40 is $2^3 3^0 5^1$, in which the exponent of p_0 is 3. Similarly, $E(1, 40) = 0$, $E(2, 40) = 1$, $E(3, 40) = 0$, etc. The value assigned to $E(i, 0)$ is unimportant; we arbitrarily choose 0.

Evidently E is just the function needed to decode Gödel numbers, for if z is the Gödel number of the sequence $x_0, \ldots, x_n$, then $E(i, z) = x_i$ for each i in the range $0 \leq i \leq n$. (If $i > n$, $E(i, z) = 0$.) The function E is called the *extraction function*, and the values $E(0, z)$, $E(1, z)$, ... are called the *components* of z.

In order to show that E is primitive recursive, we observe that for $z > 0$, $E(i, z)$ is the largest natural number x such that p_i^x divides z. Alternatively, $E(i, z)$ is the smallest natural number x such that p_i^{x+1} does not divide z. Since this last statement is true for all z, and since x will never exceed z, we can write

$$E(i, z) = \overset{z}{\mu}x[\text{div}(\text{pn}(i)^{x+1}, z) = 0] \tag{4-40}$$

where div and pn denote the "divides" and "nth prime" functions defined by Eqs. (4-24) and (4-27). Since both of these functions are primitive recursive, so is E.

When applying the extraction function to a natural number z, it is often convenient to know which is the "last" nonzero component of z. We define the one-variable function Lh so that whenever $z > 1$, $Lh(z)$ is the index (subscript) of the largest prime that divides z. Thus $Lh(2) = 0$, $Lh(3) = Lh(12) = 1$, $Lh(98) = 3$, and so on. If we also define $Lh(0)$ to be 1 and $Lh(1)$ to be 0, we can write

$$Lh(z) = \overset{z}{\mu}n\left[\prod_{i=0}^{n} \text{pn}(i)^{E(i, z)} = z\right]. \tag{4-41}$$

The function Lh is therefore primitive recursive. It is called the *length function* and the value of $Lh(z)$ is called the *length* of z.[†] This value is often abbreviated $|z|$.

Before discussing applications of our Gödel-numbering scheme, we mention two of its most common variants. Both of these schemes are designed to provide distinct representations for distinct sequences. In the first variant, the sequence $x_0, x_1, \ldots, x_n$ is represented by the natural number

$$p_0^{x_0+1}p_1^{x_1+1} \cdots p_n^{x_n+1}.$$

For example, the sequence 3, 1, 2 is represented by the number $2^4 3^2 5^3 = 18,000$. Since each member of a sequence is represented by a factor greater than 1, no two sequences can have the same representation, However, this scheme has the disadvantage that in order for a natural number z to represent a sequence, the powers of all the primes up to the largest prime that divides z must be nonzero. Thus many natural numbers, such as $10 = 2^1 3^0 5^1$ or $7 = 2^0 3^0 5^0 7^1$, do not represent any sequence at all.

[†] The "length" of a number z, as defined here, must not be confused with the length of a sequence of numbers that z might represent.

In the second variant, the sequence x_1, x_2, ..., x_n is represented by the natural number

$$p_0^n p_1^{x_1} p_2^{x_2} \cdots p_n^{x_n}$$

with the length of the sequence being encoded as the exponent of p_0. Thus the sequence 1, 3, 2 is represented by the number $2^3 3^1 5^3 7^2 = 147{,}000$, the sequence 1, 3, 2, 0 is represented by $2^4 3^1 5^3 7^2 = 294{,}000$, etc. This scheme also provides distinct representations for distinct sequences, but it too has the disadvantage that not every natural number represents a sequence. For instance, $2100 = 2^2 3^1 5^2 7^1$ is not a legal representation since the index of the largest prime divisor exceeds the value specified by the exponent of p_0. (We could agree to ignore any "extra" primes and interpret 2100 as representing the sequence 1, 2, but with this convention sequences would no longer have unique representations.)

Still other variations are possible, but we will not discuss them here. All the schemes that use products of prime powers to represent sequences are known loosely as *Gödel numberings*, and the representations themselves are known as *Gödel numbers*. Each scheme has its own minor advantages and disadvantages, and although we shall most often use the "standard" scheme in which x_0, ..., x_n is represented by $p_0^{x_0} \cdots p_n^{x_n}$, we will feel free to adopt an appropriate variation when the occasion requires it.

Course-of-Values Recursion

In a definition by primitive recursion, the "next" value of the function being defined is specified in terms of the "current" value. In many cases, it is more convenient to be able to specify the "next" value of a function in terms of the "current" value and one or more "preceding" values. For example, the so-called Fibonacci function f is specified by the equations

$$f(0) = 1$$
$$f(1) = 1 \tag{4-42}$$
$$f(y^+) = f(y) + f(y - 1) \qquad \text{for } y \geq 1$$

wherein $f(y^+)$ depends not just on $f(y)$ but on $f(y - 1)$ as well. It is easy to see that these equations describe a well-defined total function whose values are $f(0) = 1$, $f(1) = 1$, $f(2) = 2$, $f(3) = 3$, $f(4) = 5$, $f(5) = 8$, $f(6) = 13$, $f(7) = 21$, and so on.

More generally, we might wish to specify the next value of a function in terms of *all* the preceding values of that function. For example, we might wish to define the function f so that

$$f(0) = 1$$
$$f(y^+) = f(y) + 2f(y - 1) + 3f(y - 2) + \cdots + (y + 1)f(0) \tag{4-43}$$

Again it is clear that these equations describe a well-defined function and that this function is effectively computable. In particular, we can readily determine that $f(1) = 1, f(2) = 3, f(3) = 8, f(4) = 21$, and so on.

Now, what does it mean to specify the next value of a function "in terms of" all the preceding values of that function? No doubt what we have in mind is that $f(y^+)$ should be expressible as a function of $f(0), \ldots, f(y)$. But because a function has a fixed number of arguments, it would be necessary to provide a different function for each value of y^+. In most cases this would be very inconvenient. Another approach is to note that the sequence of values $f(0), \ldots, f(y)$ can be represented by a single Gödel number, which contains all the information provided by $f(0), \ldots, f(y)$. Thus a practical way of specifying $f(y^+)$ in terms of $f(0), \ldots, f(y)$ is to express $f(y^+)$ as a function of the Gödel number $\langle f(0), \ldots, f(y) \rangle$. This is in fact the approach that we will adopt.

Of course, we shall want to be able to define functions of more than one variable, and to allow the "next" value of the function being defined to depend on the current arguments of the function as well as on the preceding function values. Making these obvious generalizations, we arrive at the following format for the definition of an $(n + 1)$-variable function f, where g and h are given functions of n and $n + 2$ variables, respectively:

$$f(x_1, \ldots, x_n, 0) = g(x_1, \ldots, x_n)$$
$$f(x_1, \ldots, x_n, y^+) = h(x_1, \ldots, x_n, y, \langle f(x_1, \ldots, x_n, 0), \ldots, f(x_1, \ldots, x_n, y) \rangle)$$

$$(4\text{-}44)$$

Any function f defined in accordance with this format is said to be defined by *course-of-values recursion* from the functions g and h. As usual, the function g reduces to a constant when $n = 0$.

Example 4-8. Consider how the function f specified by Eqs. (4-43) might be defined by course-of-values recursion. To obtain the value of $f(y^+)$ from the Gödel number $z = \langle f(0), \ldots, f(y) \rangle$, it is necessary to extract the various components of z, multiply each one by the appropriate coefficient, and sum the results. If we define the two-variable function h so that

$$h(y, z) = \sum_{i=0}^{y} (y^+ \doteq i) E(i, z)$$

we find that $h(y, \langle f(0), \ldots, f(y) \rangle) = y^+ f(0) + y f(1) + \cdots + 2f(y - 1) + f(y)$. The function f can therefore be defined by course-of-values recursion from the constant 1 and the (primitive recursive) function h as follows:

$$f(0) = 1$$
$$f(y^+) = h(y, \langle f(0), \ldots, f(y) \rangle) \qquad \square$$

Course-of-values recursions are important because, as the next theorem shows, they can be used to establish primitive recursiveness.

Theorem 4-11. *Let the $(n + 1)$-variable function f be defined by course-of-values recursion from the primitive recursive functions g and h, according to the equations:*

$$f(x_1, \ldots, x_n, 0) = g(x_1, \ldots, x_n)$$

$$f(x_1, \ldots, x_n, y^+) = h(x_1, \ldots, x_n, y, \langle f(x_1, \ldots, x_n, 0), \ldots, f(x_1, \ldots, x_n, y) \rangle)$$

Then the function f is also primitive recursive.

Proof. Define the auxiliary function Ψ so that

$$\Psi(x_1, \ldots, x_n, y) = \langle f(x_1, \ldots, x_n, 0), \ldots, f(x_1, \ldots, x_n, y) \rangle.$$

Noting that

$$f(x_1, \ldots, x_n, y^+) = h(x_1, \ldots, x_n, y, \Psi(x_1, \ldots, x_n, y))$$

we can write

$$\Psi(x_1, \ldots, x_n, 0) = 2^{f(x_1, \ldots, x_n, 0)} = 2^{g(x_1, \ldots, x_n)}$$

$$\Psi(x_1, \ldots, x_n, y^+) = \Psi(x_1, \ldots, x_n, y) p_{y+1}^{f(x_1, \ldots, x_n, y^+)}$$

$$= \Psi(x_1, \ldots, x_n, y) \, \mathrm{pn}(y^+)^{h(x_1, \ldots, x_n, y, \Psi(x_1, \ldots, x_n, y))}.$$

Thus Ψ can be defined by primitive recursion in terms of g and h. Since g and h are primitive recursive, it follows that Ψ is primitive recursive. Finally, we note that

$$f(x_1, \ldots, x_n, y) = \mathrm{E}(y, \Psi(x_1, \ldots, x_n, y))$$

so that f is also primitive recursive.   □

When we rely on Theorem 4-11 to establish the primitive recursiveness of a function f, we do not usually bother to construct a formal course-of-values recursion for f. Instead we simply indicate how the value of $f(x_1, \ldots, x_n, y^+)$ can be obtained in a primitive recursive manner from $f(x_1, \ldots, x_n, 0)$ through $f(x_1, \ldots, x_n, y)$. Thus we would normally be satisfied with Eqs. (4-43) as an informal definition by course-of-values recursion. Similarly, we would accept the equations

$$f(x, 0) = g(x)$$

$$f(x, y^+) = h(x, y, \sum_{i=0}^{y} f(x, i))$$

as an informal course-of-values recursion. In each case it is left to the reader to convince himself that the equations in question can be recast in the format of Eqs. (4-44).

In order to obtain some practice in recognizing informal descriptions of course-of-values recursions, we now examine two common forms that such descriptions may take. First consider the case in which the "next" value of the function being defined is specified in terms of a single "previous" value. As long as this previous value can be selected by a primitive recursive function, the resulting definition can be recast as a course-of-values recursion and the function being defined must be primitive recursive.

Corollary 4-11.1. *Let g, h, and w be primitive recursive functions of n, n + 2, and n + 1 variables, respectively, and suppose that* $w(x_1, \ldots, x_n, y) \le y$ *for all* $x_1, \ldots, x_n, y$. *Then the function f defined by the equations*

$$f(x_1 \ldots, x_n, 0) = g(x_1, \ldots, x_n)$$
$$f(x_1, \ldots, x_n, y^+) = h(x_1, \ldots, x_n, y, f(x_1, \ldots, x_n, w(x_1, \ldots, x_n, y)))$$

(4-45)

is also primitive recursive.

Proof. Define the function h' so that

$$h'(x_1, \ldots, x_n, y, z) = h(x_1, \ldots, x_n, y, E(w(x_1, \ldots, x_n, y), z)).$$

The primitive recursiveness of h' follows immediately from that of h and w. Moreover,

$$f(x_1, \ldots, x_n, y^+) = h(x_1, \ldots, x_n, y, f(x_1, \ldots, x_n, w(x_1, \ldots, x_n, y)))$$
$$= h'(x_1, \ldots, x_n, y, \langle f(x_1, \ldots, x_n, 0), \ldots, f(x_1, \ldots, x_n, y)\rangle).$$

Thus f can be defined by course-of-values recursion from the primitive recursive functions g and h'. Theorem 4-11 then guarantees the primitive recursiveness of f.   □

The following example provides a simple illustration of the role of Corollary 4-11.1.

Example 4-9. Suppose that g and h are given primitive recursive functions and that the new function f is defined from g and h as follows:

$$f(x, 0) = g(x)$$
$$f(x, y^+) = h(x, y, f(x, \text{quo}(x, y)))$$

In this definition the value of $f(x, y^+)$ is specified in terms of x, y, and the single previous value $f(x, \text{quo}(x, y))$. Since $\text{quo}(x, y) \le y$, this definition fits the format of Corollary 4-11.1 with quo as the function w. Thus the given equations represent an informal course-of-values recursion, and the function f is primitive recursive.   □

Corollary 4-11.1 can be extended in an obvious way to the case in which $f(x_1, \ldots, x_n, y^+)$ depends on $f(x_1, \ldots, x_n, y, w_1(x_1, \ldots, x_n, y)), f(x_1, \ldots, x_n, y, w_2(x_1, \ldots, x_n, y)), \ldots, f(x_1, \ldots, x_n, y, w_k(x_1, \ldots, x_n, y))$, where k is fixed and $w_1, w_2, \ldots, w_k$ are specified primitive recursive functions. The details are straightforward and are left to the reader.

Next consider the definition scheme in which the first two values of the function being defined are given explicitly, and each new value is specified in terms of the two immediately preceding values. Such definitions can also be recast as course-of-values recursions.

Corollary 4-11.2. *Let g_0, g_1, and h be primitive recursive functions of n, n, and $n + 3$ variables, respectively. Then the function f defined by the equations*

$$f(x_1, \ldots, x_n, 0) = g_0(x_1, \ldots, x_n)$$

$$f(x_1, \ldots, x_n, 1) = g_1(x_1, \ldots, x_n) \tag{4-46}$$

$$f(x_1, \ldots, x_n, y^+)$$

$$= h(x_1, \ldots, x_n, y, f(x_1, \ldots, x_n, y), f(x_1, \ldots, x_n, y - 1)) \qquad \textit{for } y > 0$$

is also primitive recursive.

Proof. The function f can be defined by course-of-values recursion from the functions g_0 and h', where

$$h'(x_1, \ldots, x_n, y, z) = \text{eqs}(y, 0)g_1(x_1, \ldots, x_n)$$

$$+ \text{gr}(y, 0)h(x_1, \ldots, x_n, y, \text{E}(y, z), \text{E}(y \dotdiv 1, z)).$$

Since h' is obviously primitive recursive, so is f. $\square$

As a consequence of this result we see, for example, that the Fibonacci function defined by Eqs. (4-42) is primitive recursive. And as an obvious extension, we see that if the function f is defined by equations of the form

$$f(x_1, \ldots, x_n, 0) = g_0(x_1, \ldots, x_n)$$

$$\vdots$$

$$f(x_1, \ldots, x_n, k) = g_k(x_1, \ldots, x_n) \tag{4-47}$$

$$f(x_1, \ldots, x_n, y^+)$$

$$= h(x_1, \ldots, x_n, y, f(x_1, \ldots, x_n, y), \ldots, f(x_1, \ldots, x_n, y - k)) \qquad \textit{for } y \geq k$$

where $g_0, \ldots, g_k$ and h are all primitive recursive, then f must also be primitive recursive. Again the details are left to the reader.

Primitive Recursion on Several Variables

In all of the recursive definitions discussed so far, just one of the arguments of the function being defined has served as a recursion variable, and all the others have played the role of parameters. By this we mean that "new" function values have always been specified in terms of "old" function values obtained by varying just one of the arguments and leaving all the others fixed. In particular, when defining the $(n + 1)$-variable function f, we have always specified $f(m_1, \ldots, m_n, k^+)$ in terms of values of the form $f(m_1, \ldots, m_n, y)$, where $y \leq k$.

We now turn to the more general situation in which two or more arguments are used as recursion variables. As an introductory example, consider the two-variable function f whose value for the arguments y_1 and y_2 is the binomial coefficient

$$\binom{y_1 + y_2}{y_1}.$$

Although this function is easily obtained from the factorial function, it can also be defined by a simple recursive scheme. First note that

$$f(0, y_2) = \binom{y_2}{0} = 1 \tag{4-48a}$$

and

$$f(y_1, 0) = \binom{y_1}{y_1} = 1. \tag{4-48b}$$

Making use of the identity

$$\binom{y_1 + y_2 + 2}{y_1 + 1} = \binom{y_1 + y_2 + 1}{y_1} + \binom{y_1 + y_2 + 1}{y_1 + 1},$$

we can also write

$$f(y_1^+, y_2^+) = f(y_1, y_2^+) + f(y_1^+, y_2). \tag{4-48c}$$

Thus the value of f for the arguments y_1^+ and y_2^+ is expressed in terms of the values assumed by f for preceding values of *both* y_1 and y_2.

To see that Eqs. (4-48) completely specify the function f, it is appropriate to think of the various ordered pairs of the form (y_1, y_2) as representing points in one quadrant of the plane, as in Fig. 4-3. Equations (4-48a) and (4-48b) obviously specify the values of f for the points along the two coordinate axes, as indicated in the figure. Given these values, Eq. (4-48c) determines the values of f for the remainder of the quadrant. For as the arrows in the figure suggest, Eq. (4-48c) specifies the value of f at a given point in terms of the values of f at the points immediately above and immediately to the left of the given point. Thus the values in the $y_1 = 0$ row together with the first value in the $y_1 = 1$ row inductively determine the values in the remainder of the $y_1 = 1$ row. These values together with the first value in the $y_1 = 2$ row then determine all the values in the $y_1 = 2$ row, and so on.

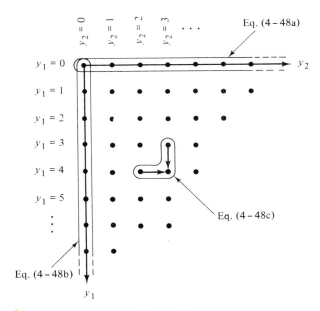

Figure 4-3

Equations (4-48) therefore provide an example of a new recursive scheme for defining functions of two variables. In its general form, this scheme defines a function f by means of the equations

$$f(0, y_2) = g_1(y_2)$$

$$f(y_1^+, 0) = g_2(y_1) \tag{4-49}$$

$$f(y_1^+, y_2^+) = h(y_1, y_2, f(y_1^+, y_2), f(y_1, y_2^+))$$

where g_1, g_2, and h are given functions. (Note that y_1^+ is used in the second equation in order to avoid a possible conflict between g_1 and g_2 when $y_1 = y_2 = 0$.) It should be clear from the reasoning presented above that these equations do in fact define a unique function f. A definition of this type will be called a *primitive recursion on two variables* or more simply a *double recursion*.[†]

It is important to understand that double recursions do not fit the format of any of the recursive definitions previously considered. Thus we have no right to presume that the function f defined by Eqs. (4-49) is necessarily primitive recursive whenever g_1, g_2, and h are primitive recursive. Nevertheless this is in fact the case, as we now set out to show.

---

† The term double recursion (or, more generally, multiple recursion) is also used in the literature to denote a more complex type of recursive definition. The form of double recursion considered here is known technically as *unnested* double recursion.

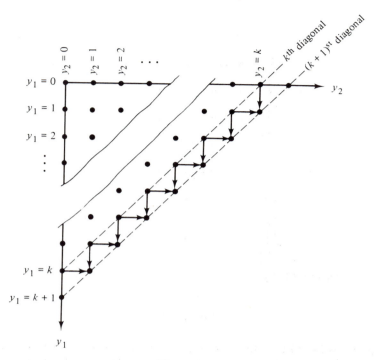

Figure 4-4

The key to understanding the argument is a diagram like that of Fig. 4-3. We define the kth *diagonal* in such a diagram to be the collection of points (y_1, y_2) for which $y_1 + y_2 = k$. (See Fig. 4-4.) Then we note that the values assumed by f along the $(k + 1)$st diagonal are completely determined by the values assumed along the kth diagonal. In particular, the top-right-most value on the $(k + 1)$st diagonal is given by the equation $f(0, k + 1) = g_1(k + 1)$; the bottom-left-most value is given by the equation $f(k^+, 0) = g_2(k)$; and the intervening values are given by the equation $f(y_1^+, k^+ - y_1^+) = h(y_1, k - y_1^+, f(y_1^+, k - y_1^+), f(y_1, k^+ - y_1^+))$, as suggested by the arrows in Fig. 4-4. Thus if the sequence of values assumed along each diagonal is represented by a single Gödel number, it should be possible to obtain the Gödel number corresponding to the $(k + 1)$st diagonal from that corresponding to the kth diagonal.

With this goal in mind, we define a one-variable auxiliary function Ψ whose value for argument z is the Gödel number associated with the zth diagonal in the diagram of Fig. 4-4. Specifically, we define Ψ so that:

$$\Psi(z) = \prod_{i=0}^{z} p_i^{f(i, z-i)}. \tag{4-50}$$

Thus $\Psi(0) = 2^{f(0,\,0)}$, $\Psi(3) = 2^{f(0,\,3)}3^{f(1,\,2)}5^{f(2,\,1)}7^{f(3,\,0)}$, and so on. To recover the values of f from Ψ, we note that (y_1, y_2) is the y_1th point on the $(y_1 + y_2)$th diagonal. Therefore

$$f(y_1, y_2) = E(y_1, \Psi(y_1 + y_2)) \tag{4-51}$$

and f will be primitive recursive as long as Ψ is.

To show that Ψ is primitive recursive, we note that

$$\Psi(0) = 2^{f(0,\,0)} = 2^{g_1(0)}$$

and

$$\Psi(z^+) = 2^{f(0,\,z^+)}\left(\prod_{i=1}^{z} p_i^{f(i,\,z^+ - i)}\right) p_{z+1}^{f(z^+,\,0)}.$$

In this last equation we make use of the convention that a repeated product has the value 1 if its upper limit is less than its lower limit. Thus for $z = 0$ the equation gives $\Psi(1) = 2^{f(0,\,1)}3^{f(1,\,0)}$, as required.

In view of the defining equations for f, we may also write

$$\Psi(z^+) = 2^{g_1(z^+)}\left(\prod_{i=1}^{z} p_i^{h(i-1,\,z-i,\,f(i,\,z-i),\,f(i-1,\,z^+ - i))}\right) p_{z+1}^{g_2(z)}.$$

If we now use Eq. (4-51) to express $f(i, z - i)$ and $f(i - 1, z^+ - i)$ in terms of $\Psi(z)$, we find that Ψ can be defined by primitive recursion as follows:

$$\Psi(0) = 2^{g_1(0)}$$

$$\Psi(z^+) = 2^{g_1(z^+)}\left(\prod_{i=1}^{z} \mathrm{pn}(i)^{h(i - 1,\, z \,\dot-\, i,\, E(i,\,\Psi(z)),\, E(i \,\dot-\, 1,\,\Psi(z)))}\right) \mathrm{pn}(z^+)^{g_2(z)}$$

Thus Ψ is primitive recursive, which in turn implies that f is primitive recursive. We have therefore established:

Theorem 4-12. *Let g_1, g_2, and h be primitive recursive functions of one, one, and four variables, respectively. Then the function f defined by double recursion from g_1, g_2, and h in the manner prescribed in Eqs. (4-49) must also be primitive recursive.*

Example 4-10. Consider the modified Cantor-numbering scheme shown in Fig. 4-5, where the natural numbers are assigned in sequence along successive L-shaped paths. If $\tau(x, y)$ denotes the number assigned to the pair (x, y), it is easy to see that $\tau(0, y) = y^2$ and $\tau(x^+, 0) = (x + 2)^2 - 1$. Moreover:

$$\tau(x^+, y^+) = \begin{cases} \tau(x, y^+) + 1 & \text{if} \quad x \le y \\ \tau(x^+, y) - 1 & \text{if} \quad x > y. \end{cases}$$

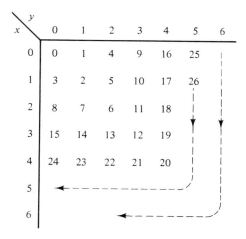

Figure 4-5

Thus τ can be defined by double recursion[†] from the functions g_1, g_2, and h, where:

$$g_1(y) = y^2$$

$$g_2(x) = (x + 2)^2 \mathbin{\dot-} 1$$

$$h(x, y, a, b) = \mathrm{ls}(x, y^+)(b + 1) + \mathrm{gr}(x, y)(a \mathbin{\dot-} 1)$$

Since g_1, g_2, and h are primitive recursive, it follows that τ is also primitive recursive.   □

The result of Theorem 4-12 can be extended in several ways. One of the simplest extensions permits the use of parameters—i.e., additional variables not involved in the actual recursion.

Corollary 4-12.1. *Let g_1, g_2, and h be primitive recursive functions of $n + 1$, $n + 1$, and $n + 4$ variables, respectively, and let f be the $(n + 2)$-variable function defined by the equations:*

$$f(x_1, \ldots, x_n, 0, y_2) = g_1(x_1, \ldots, x_n, y_2)$$

$$f(x_1, \ldots, x_n, y_1^+, 0) = g_1(x_1, \ldots, x_n, y_1)$$

$$f(x_1, \ldots, x_n, y_1^+, y_2^+)$$

$$= h(x_1, \ldots, x_n, y_1, y_2, f(x_1, \ldots, x_n, y_1^+, y_2), f(x_1, \ldots, x_n, y_1, y_2^+))$$

(4-52)

Then f is also primitive recursive.

[†] The use of subscripted y's to denote recursion variables is often unwieldy in practice. From now on, we shall freely use whatever variables seem convenient, especially in examples.

The proof is a straightforward extension of the proof of Theorem 4-12 and will not be given.

It is also possible to define a two-variable function f by a "double course-of-values recursion," in which the value of $f(a^+, b^+)$ is expressed in terms of all the function values $f(y_1, y_2)$ for which $y_1 + y_2 < a^+ + b^+$. To specify such a recursion precisely, we resort to a kind of two-level Gödel numbering. As before, let $\Psi(z)$ denote the Gödel number of the sequence $f(0, z)$, $f(1, z-1), \ldots, f(z, 0)$. Now define $\theta(z)$ to be the Gödel number of the sequence $\Psi(0), \Psi(1), \ldots, \Psi(z)$. Evidently $\theta(z)$ contains complete information about all the function values appearing on or before the zth diagonal in a diagram like that of Fig. 4-4. The values of f are easily extracted from such Gödel numbers, for

$$f(y_1, y_2) = E(y_1, \Psi(y_1 + y_2)) = E(y_1, E(y_1 + y_2, \theta(z)))$$

as long as $z \geq y_1 + y_2$.

Now for the formal definition: the two-variable function f is said to be defined by a *double course-of-values recursion* if f is specified by the equations

$$f(0, y_2) = g_1(y_2)$$
$$f(y_1^+, 0) = g_2(y_1) \qquad\qquad (4\text{-}53)$$
$$f(y_1^+, y_2^+) = h(y_1, y_2, \theta(y_1 + y_2 + 1))$$

where g_1, g_2, and h are given functions of one, one, and three variables, respectively, and $\theta(z)$ is the "second order" Gödel number defined above. As might be expected, we have:

Theorem 4-13. *Let g_1, g_2, and h be primitive recursive functions of one, one, and three variables, respectively. Then the two-variable function f defined from g_1, g_2, and h by double course-of-values recursion, as specified in Eqs. (4-53), must also be primitive recursive.*

The proof relies on arguments similar to those used in Theorems 4-11 and 4-12 and will be left as an exercise.

Example 4-11. Consider the two-variable function f, where $f(x, y)$ is the greatest common divisor of x and y when x and y are nonzero and is 0 otherwise. Elementary number theory tells us that:

$$\gcd(x, y) = \begin{cases} x & \text{if} \quad x = y \\ \gcd(x, y - x) & \text{if} \quad x < y \\ \gcd(x - y, y) & \text{if} \quad x > y \end{cases}$$

The values of f may therefore be specified by the equations:

$$f(0, y) = 0$$
$$f(x^+, 0) = 0$$
$$f(x^+, y^+) = x^+ \text{ eqs } (x^+, y^+) + f(x^+, y \doteq x) \text{ ls } (x^+, y^+)$$
$$+ f(x \doteq y, y^+) \text{ gr } (x^+, y^+)$$

These equations can be recast as a formal double course-of-values recursion. Again using θ to denote the second-order Gödel-numbering function referred to in Eqs. (4-53), we may write:

$$f(0, y) = 0 = g_1(y)$$
$$f(x^+, 0) = 0 = g_2(x)$$
$$f(x^+, y^+) = x^+ \text{ eqs } (x^+, y^+) + E(x^+, E(y^+, \theta(x + y + 1))) \text{ ls } (x^+, y^+)$$
$$+ E(x \doteq y, E(x^+, \theta(x + y + 1))) \text{ gr } (x^+, y^+)$$
$$= h(x, y, \theta(x + y + 1))$$

Since these equations meet the requirements of Theorem 4-13, we conclude that the function f is primitive recursive. ☐

The definition scheme of Eqs. (4-53) can be extended to include the use of parameters or the use of recursion on more than two variables. The resulting schemes are readily shown to yield primitive recursive functions whenever the functions to which they are applied are primitive recursive. All such definition schemes will be referred to as *multiple course-of-values recursions*.

Of course in practice we do not usually bother to express the specification of a function in the precise format required by Eqs. (4-53) or their generalizations. We will often rely on a relatively informal description of the manner in which a "new" function value is to be obtained from various "old" values. Thus in the case of Example 4-11, we would normally accept the equations

$$f(0, y) = 0$$
$$f(x^+, 0) = 0$$
$$f(x^+, y^+) = x^+ \text{ eqs } (x^+, y^+) + f(x^+, y \doteq x) \text{ ls}(x^+, y^+)$$
$$+ f(x \doteq y, y^+) \text{ gr}(x^+, y^+)$$

as an adequate course-of-values specification of the function f.

In summary, we see that Cantor and Gödel numbering schemes can be of great help in showing that certain functions are primitive recursive. We turn now to one more fundamental concept that will also prove useful in establishing the primitive recursiveness of various functions.

4.3 PRIMITIVE RECURSIVE PREDICATES

In this section we introduce the notion of a predicate as a way to describe certain relationships or conditions. Closely related to functions, predicates can provide a convenient way of describing and defining functions, especially primitive recursive functions. Extensive use of predicates will be made in Chapter 5.

Basic Definitions

It is often convenient to define a function in terms of one or more conditions on the arguments of that function. Thus when we define the function f so that

$$f(x, y) = \begin{cases} x - y & \text{if} & y < x \\ 2x & \text{if} & y = x \\ y & \text{if} & y > x \end{cases}$$

we are making use of the conditions "y is less than x," "y equals x," and "y is greater than x." Similarly, when we define the function g so that $g(x)$ is the least y such that $y^2 \geq x$, we are making use of the condition "the square of y is greater than or equal to x."

Note that each of the phrases, "y is less than x," "y equals x," etc., becomes a statement (i.e., an expression that is either true or false) when specific numbers are substituted for x and y. Such phrases are commonly called *predicates*. Thus "x is an even number" is a predicate, while "the even number x" is not. The letters x and y that play the role of variables in a predicate like "x equals y" are called the *arguments* of the predicate. A predicate that has n distinct arguments is called an *n-place predicate* or an *n-ary predicate*. Thus "x is an even number" is a one-place predicate, "x equals y" is a two-place predicate, and "x is less than y and z is odd" is a three-place predicate.

When we use a predicate like "x equals y," we usually have in mind a particular set of elements that may be substituted for the arguments x and y. This set is called the *domain* of the predicate in question. For instance, the predicate "x is less than y" might have the set of real numbers as its domain, and "x is a parent of y" might have the set of human beings as its domain. Actually, we cannot really specify a predicate precisely unless we also specify its domain. For in order to define a predicate such as "x is less than y" we must first define what "less than" means and, in order to do that, we must specify the set of objects for which "less than" is defined. Thus a predicate and its domain are inseparable. A predicate whose domain is D is referred to as a predicate *on D*.

From the preceding remarks, we see that an n-place predicate on the domain D can be thought of as mapping each n-tuple of elements of D into a statement. For instance, the predicate "x is less than y" on the domain of natural numbers can be thought of as mapping the ordered pair $(3, 7)$ into the statement "3 is less than 7," the pair (14, 5) into the statement "14 is less than 5," and so on. This point of view is sufficiently useful that we now adopt it officially, and formally define an *n-place predicate on the domain D* to be a function that maps each element of D^n into a statement.

Predicates are typically represented by letters such as P, Q, R, etc. If P represents an n-place predicate, we use $P(k_1, \ldots, k_n)$ to denote the value of P for the argument values $k_1, \ldots, k_n$—that is, the statement that P assigns to the n-tuple $(k_1, \ldots, k_n)$. For example, if P stands for the predicate "x_1 is less than x_2" on the domain N, then $P(3, 8)$ denotes the statement "3 is less than 8."[†] In general, we use $P(x_1, \ldots, x_n)$ to denote the statement that results from an arbitrary choice of the argument values $x_1, \ldots, x_n$. If the statement $P(x_1, \ldots, x_n)$ is true, we say that P *holds* for the choice of arguments $x_1, \ldots, x_n$, and write "$P(x_1, \ldots, x_n)$ holds" or very often simply "$P(x_1, \ldots, x_n)$."

We are often less interested in the detailed nature of a statement $P(x_1, \ldots, x_n)$ than in the simple truth or falsity of that statement. If P is an n-place predicate on the domain D, we define the function χ_P to be the function from D^n to $\{0, 1\}$ such that

$$\chi_P(x_1, \ldots, x_n) = \begin{cases} 1 & \text{if} \quad P(x_1, \ldots, x_n) \text{ holds (is true)} \\ 0 & \text{if} \quad P(x_1, \ldots, x_n) \text{ does not hold (is false).} \end{cases} \quad (4\text{-}54)$$

This function χ_P is called the *characteristic function*[‡] of the predicate P. For many purposes, characteristic functions provide the most concise and convenient descriptions of predicates.

Since expressions like $P(x_1, \ldots, x_n)$ and $Q(y_1, \ldots, y_n)$ represent statements, the usual logical connectives can be applied to them. Thus we write

$$\neg P(x_1, \ldots, x_n)$$

to denote the statement that is true iff $P(x_1, \ldots, x_n)$ is false. Similarly, we write

$$P(x_1, \ldots, x_n) \vee Q(y_1, \ldots, y_m)$$

and

$$P(x_1, \ldots, x_n) \wedge Q(y_1, \ldots, y_m)$$

to denote, respectively, the statement that is true iff at least one of $P(x_1, \ldots, x_n)$ or $Q(y_1, \ldots, y_m)$ holds and the statement that is true iff both $P(x_1, \ldots, x_n)$ and $Q(y_1, \ldots, y_m)$ hold. The use of the connectives $\vee$ and $\wedge$ is extended to more than two statements in the obvious way.

The symbol $\Leftrightarrow$ is commonly used to indicate that two predicates hold under precisely the same circumstances. Thus if P and Q denote three-place predicates,

---

[†] Note that in order to interpret an expression like $P(3, 8)$ it is necessary to know how the arguments of P are ordered. This ordering is usually implicit in the names assigned to the arguments, as in the case of "x_1 is less than x_2," "x is greater than y," etc. When this is not the case, or when the obvious ordering is not to be used, the intended ordering must be specified explicitly.

[‡] This definition is not universally followed in the literature; sometimes the roles of 0 and 1 are interchanged.

we write

$$P(x_1, x_2, x_3) \Leftrightarrow Q(x_1, x_2, x_3)$$

to indicate that P holds for a given ordered triple (x_1, x_2, x_3) iff Q holds for the same triple. Similarly, we write

$$R(x_1, x_2, x_3) \Leftrightarrow Q(x_3, x_1, x_2)$$

to indicate that the predicate R holds for the triple (x_1, x_2, x_3) iff Q holds for the triple (x_3, x_1, x_2).

The symbol $\Leftrightarrow$ plays an important role in the definition of new predicates from given predicates, especially when used in conjunction with the connectives $\neg$, $\wedge$, and $\vee$.

Example 4-12. Let L and E denote the two-place predicates "x_1 is less than or equal to x_2" and "x_1 equals x_2," respectively, on the domain of natural numbers. Thus $L(x_1, x_2)$ holds iff $x_1 \leq x_2$ and $E(x_1, x_2)$ holds iff $x_1 = x_2$. Now define the new two-place predicate G so that

$$G(x_1, x_2) \Leftrightarrow L(x_2, x_1).$$

Evidently G is the predicate "x_1 is greater than or equal to x_2." Similarly, the two-place predicate M, defined so that

$$M(x_1, x_2) \Leftrightarrow L(x_1, x_2) \wedge (\neg E(x_1, x_2)),$$

is the predicate "x_1 is strictly less than x_2."   □

From now on, we shall be concerned exclusively with predicates whose domain is the set of natural numbers. Such predicates are called *number-theoretic predicates*. Thus from now on when we speak of the predicate "x_1 is less than x_2" we assume that only natural numbers are to be substituted for x_1 and x_2, and that "less than" refers to the usual ordering relation for the natural numbers.

Whenever it is practical, we will use established infix notation for describing number-theoretic predicates. Thus we will write "$x < y$" for "x is less than y," "$x \mid y$" for "x divides y," and so on. We will also make use of established notation for the familiar arithmetic operations, writing "$x + y = z$" for "the sum of x and y equals z." In particular, we will use such familiar notation in the definitions of simple number-theoretic predicates, writing

$$P(x, y, z) \Leftrightarrow x + y = z$$

in order to define P to be the three-place predicate that holds for the numbers x, y, z iff the sum of x and y equals z.

It is obvious from the definitions that the characteristic function of an n-place number-theoretic predicate is an n-variable number-theoretic function. Number-theoretic predicates may therefore be classified according to the number-theoretic properties of their characteristic functions. In the remainder of this chapter we are concerned primarily with predicates whose characteristic functions are primitive recursive. Such predicates are called *primitive recursive predicates*. As we shall see, primitive recursive predicates play an important part in the study of computability.

Example 4-13. The number-theoretic predicates "$x = y$" and "$x < y$" are primitive recursive because their characteristic functions are the functions eqs and ls, which have been shown to be primitive recursive. The three-place predicate P that holds for x, y, z iff z equals the larger of x and y is also primitive recursive, because its characteristic function can be expressed in the form

$$\chi_P(x, y, z) = \text{eqs}((x \mathbin{\dot-} y) + y, z)$$

and hence is primitive recursive. ☐

The Derivation of Primitive Recursive Predicates

We now investigate some techniques for obtaining new primitive recursive predicates from old ones. The first of these techniques corresponds to the composition of functions. Let Q be a given m-place predicate and let $g_1, \ldots, g_m$ be given n-variable functions. Then the n-place predicate P defined so that

$$P(x_1, \ldots, x_n) \Leftrightarrow Q(g_1(x_1, \ldots, x_n), \ldots, g_m(x_1, \ldots, x_n)) \qquad (4\text{-}55)$$

is said to be obtained by *substituting* the functions $g_1, \ldots, g_m$ in the predicate Q. This new predicate is sometimes denoted $Q \circ (g_1, \ldots, g_m)$.

Example 4-14. Let Q be the two-place predicate "$y_1 = y_2$" and let g_1 and g_2 be the three-variable functions defined as follows:

$$g_1(x_1, x_2, x_3) = x_1 + x_2$$
$$g_2(x_1, x_2, x_3) = x_3$$

Then substitution of g_1 and g_2 in Q yields the three-place predicate P, where

$$P(x_1, x_2, x_3) \Leftrightarrow x_1 + x_2 = x_3.$$ ☐

Our next theorem establishes the fact that substitution preserves primitive recursiveness.

Theorem 4-14. *Let Q be an m-place primitive recursive predicate and let $g_1, \ldots, g_m$ be n-variable primitive recursive functions. If P is the n-place predicate defined so that*

$$P(x_1, \ldots, x_n) \Leftrightarrow Q(g_1(x_1, \ldots, x_n), \ldots, g_m(x_1, \ldots, x_n))$$

then P is also primitive recursive.

Proof. Let χ_P and χ_Q denote the characteristic functions of P and Q, respectively. Evidently χ_P is just the composite function $\chi_Q \circ (g_1, \ldots, g_m)$. Since χ_Q

and $g_1, \ldots, g_m$ are known to be primitive recursive, it follows that χ_P must also be primitive recursive. And this in turn means that P is primitive recursive. $\square$

An important consequence of this theorem is the fact that the arguments of a primitive recursive predicate can be permuted, or dummy arguments added, without loss of primitive recursiveness.

Example 4-15. Let Q be a given two-place primitive recursive predicate. Define the two-place predicate R_1 and the three-place predicate R_2 so that:

$$R_1(x_1, x_2) \Leftrightarrow Q(x_2, x_1)$$

$$R_2(x_1, x_2, x_3) \Leftrightarrow Q(x_1, x_2)$$

Then both R_1 and R_2 are primitive recursive, since each one can be obtained from Q by substitution of appropriate projection functions. Specifically:

$$R_1 = Q \circ (P_2^{(2)}, P_1^{(2)})$$

and

$$R_2 = Q \circ (P_1^{(3)}, P_2^{(3)}).$$

$\square$

In practice, we do not usually bother to mention the use of substitution explicitly, any more than we usually mention the use of functional composition. If we wish to define the primitive recursive predicate R_1 of Example 4-15, we will simply write

$$R_1(x_1, x_2) \Leftrightarrow Q(x_2, x_1).$$

It will be left up to the reader to recognize that R_1 is obtained from Q by substitution and that the primitive recursiveness of R_1 follows from that of Q. Similarly, we will take

$$P(x_1, x_2, x_3) \Leftrightarrow x_1 + x_2 = x_3$$

to be an adequate informal definition of the predicate of Example 4-14, and will assume the reader understands that the primitive recursiveness of P follows from the primitive recursiveness of the equality predicate "$y_1 = y_2$," the addition function, and the appropriate projection functions.

The logical connectives $\neg$, $\wedge$, and $\vee$ provide a second way of obtaining new primitive recursive predicates from old ones. Consideration of the pertinent characteristic functions leads easily to:

Theorem 4-15. Let Q_1 and Q_2 be given n-place primitive recursive predicates, and let R_1, R_2, and R_3 be the n-place predicates defined so that:

$$R_1(x_1, \ldots, x_n) \Leftrightarrow \neg Q_1(x_1, \ldots, x_n)$$

$$R_2(x_1, \ldots, x_n) \Leftrightarrow Q_1(x_1, \ldots, x_n) \vee Q_2(x_1, \ldots, x_n)$$

$$R_3(x_1, \ldots, x_n) \Leftrightarrow Q_1(x_1, \ldots, x_n) \wedge Q_2(x_1, \ldots, x_n)$$

Then R_1, R_2, and R_3 are also primitive recursive.

Note that Theorem 4-15 is applicable only when the predicates in question all have the same arguments. But since the substitution of various projection functions can always be used to permute the arguments of a predicate or to add dummy arguments, the conclusion of the theorem is readily extended to apply to other definitions involving the logical connectives.

Example 4-16. Suppose that Q_1 and Q_2 are given primitive recursive predicates of two and three arguments, respectively, and let the four-place predicate R be defined so that

$$R(w, x, y, z) \Leftrightarrow Q_1(y, w) \wedge Q_2(z, w, x).$$

To see that R is also primitive recursive, we note that the four-place predicates Q_1' and Q_2', where

$$Q_1'(w, x, y, z) \Leftrightarrow Q_1(y, w)$$

$$Q_2'(w, x, y, z) \Leftrightarrow Q_2(z, w, x)$$

are primitive recursive by virtue of Theorem 4-14. The primitive recursiveness of R then follows from Theorem 4-15 and the fact that

$$R(w, x, y, z) \Leftrightarrow Q_1'(w, x, y, z) \wedge Q_2'(w, x, y, z). \qquad \square$$

In practice, we will not bother to give detailed derivations of primitive recursive predicates that are obtained from other primitive recursive predicates through the use of logical connectives. Thus if Q_1 and Q_2 are known to be primitive recursive, we will accept the defining relationship

$$R(w, x, y, z) \Leftrightarrow Q_1(y, w) \wedge Q_2(z, w, x)$$

as an adequate informal justification for the contention that the predicate R is primitive recursive.

A third way of constructing new primitive recursive predicates from old ones involves a process that corresponds to the use of bounded sums and products. If P is a given $(n + 1)$-place predicate and $x_1, \ldots, x_n$ are given natural numbers, we use the expression

$$(\overset{y}{\exists} i)[P(x_1, \ldots, x_n, i)]$$

to denote the statement that is true iff $P(x_1, \ldots, x_n, i)$ holds for *at least one* natural number i in the range $0 \leq i \leq y$. Thus

$$(\overset{y}{\exists} i)[P(x_1, \ldots, x_n, i)]$$

can be thought of as an abbreviation for

$$P(x_1, \ldots, x_n, 0) \vee P(x_1, \ldots, x_n, 1) \vee \cdots \vee P(x_1, \ldots, x_n, y).$$

Since in general the truth or falsity of

$$(\overset{y}{\exists}i)[P(x_1, \ldots, x_n, i)]$$

depends on the choice of $x_1, \ldots, x_n$, and y, this expression can be used to define a new $(n + 1)$-place predicate Q, where:

$$Q(x_1, \ldots, x_n, y) \Leftrightarrow (\overset{y}{\exists}i)[P(x_1, \ldots, x_n, i)] \qquad (4\text{-}56)$$

This predicate Q is said to be obtained from P by *bounded existential quantification*. The value y is referred to as the (upper) bound, or *limit*, of the quantification.

In an analogous way, we use the expression $(\overset{y}{\forall}i)[P(x_1, \ldots, x_n, i)]$ to denote the statement that is true iff $P(x_1, \ldots, x_n, i)$ holds for *every* natural number i in the range $0 \le i \le y$. Thus

$$(\overset{y}{\forall}i)[P(x_1, \ldots, x_n, i)]$$

is an abbreviation for

$$P(x_1, \ldots, x_n, 0) \wedge P(x_1, \ldots, x_n, 1) \wedge \cdots \wedge P(x_1, \ldots, x_n, y).$$

This expression can be used to define a new $(n + 1)$-place predicate R, where

$$R(x_1, \ldots, x_n, y) \Leftrightarrow (\overset{y}{\forall}i)[P(x_1, \ldots, x_n, i)]. \qquad (4\text{-}57)$$

The predicate R is said to be obtained from P by *bounded universal quantification*.

Theorem 4-16. *Let P be an $(n + 1)$-place primitive recursive predicate. Then the $(n + 1)$-place predicates Q and R, defined so that*

$$Q(x_1, \ldots, x_n, y) \Leftrightarrow (\overset{y}{\exists}i)[P(x_1, \ldots, x_n, i)]$$

$$R(x_1, \ldots, x_n, y) \Leftrightarrow (\overset{y}{\forall}i)[P(x_1, \ldots, x_n, i)]$$

are also primitive recursive.

Proof. Let χ_P be the characteristic function of P. Then the characteristic functions of Q and R are given by the equations:

$$\chi_Q(x_1, \ldots, x_n, y) = \mathrm{sg}\left(\sum_{i=0}^{y} \chi_P(x_1, \ldots, x_n, i)\right)$$

$$\chi_R(x_1, \ldots, x_n, y) = \prod_{i=0}^{y} \chi_P(x_1, \ldots, x_n, i)$$

Since χ_P is known to be primitive recursive, it follows that χ_Q and χ_R are also primitive recursive, and hence that Q and R are primitive recursive.   □

As in the case of bounded sums and products, the limits of a bounded quantification can be expressed as functions of the arguments of the predicate being defined. If w_1 and w_2 are given n-variable functions, we write

$$Q(x_1, \ldots, x_n) \Leftrightarrow \overset{w_2(x_1, \ldots, x_n)}{\underset{i = w_1(x_1, \ldots, x_n)}{(\exists i)}} [P(x_1, \ldots, x_n, i)] \qquad (4\text{-}58)$$

to define the n-place predicate Q that holds for $x_1, \ldots, x_n$ iff $P(x_1, \ldots, x_n, i)$ holds for at least one value of i in the range $w_1(x_1, \ldots, x_n) \le i \le w_2(x_1, \ldots, x_n)$. By convention, $Q(x_1, \ldots, x_n)$ is *not to hold* if $w_2(x_1, \ldots, x_n) < w_1(x_1, \ldots, x_n)$. In an analogous way we write

$$R(x_1, \ldots, x_n) \Leftrightarrow \overset{w_2(x_1, \ldots, x_n)}{\underset{i = w_1(x_1, \ldots, x_n)}{(\forall i)}} [P(x_1, \ldots, x_n, i)] \qquad (4\text{-}59)$$

to define the n-place predicate R that holds for $x_1, \ldots, x_n$ iff $P(x_1, \ldots, x_n, i)$ holds for all values of i in the range $w_1(x_1, \ldots, x_n) \le i \le w_2(x_1, \ldots, x_n)$. By convention, $R(x_1, \ldots, x_n)$ is defined *to hold* if $w_2(x_1, \ldots, x_n) < w_1(x_1, \ldots, x_n)$.

Paralleling the development of bounded sums and products, we now have:

Corollary 4-16.1. *Let P be an $(n + 1)$-place primitive recursive predicate and let w_1 and w_2 be n-variable primitive recursive functions. Then the n-place predicates Q and R, defined so that*

$$Q(x_1, \ldots, x_n) \Leftrightarrow \overset{w_2(x_1, \ldots, x_n)}{\underset{i = w_1(x_1, \ldots, x_n)}{(\exists i)}} [P(x_1, \ldots, x_n, i)]$$

and

$$R(x_1, \ldots, x_n) \Leftrightarrow \overset{w_2(x_1, \ldots, x_n)}{\underset{i = w_1(x_1, \ldots, x_n)}{(\forall i)}} [P(x_1, \ldots, x_n, i)]$$

are also primitive recursive.

The proof is easy and will be omitted.

Example 4-17. Consider the Gödel-numbering scheme in which the sequence $x_0, x_1, \ldots, x_n$ is represented by the number

$$2^{x_0 + 1} 3^{x_1 + 1} \cdots p_n^{x_n + 1}$$

and let R be the one-place predicate that holds for a given natural number z iff z is the Gödel number of some sequence. Since $R(z)$ is to hold iff the exponent of every prime factor of z—up to the largest prime that divides z—exceeds zero, we can write

$$R(z) \Leftrightarrow \left(\overset{\text{Lh } (z)}{\forall i} \right) [E \ (i, z) > 0].$$

Since the predicate "$E(i, z) > 0$" is primitive recursive (it is obtained from the predicate "$u > v$" by substitution of the extraction function E and the constant function $C_0^{(2)}$), and since Lh is a primitive recursive function, it follows that the predicate R is also primitive recursive.   □

Simple Applications

We now establish the primitive recursiveness of a number of useful predicates. Although in each case this could be done directly from the appropriate characteristic function, we shall for the most part rely on the indirect techniques discussed above, in order to gain some practice in using those techniques. Our starting point is the equality predicate, already seen to be primitive recursive by virtue of its characteristic function.

Comparison Predicates. There are six basic two-place comparison predicates: *equality, inequality, less than, less than or equal, greater than,* and *greater than or equal.* All six are defined in the obvious way and expressed in terms of the familiar infix symbols $=$, $\neq$, $<$, $\leq$, $>$, and $\geq$. As already noted, the equality predicate is primitive recursive because its characteristic function is the primitive recursive function eqs defined in Eq. (4-15). The primitive recursiveness of the inequality predicate follows immediately from that of equality, since

$$x \neq y \Leftrightarrow \neg (x = y).$$

To establish the primitive recursiveness of the less-than-or-equal predicate, we note that $x \leq y$ iff there exists a natural number z such that $x + z = y$. Since z need never exceed y, we can write

$$x \leq y \Leftrightarrow (\overset{y}{\exists z})[x + z = y].$$

The primitive recursiveness of the less-than-or-equal predicate now follows from Corollary 4-16.1, for the three-place predicate described by the expression "$x + z = y$" is obtained from the equality predicate by substitution of primitive recursive functions (see Example 4-14) and is therefore primitive recursive.

The remaining comparison predicates are easily derived from equality, inequality, and less than or equal. The less-than predicate can be expressed in the form

$$x < y \quad \Leftrightarrow \quad (x \leq y) \wedge (\neg (x = y))$$

and so is primitive recursive. The greater-than predicate is obtained from less-than by substitution:

$$x > y \Leftrightarrow y < x$$

Finally, the greater-than-or-equal predicate can be expressed in the form

$$x \geq y \Leftrightarrow (x > y) \vee (x = y)$$

and so must be primitive recursive.

Divisibility Predicates.   We next consider a series of predicates related to division and divisibility. The first of these is the two-place predicate *divides*, defined to hold for the pair (x, y) iff x divides y. This predicate will be represented by the usual infix symbol $|$. The next two predicates are the one-place *even* and *odd* predicates, denoted Ev and Od and defined to hold for the natural number x iff x is even or odd, respectively. The fourth is the one-place *prime* predicate, denoted Pr and defined to hold for the natural number x iff x is prime.

The divides predicate can be derived by bounded quantification, in much the same way as the less-than-or-equal predicate:

$$x \,|\, y \Leftrightarrow \overset{y}{(\exists m)}[m \cdot x = y].$$

Thus the divides predicate is primitive recursive. The even and odd predicates are now easily obtained by substitution and the use of the logical connective $\neg$:

$$\mathrm{Ev}(x) \Leftrightarrow 2 \,|\, x$$

$$\mathrm{Od}(x) \Leftrightarrow \neg\, \mathrm{Ev}(x)$$

Finally, the prime predicate can be shown to be primitive recursive through the use of bounded quantification and various logical connectives:

$$\mathrm{Pr}(x) \Leftrightarrow (x \geq 2) \wedge \neg\, \overset{x}{(\exists z)}[z \neq 1 \wedge z \neq x \wedge z \,|\, x]$$

Now that we have some simple predicates to work with, we can illustrate the usefulness of primitive recursive predicates in the derivation of various primitive recursive functions. We begin with the situation in which a function is defined by specifying the value that it is to assume in each of several cases.

Theorem 4-17. *Let $P_1, \ldots, P_m$ be n-place primitive recursive predicates, and let $g_1, \ldots, g_m, g_{m+1}$ be n-variable primitive recursive functions. Assume that $P_1, \ldots, P_m$ are mutually exclusive—i.e., that no two of them hold for the same choice of argument values. Then the n-variable function f, defined so that*

$$f(x_1, \ldots, x_n) = \begin{cases} g_1(x_1, \ldots, x_n) & \text{if} \quad P_1(x_1, \ldots, x_n) \\ \quad\vdots & \qquad\vdots \\ g_m(x_1, \ldots, x_n) & \text{if} \quad P_m(x_1, \ldots, x_n) \\ g_{m+1}(x_1, \ldots, x_n) & \text{otherwise} \end{cases} \qquad (4\text{-}60)$$

is primitive recursive.

Proof. Note that the value of f can be expressed in the form

$$f(x_1, \ldots, x_n) = g_1(x_1, \ldots, x_n)\chi_1(x_1, \ldots, x_n)$$
$$+ \cdots + g_m(x_1, \ldots, x_n)\chi_m(x_1, \ldots, x_n)$$
$$+ g_{m+1}(x_1, \ldots, x_n) \overline{\text{sg}} \ (\chi_1(x_1, \ldots, x_n) + \cdots + \chi_m(x_1, \ldots, x_n))$$

where χ_i is the characteristic function of P_i. Since the functions $g_1, \ldots, g_{m+1}$ and $\chi_1, \ldots, \chi_m$ are all primitive recursive, it follows that f is primitive recursive. □

Functions defined in the manner of Eq. (4-60) are said to be *defined by cases*, or by *parts*. The use of such a definition often simplifies the job of establishing the primitive recursiveness of a function.

Example 4-18

a) The function g introduced in Example 4-4 is most naturally defined by parts, as follows:

$$g(x, y) = \begin{cases} 3 & \text{if} \quad x = 1 \wedge y = 7 \\ 5 & \text{if} \quad x = 2 \wedge y = 0 \\ 1 & \text{if} \quad x = 4 \wedge y = 4 \\ 0 & \text{otherwise} \end{cases}$$

The primitive recursiveness of the constant functions and of the predicates "$x = 1 \wedge y = 7$," "$x = 2 \wedge y = 0$," and "$x = 4 \wedge y = 4$" ensures that g is primitive recursive.

b) The absolute difference function can be defined by parts as follows:

$$|x - y| = \begin{cases} x \dot- y & \text{if} \quad x \geq y \\ y \dot- x & \text{otherwise} \end{cases}$$

The primitive recursiveness of proper subtraction and of the predicate "$x \geq y$" ensures that the absolute difference function is primitive recursive. □

Of course, a definition by parts can be used in conjunction with other defining schemes. In particular, a definition by parts can be used within a primitive recursion or a course-of-values recursion.

Example 4-19. The remainder function introduced in Section 4.1 can be defined by primitive recursion as follows:

$$\text{rem} \ (x, 0) = 0$$

$$\text{rem} \ (x, y^+) = \begin{cases} y^+ & \text{if} \quad x = 0 \\ \text{rem} \ (x, y) + 1 & \text{if} \quad (x \neq 0) \wedge (\text{rem} \ (x, y) + 1 < x) \\ 0 & \text{otherwise} \end{cases}$$

To see that these equations do in fact constitute a primitive recursion, we note that the second equation provides a definition by parts of rem (x, y^+) in terms of x, y, and rem (x, y). Thus

$$\text{rem}(x, y^+) = h(x, y, \text{rem}(x, y))$$

where:

$$h(a, b, c) = \begin{cases} b^+ & \text{if} \quad a = 0 \\ c + 1 & \text{if} \quad (a \neq 0) \wedge (c + 1 < a) \\ 0 & \text{otherwise} \end{cases}$$

Since h is obviously a primitive recursive function, so is the remainder function. $\qquad\square$

The use of predicates can also simplify the definition of functions by bounded minimalization. Instead of defining the value of the new function to be the least natural number that causes a given function to assume the value 0, we can think of defining it to be the least natural number that causes a given predicate to hold. If P is an $(n + 1)$-place predicate and $x_1, \ldots, x_n$ are given natural numbers, we use the expression

$$\overset{y}{\mu z}[P(x_1, \ldots, x_n, z)]$$

to denote the least value of z in the range $0 \leq z \leq y$ for which $P(x_1, \ldots, x_n, z)$ holds, or $y + 1$ if there is no such z in the range $0 \leq z \leq y$. The $(n + 1)$-variable function f, where

$$f(x_1, \ldots, x_n, y) = \overset{y}{\mu z}[P(x_1, \ldots, x_n, z)] \tag{4-61}$$

is said to be defined by *bounded minimalization from the predicate P*.

Definitions by bounded minimalization from predicates are readily extended to include the use of computed bounds on the minimalization. All such minimalizations from predicates necessarily yield primitive recursive functions as long as the predicates to which they are applied are primitive recursive.

Theorem 4-18. *Let P be an $(n + 1)$-place primitive recursive predicate, and let w be an n-variable primitive recursive function. Then the n-variable function f defined so that*

$$f(x_1, \ldots, x_n) = \overset{w(x_1, \ldots, x_n)}{\mu z} [P(x_1, \ldots, x_n, z)]$$

is also primitive recursive.

Proof. The function f can be expressed in terms of the characteristic function of P as follows:

$$f(x_1, \ldots, x_n) = \overset{w(x_1, \ldots, x_n)}{\mu z} [\chi_P(x_1, \ldots, x_n, z) = 1]$$

Since χ_P is known to be primitive recursive, f must also be primitive recursive. □

Although bounded minimalization from predicates is essentially equivalent to bounded minimalization from functions, the use of a predicate often leads to a more natural and transparent description of the function being defined.

Example 4-20. Consider the quotient function, which was defined in Eq. (4-22) as follows

$$\text{quo}(x, y) = \text{sg}(x)(\overset{y}{\mu}z[\text{gr}(x(z + 1), y) = 1]).$$

Recognizing that the minimalization is designed to select the smallest value of z such that $x(z + 1)$ exceeds y, we can write

$$\text{quo}(x, y) = \text{sg}(x)(\overset{y}{\mu}z[x(z + 1) > y]).$$

This last expression undoubtedly provides a more understandable description of the quotient function than the original definition of Eq. (4-22). Yet because the three-place predicate represented by "$x(z + 1) > y$" is primitive recursive, the new expression suffices to establish the primitive recursiveness of the quotient function. □

Thus the use of primitive recursive predicates can simplify the derivation of primitive recursive functions. We will rely on this fact in the next chapter when we examine various extensions of the class of primitive recursive functions.

4.4 SUMMARY

A function f is said to be defined by primitive recursion from the functions g and h if g specifies the value of $f(x_1, \ldots, x_n, 0)$ and h specifies the value of $f(x_1, \ldots, x_n, y + 1)$ in terms of $x_1, \ldots, x_n, y$, and $f(x_1, \ldots, x_n, y)$. Primitive recursion provides one of the simplest and most important ways of defining number-theoretic functions. It leads naturally to the class of primitive recursive functions, which consists of all those functions that can be obtained from the constant, successor, and projection functions by means of functional composition and primitive recursion. Every primitive recursive function is necessarily both total and Turing computable. Although the primitive recursive functions do not include all the Turing-computable functions, they do include most of the total, effectively computable functions of practical interest.

There are a number of definition schemes that, although not representing definitions by primitive recursion, necessarily yield primitive recursive functions when applied to primitive recursive functions. Among the simplest of these schemes are definitions by bounded sums or products, in which the value of a new function is defined to be the sum or product of successive values of a given function. Also included are definitions by bounded minimalization, in which the value of a new function is defined to be the smallest argument value—up to a designated limit—for which a given function assumes the value zero. Use of such definition schemes can simplify the job of showing that certain functions are primitive recursive. In particular, these schemes make it easy to establish the primitive recursiveness of a variety of comparison functions and a number of useful functions associated with division and divisibility.

It is often necessary in the study of computability to be able to represent sequences of natural numbers by single numbers. The Cantor-numbering scheme provides a simple way of encoding ordered pairs of natural numbers into single natural numbers and can be extended to provide a similar encoding for n-tuples of any fixed length n. In each case the necessary encoding and decoding can be effected by primitive recursive functions. Cantor numbering has many applications. It is used here to show that two functions defined by simultaneous recursion from given primitive recursive functions are necessarily primitive recursive. This is done by combining the values of the two functions into the values of a single auxiliary function and showing that the auxiliary function is primitive recursive. The values of the desired functions can then be obtained by "decoding" the values of the auxiliary function.

Gödel-numbering schemes make use of the uniqueness of prime power decompositions to provide representations for arbitrary sequences of natural numbers. As long as a sequence can be produced as successive values of a primitive recursive function, the encoding and decoding of the corresponding Gödel number representation can also be effected by primitive recursive functions. Using Gödel numbers as representations of sequences makes it possible to specify a new function value in terms of an arbitrary number of preceding values and to show that such specifications yield primitive recursive functions. In this way it is possible to prove that a variety of recursive definition schemes—including course-of-values recursions and recursions on two or more variables—necessarily yield primitive recursive functions when they are applied to primitive recursive functions.

The definition of a function is often phrased in terms of conditions on the arguments of the function. Such conditions can be represented by predicates, which are formally functions whose values are statements. A predicate can in turn be represented by its characteristic function, whose values specify the situations in which the condition expressed by the predicate holds. Predicates whose characteristic functions are primitive recursive are called primitive recursive predicates. Predicates are commonly used in definitions by parts—in which the values of a function are specified separately for each of several cases—and in definitions by bounded minimalization—in which the value of a function is defined to be the smallest argument value for which a given condition holds. In each case the use of primitive recursive predicates ensures the primitive recursiveness of the function being defined.

Primitive recursive predicates can be built up by means of several basic constructions. These include substitution of primitive recursive functions for the arguments of primitive recursive predicates; combining primitive recursive predicates through the use of logical connectives; and applying bounded universal or existential quantifiers to primitive recursive predicates. These constructions make it possible to establish the primitive recursiveness of a wide variety of predicates, including a number of useful comparison and divisibility predicates. As will be apparent in later chapters, the ability to show that certain predicates are primitive recursive plays an important part in the study of computation.

4.5 PROBLEMS

Section 4.1

4.1.1   The functions Ψ and θ are total number-theoretic functions of one and two variables, respectively. Which of the following represent definitions of a new function f by primitive recursion?

a) $\begin{aligned} f(x, 0) &= 17 \\ f(x, y^+) &= f(0, \theta(x, y)) \end{aligned}$

b) $\begin{aligned} f(x, 0) &= \theta(0, x) \\ f(x, y^+) &= \theta(f(x, y), y + 1) \end{aligned}$

c) $\begin{aligned} f(x, 0) &= \Psi(x) \\ f(x, y^+) &= f(x, y) + \theta(y, x) \end{aligned}$

d) $\begin{aligned} f(0) &= 35 \\ f(y^+) &= \Psi(f(0) + f(1) + \cdots + f(y)) \end{aligned}$

For each definition that does represent a primitive recursion, specify the associated functions g and h from which f is obtained by primitive recursion.

4.1.2   Show that each of the functions described below is primitive recursive. You may use any of the primitive recursive functions mentioned in Eqs. (4-4) through (4-15), but not bounded sums, products, or minimalization.

a) $\max(x, y) = \begin{cases} x & \text{if} \quad x \geq y \\ y & \text{if} \quad x < y \end{cases}$

b) $\min(x, y) = \begin{cases} x & \text{if} \quad x \leq y \\ y & \text{if} \quad x > y \end{cases}$

c) $\mathrm{ev}(x) = \begin{cases} 1 & \text{if} \quad x \text{ is even} \\ 0 & \text{if} \quad x \text{ is odd} \end{cases}$ (*continued*)

d) $hf(x) = \left\lfloor \dfrac{x}{2} \right\rfloor$

e) $sqrt(x) = \lfloor \sqrt{x} \rfloor$

f) $psq(x) = \begin{cases} 1 & \text{if} \quad x \text{ is a perfect square} \\ 0 & \text{otherwise} \end{cases}$

4.1.3   Let Ψ and θ be designated primitive recursive functions of one and two variables, respectively. Show that each of the functions described below is also primitive recursive.

a) The one-variable function f_1, where $f_1(0) = \Psi(0) + 1$, $f_1(1) = \Psi(\Psi(1) + 1) + 1$, and in general

$$f_1(x) = \underbrace{\Psi(\Psi(\cdots(\Psi(x) + 1)\cdots) + 1) + 1.}_{x + 1}$$

b) The two-variable function f_2, where $f_2(x, 0) = \theta(x, 0)$, $f_2(x, 1) = \theta(\theta(x, 1), 0)$, and in general $f_2(x, y) = \theta(\theta(\theta(\cdots\theta(\theta(x, y), y - 1) \cdots, 2),1), 0)$.

4.1.4   Let g and θ be specified primitive recursive functions of one and two variables, respectively. Prove that each of the two-variable functions defined below is also primitive recursive.

a)
$$f(x, 0) = g(x)$$
$$f(x, y^+) = \theta\left(x, \sum_{i=0}^{y} f(x, i)\right)$$

b)
$$f(x, 0) = g(x)$$
$$f(x, y^+) = \theta(f(x^+, y), y^+)$$

Hint: In each case introduce an auxiliary function that can be defined by primitive recursion.

4.1.5   Use definitions by bounded sums and/or products to establish the primitiveness of each of the functions described below. Assume that g is a specified one-variable primitive recursive function.

a) $f(y) = $ the number of values of i in the range $0 \le i \le y$ for which $g(i) > 3$

b) $f(x, y) = \begin{cases} 1 & \text{if } g(i + 1) > g(i) \text{ for all values of } i \text{ in the range} \\ & x \le i \le y \\ 0 & \text{otherwise} \end{cases}$

c) $f(w, x, y) = \begin{cases} 1 & \text{if } x \le y \text{ and } w \text{ is the largest of } g(x), \\ & g(x + 1), \ldots, g(y) \\ 0 & \text{otherwise} \end{cases}$

d) $f(y) = $ the smallest value of i, up to y, for which $g(i) = g(i + 1) = 0$, or $y + 1$ if no such value exists

4.1.6 Let g be a given primitive recursive function of $n + 1$ variables. The new function f is defined from g so that

$$f(x_1, \ldots, x_n, y) = \max_{i=0}^{y} \left(g(x_1, \ldots, x_n, i) \right)$$

$$= \max \left(g(x_1, \ldots, x_n, 0), \ldots, g(x_1, \ldots, x_n, y) \right).$$

Thus f may be thought of as being obtained from g by a "bounded maximum" operation.

    a) Prove that the function f is primitive recursive.

    b) Generalize the given definition scheme so as to permit the upper and lower limits of i to be arbitrary primitive recursive functions of the arguments of f. (Use the convention that the value of f is to be 0 if the lower bound exceeds the upper bound.) Show that the resulting function f is still primitive recursive.

4.1.7 Use bounded minimalization to show that each of the functions described below is primitive recursive.

a) $\max(x, y) = \begin{cases} x & \text{if} & x \geq y \\ y & \text{if} & x < y \end{cases}$

b) $\min(x\ y) = \begin{cases} x & \text{if} & x \leq y \\ y & \text{if} & x > y \end{cases}$

c) $\text{hf}(x) = \left\lfloor \dfrac{x}{2} \right\rfloor$

d) $\text{sqrt}(x) = \lfloor \sqrt{x} \rfloor$

4.1.8 Let the one-variable function h be defined so that $h(n)$ is the $(n + 1)$st significant digit in the decimal representation of $\sqrt{2}$. Thus $h(0) = 1$, $h(1) = 4$, $h(2) = 1$, $h(3) = 4$, $h(4) = 2$, and so on. Show that h is primitive recursive.

Hint: Consider the auxiliary function g, where $g(n)$ is the natural number represented by the first $n + 1$ significant digits in the representation of $\sqrt{2}$. Thus $g(0) = 1$, $g(1) = 14$, $g(2) = 141$, $g(3) = 1414$, $g(4) = 14142$, etc. Use bounded minimalization to show that g is primitive recursive.

4.1.9 Show that each of the functions described below is primitive recursive. Any of the functions presented in Section 4.1 may be used as auxiliary functions, except for exponentiation, which should not be used. Results of earlier parts of the problem may be used in later parts.

a) $\text{shr}(x, n) = \left\lfloor \dfrac{x}{2^n} \right\rfloor$ b) $\text{lg}(x) = \begin{cases} 0 & \text{if} & x = 0 \\ \lfloor \log_2 (x) \rfloor + 1 & \text{if} & x > 0 \end{cases}$ *(continued)*

c) $\mathrm{dig}(x, n) = $ the nth digit in the binary representation of x,
counting from the right and starting with 0

Thus $\mathrm{dig}(13, 0) = 1$, $\mathrm{dig}(13, 1) = 0$, $\mathrm{dig}(13, 2) = 1$, $\mathrm{dig}(13, 3) = 1$, $\mathrm{dig}(13, 4) = 0$, etc.

d) $\mathrm{wgt}(x) = $ the number of 1's in the binary representation of x

e) $\mathrm{rev}(x) = $ the number whose binary representation is the reverse of that of x. (Leading 0's are to be ignored.)

Thus $\mathrm{rev}(6) = \mathrm{rev}\ (3) = 3$, $\mathrm{rev}(13) = 11$, $\mathrm{rev}(8) = \mathrm{rev}(4) = \mathrm{rev}(2) = \mathrm{rev}(1) = 1$, $\mathrm{rev}(25) = 19$, etc.

4.1.10   The evaluation of a total n-variable function f will be said to be *bounded* by the n-variable function $\hat{f}$ if there exists a Turing machine that evaluates f in standard form and that requires at most $\hat{f}(x_1, \ldots, x_n)$ steps to compute $f(x_1, \ldots, x_n)$. The purpose of this problem is to show that the evaluation of every primitive recursive function is bounded by another primitive recursive function.

a) Exhibit primitive recursive functions that bound the evaluation of the various constant, projection, and successor functions.

b) Let $g_1, \ldots, g_m$ be n-variable primitive recursive functions, and let h be an m-variable primitive recursive function. Assume that the evaluations of $g_1, \ldots, g_m$ and h are bounded by the primitive recursive functions $\hat{g}_1, \ldots, \hat{g}_m$, and $\hat{h}$, respectively. Let f denote the n-variable function defined from $g_1, \ldots, g_m$ and h by functional composition. Show that the evaluation of f is bounded by a primitive recursive function $\hat{f}$.

c) Let g and h be n- and $(n + 2)$-variable primitive recursive functions. Assume that the evaluations of g and h are bounded by the primitive recursive functions $\hat{g}$ and $\hat{h}$. Let f denote the $(n + 1)$-variable function defined from g and h by primitive recursion. Show that the evaluation of f is bounded by a primitive recursive function $\hat{f}$.

d) Prove that the evaluation of every primitive recursive function f is bounded by a primitive recursive function $\hat{f}$.

Section 4.2

4.2.1   Let π denote the Cantor-numbering function specified in Fig. 4-2, and let σ_1 and σ_2 denote the associated decoding functions.

a) Show that $\pi(0, y) = \dfrac{y(y + 1)}{2}$.

b) Show that $\pi(x, y) = \frac{1}{2}(x^2 + 2xy + y^2 + 3x + y)$.

c) Consider the function σ such that $\sigma(z) = \sigma_1(z) + \sigma_2(z)$. (Thus $\sigma(z)$ is the number of the diagonal in Fig. 4-2 on which the natural number z lies.)

Establish the primitive recursiveness of σ from that of π by means of an appropriate primitive recursion. Do not use minimalization.

    d) Use the result of part (c) to show that the functions σ_1 and σ_2 are primitive recursive. Do not use minimalization.

4.2.2   In this problem we consider the possibility of devising a single Cantor-numbering scheme for *all* ordered tuples of natural numbers.

    a) Let τ denote the mapping such that, for each n,

$$\tau(x_1, \ldots, x_n) = \pi^{n+1}(x_n, \ldots, x_2, x_1, n-1).$$

Show that this mapping is a bijection from $N \cup N^2 \cup N^3 \cup \cdots$ to N, and hence can be used to assign "Cantor numbers" to all tuples of natural numbers. Give the tuples represented by the Cantor numbers 0, 1, 2, 3, 4, 5, and 32.

    b) Let the "decoding function" β be defined so that if $z = \tau(x_1, \ldots, x_n)$ then:

$$\beta(z, i) = \begin{cases} n & \text{if} & i = 0 \\ x_i & \text{if} & 1 \le i \le n \\ 0 & \text{if} & i > n. \end{cases}$$

Show that β is primitive recursive.

    c) Suppose that f is a given one-variable primitive recursive function. Define the function $\hat{f}$ so that $\hat{f}(n) = \tau(f(0), \ldots, f(n))$. Show that $\hat{f}$ is also primitive recursive.

4.2.3   Give a definition by simultaneous recursion for each of the following pairs of functions.

    a) f_1 and f_2, where f_1 is the Fibonacci function defined by Eqs. (4-42) and:

$$f_2(x) = \begin{cases} 0 & \text{if} & x = 0 \\ f_1(x-1) & \text{if} & x > 0 \end{cases}$$

    b) σ_1 and σ_2, the Cantor-number decoding functions.

    c) s and r, where $s(x) = \lfloor \sqrt{x} \rfloor$ and $r(x) = x - (s(x))^2$.

4.2.4   Establish the primitive recursiveness of each of the following one-variable functions by finding a second function with which the given function can be defined by simultaneous recursion. In each case tabulate the first 25 values of the new function and provide the definition by simultaneous recursion.

(a)

x	0	1	2	3	4	5	6	7	8	9	10	11	12	13	14	15	16	17	18	19	20	21	22	23	24	...
$f_1(x)$	0	1	0	2	1	0	3	2	1	0	4	3	2	1	0	5	4	3	2	1	0	6	5	4	3	...

(b)

x	0	1	2	3	4	5	6	7	8	9	10	11	12	13	14	15	16	17	18	19	20	21	22	23	24	...
$f_2(x)$	0	1	1	2	2	2	2	3	3	3	3	3	3	3	3	4	4	4	4	4	4	4	4	4	4	...

$\underbrace{\quad}_{1}\ \underbrace{\quad}_{2}\ \underbrace{\quad}_{4}\ \underbrace{\qquad\quad}_{8}\ \underbrace{\qquad\qquad}_{16}$

(c)

x	0	1	2	3	4	5	6	7	8	9	10	11	12	13	14	15	16	17	18	19	20	21	22	23	24	...
$f_3(x)$	0	1	1	0	2	2	2	1	0	3	3	3	3	2	1	0	4	4	4	4	4	3	2	1	0	...

$\underbrace{\quad}_{2}\ \underbrace{\quad}_{3}\ \underbrace{\qquad}_{4}\ \underbrace{\qquad\quad}_{5}$

4.2.5   In this problem we use Cantor numbering as a way of representing two-variable functions by one-variable functions. Specifically, with each two-variable number-theoretic function g we associate the one-variable function $\hat{g}$ such that $\hat{g}(\pi(x, y)) = g(x, y)$.

   a) Show that $\hat{g}$ is primitive recursive iff g is.

   b) Suppose that $g = g_0 \circ (g_1, g_2)$, where g_0, g_1, and g_2 are all two-variable functions. Express $\hat{g}$ in terms of $\hat{g}_0$, $\hat{g}_1$, and $\hat{g}_2$.

   c) Let h be a given one-variable primitive recursive function, and consider the two-variable function f defined as follows.

$$f(0, 0) = c$$
$$f(0, y^+) = h(f(y, 0))$$
$$f(x^+, y) = h(f(x, y^+))$$

Show that $\hat{f}$, and therefore f, is primitive recursive.

4.2.6   The $(n + 1)$-variable function f is said to be defined from the n-variable function g and the one-variable function h by *iteration* if

$$f(x_1, \ldots, x_n, 0) = g(x_1, \ldots, x_n)$$
$$f(x_1, \ldots, x_n, y^+) = h(f(x_1, \ldots, x_n, y)).$$

Thus iteration is a special form of primitive recursion.

   a) Suppose that the $(n + 1)$-variable function φ is obtained from the n-variable function γ and the $(n + 2)$-variable function θ by primitive recursion.

Using an appropriate Cantor-numbering scheme, show that φ can be obtained from γ and θ by composition and iteration.

b) Show that a function is primitive recursive iff it can be obtained from the constant, projection, and successor functions, together with π^2, σ_1^2, and σ_2^2, by means of functional composition and iteration.

4.2.7 Consider the alternative Gödel-numbering scheme in which the sequence $x_0, \ldots, x_n$ is represented by the natural number

$$p_0^{x_0+1} p_1^{x_1+1} \cdots p_n^{x_n+1}.$$

Under this scheme, distinct sequences are represented by distinct numbers, but not every natural number represents a sequence.

a) What is the Gödel number of the sequence 3, 2, 0, 1? Of what sequence is 404,250 the Gödel number?

b) Which of the following numbers represent sequences?

    i) 9   ii) 150
    iii) 990   iv) 16

c) Show that the function gn. where

$$gn(x) = \begin{cases} 1 & \text{if } x \text{ is the Gödel number of a sequence} \\ 0 & \text{if } x \text{ is not the Gödel number of a sequence} \end{cases}$$

is primitive recursive.

d) Show that the function Lgn, where

$$Lgn(x) = \begin{cases} n+1 & \text{if } x \text{ is the Gödel number of a sequence } x_0, \ldots, x_n \\ 0 & \text{if } x \text{ is not the Gödel number of a sequence} \end{cases}$$

is primitive recursive.

4.2.8 Consider the set $\mathcal{F}$ consisting of all the polynomial functions of one variable—i.e., the set of all functions of the form

$$f(x) = c_0 + c_1 x + \cdots + c_n x^n$$

where $c_0, \ldots, c_n$ are natural numbers. Let the polynomial function whose coefficients are $c_0, \ldots, c_n$ be assigned the index $u = 2^{c_0} 3^{c_1} \cdots p_n^{c_n}$ and denoted F_u.

a) Show that the two-argument universal function for the class $\mathcal{F}$ is primitive recursive.

b) Show that there exist primitive recursive functions a and m such that
$$F_{a(u,\,v)}(x) = F_u(x) + F_v(x)$$
and
$$F_{m(u,\,v)}(x) = (F_u(x))(F_v(x)).$$

4.2.9 Recast each of the following definitions as a formal course-of-values recursion. That is, specify functions g and h from which the desired function f can be obtained in the format of Eqs. (4-44). Assume that Ψ and θ are given functions of one and two variables, respectively.

a) $f(0) = 7$
$$f(y^+) = \theta\left(f\left(\left\lfloor\frac{y}{2}\right\rfloor\right), f\left(\left\lfloor\frac{y}{3}\right\rfloor\right)\right)$$

b) $f(x, 0) = \Psi(x)$
$$f(x, y^+) = \theta\left(x, \sum_{i=0}^{y} f(x, i)\right)$$

c) $f(x, 0) = x^2$
$$f(x, y^+) = \Psi(\max(f(x, 0), \ldots, f(x, y))) \qquad \text{if} \quad y < x$$
$$f(x, y^+) = \Psi(\max(f(x, y - x), \ldots, f(x, y))) \qquad \text{if} \quad y \geq x.$$

4.2.10 Show that each of the functions described below is primitive recursive by providing an appropriate course-of-values recursion. Restrict your auxiliary functions to those presented in Eqs. (4-4) through (4-15). Do not use bounded sums, products, or minimalization.

a) $\mathrm{sv}(x) = \left\lfloor\dfrac{x}{7}\right\rfloor$

b) $\mathrm{rv}(x) =$ the remainder of x modulo 7

c) $\mathrm{sd}(x) =$ the number of digits in the base 7 representation of x (Let sd $(0) = 0$)

d) $\mathrm{ps}(x) = \begin{cases} 1 & \text{if } x \text{ is a power of } 7 \\ 0 & \text{otherwise} \end{cases}$

4.2.11 The functions g_0, g_1, g_2, h_1, h_2, and h_3 are known to be primitive recursive. Prove that each of the functions $f_1, f_2,$ and f_3 defined below is also primitive recursive. In each case this can be done by showing that the Gödel number of an appropriate sequence of function values can be obtained by a primitive recursion.

a) $f_1(x, 0) = g_0(x)$

$f_1(x, y^+) = h_1\left(\sum_{j=0}^{x} f_1(j, y) \right)$

Hint: Consider the quantity $\langle f_1(0, y), \ldots, f_1(x, y) \rangle$.

b) $f_2(0, y) = g_1(y)$

$f_2(x^+, 0) = g_2(x)$

$f_2(x^+, y^+) = h_2\left(f_2(x, y^+), f_2\left(x, \left\lfloor \frac{y^+}{2} \right\rfloor \right) \right)$

c) $f_3(x, 0) = g_1(x)$

$f_3(0, y^+) = g_2(y)$

$f_3(x^+, y^+) = h(f_3(x, y))$

4.2.12 a) Prove Theorem 4-13.

b) Show that the two-variable function f defined below is primitive recursive by providing a formal double course-of-values recursion for it. In other words, show that f can be specified in the form of Eqs. (4-53), and establish the primitive recursiveness of the appropriate functions g_1, g_2, and h.

$$f(0, y) = 2y + 3$$

$$f(x, 0) = 2x + 3$$

$$f(x^+, y^+) = xf\left(x \dot- \left\lfloor \frac{y}{3} \right\rfloor, \left\lfloor \frac{y}{3} \right\rfloor \right) + yf\left(\left\lfloor \frac{x}{3} \right\rfloor, y \dot- \left\lfloor \frac{x}{3} \right\rfloor \right)$$

4.2.13 a) Devise a double Gödel-numbering scheme for representing the values of a three-variable function f. In particular, devise a scheme in which $\Psi(w, z)$ represents the values of $f(x, y, z)$ for which $x + y = w$ and $\theta(v)$ represents the values of $f(x, y, z)$ for which $x + y + z = v$. Show that θ is primitive recursive iff f is.

b) Consider the three-variable function f defined as follows.

$$f(0, y, z) = g_1(y, z)$$

$$f(x^+, 0, z) = g_2(x, z)$$

$$f(x^+, y^+, 0) = g_3(x, y)$$

$$f(x^+, y^+, z^+) = h(f(x, y^+, z^+), f(x^+, y, z^+), f(x^+, y^+, z))$$

Assume that g_1, g_2, g_3, and h are all primitive recursive. Use the scheme of part (a) to prove that f is primitive recursive.

Section 4.3

4.3.1 Consider the three-place predicate Q, where $Q(x, y, z)$ holds iff $x \le y$ and z is the square of some number n in the range $x \le n \le y$.

a) Show, by recourse to the definition of a primitive recursive predicate, that Q is primitive recursive.

b) Use appropriate substitutions in Q to establish the primitive recursiveness of each of the predicates R_1 through R_4 defined below.

i) $R_1(x, y) \Leftrightarrow x = y^2$ iii) $R_3(x) \Leftrightarrow x$ is a perfect square
ii) $R_2(x) \Leftrightarrow x = 7$ iv) $R_4(x, y) \Leftrightarrow x \le y$

4.3.2 Assume that f is a total one-variable function with the property that $f(x) < 7x^2$ for all x. Let P denote the two-place predicate such that $P(x, y) \Leftrightarrow f(x) = y$. Prove that P is a primitive recursive predicate iff f is a primitive recursive function.

4.3.3 Let P be any one-place primitive recursive predicate and let χ_P be its characteristic function. Express each of the following predicates and functions in terms of P and χ_P using bounded minimalization but not bounded quantifiers, bounded sums, or bounded products.

a) The predicate Q, where $Q(n) \Leftrightarrow (\overset{n}{\forall} i)[P(i)]$
b) χ_Q, the characteristic function of Q
c) The predicate R, where $R(n) \Leftrightarrow (\overset{n}{\exists} i)[P(i)]$
d) χ_R, the characteristic function of R

4.3.4 In this problem we adopt the Gödel-numbering scheme in which the sequence $x_0, x_1, \ldots, x_n$ is represented by the number $2^{x_0+1} 3^{x_1+1} \cdots p_n^{x_n+1}$. Use bounded quantification (and appropriate logical connectives) to show that each of the predicates defined below is primitive recursive.

a) $V(z) \Leftrightarrow z$ is the Gödel number of some sequence
b) $M(x, z) \Leftrightarrow z$ is the Gödel number of a sequence in which x is a maximal element
c) $P(z) \Leftrightarrow z$ is the Gödel number of a palindrome (a sequence that reads the same backward as forward)
d) $R(z) \Leftrightarrow z$ is the Gödel number of a sequence in which some number appears more than once

4.3.5 Show that each of the predicates described below is primitive recursive.

     a) $ME(x, y, z) \Leftrightarrow z$ is nonzero and $x \equiv y$ modulo z

     b) $RP(x, y) \Leftrightarrow x$ and y are both nonzero, and are relatively prime

     c) $LCM(x, y, z) \Leftrightarrow z$ is the least common multiple of x and y

     d) $GCD(x, y, z) \Leftrightarrow z$ is the greatest common divisor of x and y

4.3.6 The set A is defined inductively as follows: (i) The number 0 is a member of A; (ii) If the natural number n is a member of A, so is the number $2 \cdot 3^n$. The one-place predicate Q is defined so that $Q(x) \Leftrightarrow x \in A$. Show that Q is primitive recursive.

4.3.7 Making optimal use of the primitive recursive predicates presented in the text, show that each of the functions described below is primitive recursive.

     a) $f(x, y, m) = $ the sum of x and y modulo m, if x and y are both less than m, and 0 otherwise

     b) $f(x) = $ the least common multiple of x and 21

     c) $f(x) = $ the number of 1's in the ternary representation of x

     d) $f(x) = $ the largest member of the sequence whose Gödel number is x (or 0 if $x = 0$)

     e) $f(x, n) = $ the $(n + 1)$st value of i in the range $0 \le i \le x$ for which $g(i) = 0$, or $x + 1$ if there do not exist $n + 1$ such values. (Assume that g is a specified primitive recursive function.)

Chapter 5

Recursive Functions

The preceding chapter introduced a variety of recursive schemes that can be used to specify primitive recursive functions. But not all the functions that can be specified by recursive schemes turn out to be primitive recursive. In the first section of this chapter we consider an important example of a recursively defined function that is not primitive recursive. Motivated by this example, we then consider ways in which the class of primitive recursive functions can be enlarged without sacrificing effective computability. We begin in Section 5.2 by removing the upper bound from the minimalization operator. This leads to the class of μ-recursive functions, which will be shown to be identical to the class of Turing-computable functions. In Section 5.3 we attempt to make a further extension by permitting the use of a very broad class of recursive definition schemes. However, the resulting class of general recursive functions turns out to be identical to the class of μ-recursive functions. This fact, together with our previous results about Turing machines, provides the basis for the commonly accepted formalization of the notion of effective computability, which will be discussed in Chapter 6.

5.1 ACKERMANN'S FUNCTION

The definition schemes of Chapter 4 provide a great deal of freedom in the way in which new functions can be derived from given functions. But these schemes do not exhaust the ways in which functions can be defined recursively—defined, that is, by specifying "new" function values in terms of "previous" function values. There are many recursively defined and effectively computable functions that cannot be made to fit the schemes of Chapter 4 and that are consequently not primitive recursive. Among the best known of these functions is one named for W. Ackermann. As we shall see, Ackermann's function is both Turing computable and recursively definable, but not primitive recursive.

Introduction

One explanation for the fact that Ackermann's function is not primitive recursive is that it grows too fast to be primitive recursive. Since this point of view is instructive, let us explore it briefly on an informal basis before we introduce Ackermann's function itself. As we shall see, this informal discussion will lead quite naturally to the definition of Ackermann's function and will help explain the formal proof that Ackermann's function is not primitive recursive.

We begin by considering some one-variable counterparts of the familiar addition, multiplication, and exponentiation functions. Specifically, let β_1, β_2, and β_3 be the one-variable functions defined as follows:

$$\beta_1(x) = 2 + x \qquad \text{(addition of two)}$$
$$\beta_2(x) = 2 \cdot x \qquad \text{(multiplication by two)}$$
$$\beta_3(x) = 2^x \qquad \text{(exponentiation of two)}$$

Since addition, multiplication, and exponentiation can be obtained from the successor function by successive primitive recursions, we expect to find that β_1, β_2, and β_3 can be obtained in a similar way. If we let β_0 denote the successor function, we can write:

$$\beta_1(0) = 2; \qquad \beta_1(x^+) = \beta_0(\beta_1(x))$$
$$\beta_2(0) = 0; \qquad \beta_2(x^+) = \beta_1(\beta_2(x)) \qquad\qquad \text{(5-1a)}$$
$$\beta_3(0) = 1; \qquad \beta_3(x^+) = \beta_2(\beta_3(x))$$

The regularity exhibited by the second members of these pairs of defining equations suggests the possibility of extending the series to include functions β_4, β_5, β_6, etc., each defined in a similar way from its predecessor.

Following this suggestion, we define the function β_4 by primitive recursion from β_3, arbitrarily choosing the value of $\beta_4(0)$ to be 1. Thus

$$\beta_4(0) = 1; \qquad \beta_4(x^+) = \beta_3(\beta_4(x))$$

As a result of this definition, $\beta_4(1) = 2^{\beta_4(0)} = 2$, $\beta_4(2) = 2^{\beta_4(1)} = 2^2$, $\beta_4(3) = 2^{\beta_4(2)} = 2^{2^2}$, and so on. In general

$$\beta_4(x) = 2^{2^{\cdot^{\cdot^{2}}}} \Big\} x$$

where the total number of 2's is x. Thus the value of β_4 increases quite rapidly with x—certainly more rapidly than the value of β_3. (Note, for instance, that the value of $\beta_4(5)$ exceeds $10^{10,000}$.) We describe this situation by saying that the "growth rate" of β_4 exceeds that of β_3.

Functions β_5, β_6, β_7, etc. can be defined in the same way. In general, the function β_{n+1} is defined by primitive recursion from β_n as follows:

$$\beta_{n+1}(0) = 1; \qquad \beta_{n+1}(x^+) = \beta_n(\beta_{n+1}(x)) \qquad \text{(5-1b)}$$

Figure 5-1 shows the first few values of the functions β_0 through β_6. Note that the value of $\beta_5(x)$ can be expressed in the form

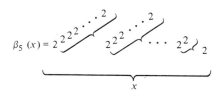

where the number of "levels" of exponentiation is x. However, the growth rates of the succeeding functions in the series increase so rapidly with n that it is very difficult even to describe the values of these functions. The growth rate of the function β_{10} (not to mention β_{100} or β_{1000}) is enough to strain the imagination.

Remember that each of the functions β_1, β_2, β_3, ... is primitive recursive, since each one is obtained from its predecessor by primitive recursion. Thus

x	0	1	2	3	4	5
β_0	1	2	3	4	5	6
β_1	2	3	4	5	6	7
β_2	0	2	4	6	8	10
β_3	1	2	4	8	16	32
β_4	1	2	4	16	32768	2^{32768}
β_5	1	2	4	32768	$2^{2^{\cdot^{\cdot^{2}}}} 32768$	
β_6	1	2	4	$2^{2^{\cdot^{\cdot^{2}}}} 32768$	$\beta_5\left(2^{2^{\cdot^{\cdot^{2}}}} 32768\right)$	

Figure 5-1

primitive recursive functions are capable of exhibiting extraordinarily large growth rates. Yet we can easily define a (nonprimitive recursive) function β that grows faster than any one of the functions $\beta_0, \beta_1, \ldots, \beta_n, \ldots$! All we need do is define β so that

$$\beta(x) = \beta_x(x). \tag{5-2}$$

Thus $\beta(0) = \beta_0(0) = 1$, $\beta(1) = \beta_1(1) = 3$, $\beta(2) = \beta_2(2) = 4$, $\beta(3) = \beta_3(3) = 8$, $\beta(4) = \beta_4(4) = 32768$, and so on.

To see informally why β grows faster than any of the functions $\beta_0, \beta_1, \ldots$, we first note that, if $m > n$, then $\beta_m(x) > \beta_n(x)$ for all x greater than three. (We do not attempt to prove this fact, but rely on Fig. 5-1 to establish its credibility.) Now compare the values of β with those of β_k, where β_k is any member of the series $\beta_0, \beta_1, \beta_2, \ldots$. As long as x exceeds both k and 3, we have $\beta(x) = \beta_x(x) > \beta_k(x)$. Thus the value of β eventually surpasses that of β_k. Indeed, the greater the value of x, the greater the disparity between $\beta(x)$ and $\beta_k(x)$. Since this is true for each choice of k, we conclude that β grows faster than any of the primitive recursive functions in the series $\beta_0, \beta_1, \beta_2, \ldots$.

Of course, we have not shown that β grows faster than *every* primitive recursive function. We must consider the possibility that some other primitive recursive function, not belonging to the series $\beta_0, \beta_1, \beta_2, \ldots$, has a growth rate exceeding even that of β. Now, how might such a fast-growing function be obtained from a sequence of compositions and primitive recursions? A little thought suggests that primitive recursion is much more effective at increasing growth rates than is composition, since recursion in essence "compounds" the growth rate of the function being defined at each new argument value. Thus for each primitive recursion function f obtained via a sequence of compositions and primitive recursions, we would expect to find another function obtained via primitive recursions alone whose growth rate equals or exceeds that of f. And since the functions $\beta_0, \beta_1, \beta_2, \ldots$ seem to make optimum use of primitive recursions for producing large growth rates, we might expect that the growth rate of each primitive recursive function f is equaled by that of some member of the series $\beta_0, \beta_1, \beta_2, \ldots$. This is in fact the case, and as a consequence the function β does grow faster than every primitive recursive function.

Ackermann's Function

The preceding informal discussion motivates our definition of Ackermann's function, which is closely related to the function β. We begin by defining a series of one-variable functions $\alpha_0, \alpha_1, \alpha_2, \ldots, \alpha_n, \ldots$, where α_0 is the successor function and α_{n+1} is obtained from α_n by primitive recursion, in essentially the same way that β_{n+1} was obtained from β_n. But in order to simplify the arguments that follow, we choose the values of $\alpha_1(0), \alpha_2(0), \ldots$ in a uniform way.

y	0	1	2	3	4	5
α_0	1	2	3	4	5	6
α_1	2	3	4	5	6	7
α_2	3	5	7	9	11	13
α_3	5	13	29	61	125	253
α_4	13	32765	$\cdots$			
α_5	32765	$\cdots$				

Figure 5-2

Specifically, we define α_{n+1} by the equations:

$$\alpha_{n+1}(0) = \alpha_n(1)$$
$$\alpha_{n+1}(y^+) = \alpha_n(\alpha_{n+1}(y)) \tag{5-3}$$

Figure 5-2 shows the first few values of the functions α_0 through α_5.

Rather than now defining a one-variable function α such that $\alpha(x) = \alpha_x(x)$, we propose to define a two-variable function A such that $A(x, y) = \alpha_x(y)$. Making use of Eqs. (5-3) and the fact that α_0 is the successor function, we see that A must satisfy the equations:

$$A(0, y) = y^+ \tag{5-4a}$$

$$A(x^+, 0) = A(x, 1) \tag{5-4b}$$

$$A(x^+, y^+) = A(x, A(x^+, y)) \tag{5-4c}$$

We take these equations to be the formal definition of the function A, which we will call *Ackermann's function.*[†]

We will not bother to prove formally that Eqs. (5-4) uniquely define a function, since the informal derivation of A from the functions $\alpha_0, \alpha_1, \alpha_2, \ldots$ makes this point intuitively clear. But it is worth noting that A is effectively computable and, in particular, that the value of $A(x, y)$ can be obtained by a finite number of applications of Eqs. (5-4). Although we postpone a formal

---

[†] In the literature, the name "Ackermann's function" is sometimes applied to certain variants of A, such as the one-variable function A_* defined so that $A_*(x) = A(x, x)$.

proof of this fact to Section 5.2, the argument, which proceeds by induction on x, is easily outlined. Certainly $A(0, y)$ can be effectively evaluated for all y. Assume that for some fixed m, $A(m, y)$ can be effectively evaluated for all y. Since repeated applications of Eqs. (5-4c) and (5-4b) allow us to express $A(m^+, y)$ solely in terms of (many) values of the form $A(m, y)$, it follows that $A(m^+, y)$ can also be effectively evaluated. Thus A is effectively computable.

To show that A is not primitive recursive, we need several preliminary results concerning the values of A. The proofs of these results are routine and can be found at the end of this section.

Lemma 5-1. $A(x, y) > y$.

Lemma 5-2. $A(x, y + 1) > A(x, y)$.

Corollary 5-2.1. If $y_2 > y_1$, then $A(x, y_2) > A(x, y_1)$.

Lemma 5-3. $A(x + 1, y) \geq A(x, y + 1)$.

Corollary 5-3.1. $A(x, y) > x$.

Corollary 5-3.2. If $x_2 > x_1$, then $A(x_2, y) > A(x_1, y)$.

Lemma 5-4. $A(x + 2, y) > A(x, 2y)$.

Thus the value of A is greater than that of either of its arguments, A is strictly monotonic in each argument, and the value of x has a greater influence on the value of A than does the value of y.

As suggested earlier, we will prove that Ackermann's function is not primitive recursive by showing that it "grows faster" than any primitive recursive function. This means that we need a precise way of comparing the "growth rate" of the two-variable function A with that of an arbitrary n-variable function. What we shall attempt to do is show that for any given n-variable primitive recursive function f, there exists a natural number k such that

$$A(k, \max(x_1, \ldots, x_n)) > f(x_1, \ldots, x_n) \tag{5-5}$$

for all values of $x_1, \ldots, x_n$.

The proof is accomplished by induction on the number of compositions and primitive recursions needed to define the function f. In order to carry out the induction step of the proof, we need two auxiliary results. The first of these results ensures that if the functions $g_1, \ldots, g_m$ and h satisfy Eq. (5-5), so does the function f obtained from $g_1, \ldots, g_m$ and h by functional composition.

Lemma 5-5. *Let the n-variable function $f = h \circ (g_1, \ldots, g_m)$ be obtained from the functions $g_1, \ldots, g_m$ and h by composition. Assume the existence of natural numbers $k_1, \ldots, k_m$, and k_0 such that*

$$A(k_i, \max(x_1, \ldots, x_n)) > g_i(x_1, \ldots, x_n) \qquad for \qquad 1 \leq i \leq m$$

and

$$A(k_0, \max(y_1, \ldots, y_m)) > h(y_1, \ldots, y_m)$$

for all $x_1, \ldots, x_n$ *and* $y_1, \ldots, y_m$. *Define* k *to be the natural number* $\max(k_0, k_1, \ldots, k_m) + 2$. *Then* $A(k, \max(x_1, \ldots, x_n)) > f(x_1, \ldots, x_n)$ *for all* $x_1, \ldots, x_n$.

Proof. For convenience, let $\max(x_1, \ldots, x_n)$ be denoted $\hat{x}$. Since $k \geq 2$, we can apply Lemma 5-3 and the definition of A to obtain

$$A(k, \hat{x}) \geq A(k - 1, \hat{x} + 1) = A(k - 2, A(k - 1, \hat{x})).$$

Since $k - 1$ is greater than each k_i, it follows from the monotonicity of A and the assumed property of k_i that

$$A(k - 1, \hat{x}) > A(k_i, \hat{x}) > g_i(x_1, \ldots, x_n)$$

for $i = 1, \ldots, m$. Again because of the monotonicity of A, this means that $A(k, \hat{x}) > A(k - 2, g_i(x_1, \ldots, x_n))$ and hence that

$$A(k, \hat{x}) > A(k - 2, \max(g_1(x_1, \ldots, x_n), \ldots, g_m(x_1, \ldots, x_n))).$$

Since $k - 2 \geq k_0$, it then follows that

$$A(k, \hat{x}) > A(k_0, \max(g_1(x_1, \ldots, x_n), \ldots, g_m(x_1, \ldots, x_n)))$$
$$> h(g_1(x_1, \ldots, x_n), \ldots, g_m(x_1, \ldots, x_n))$$
$$= f(x_1, \ldots, x_n). \qquad \square$$

The analogous result for primitive recursion is given by:

Lemma 5-6. *Let the* $(n + 1)$-*variable function* f *be defined by primitive recursion from the* n-*variable function* g *and the* $(n + 2)$-*variable function* h, *so that*

$$f(x_1, \ldots, x_n, 0) = g(x_1, \ldots, x_n)$$
$$f(x_1, \ldots, x_n, y^+) = h(x_1, \ldots, x_n, y, f(x_1, \ldots, x_n, y)).$$

Assume the existence of natural numbers k_g *and* k_h *such that*

$$A(k_g, \max(x_1, \ldots, x_n)) > g(x_1, \ldots, x_n)$$

and

$$A(k_h, \max(x_1, \ldots, x_n, y, z)) > h(x_1, \ldots, x_n, y, z)$$

for all $x_1, \ldots, x_n$, y, *and* z. *Define* k *to be the natural number* $\max(k_g, k_h) + 3$. *Then* $A(k, \max(x_1, \ldots, x_n, y)) > f(x_1, \ldots, x_n, y)$ *for all* $x_1, \ldots, x_n, y$.

Proof. Again let $\hat{x}$ denote the quantity $\max(x_1, \ldots, x_n)$. We first establish the fact that $A(k - 2, \hat{x} + y) > f(x_1, \ldots, x_n, y)$. The argument proceeds by induction on y.

> *Basis.* Since $k - 2 > k_g$, it follows that $A(k - 2, \hat{x} + 0) = A(k - 2, \hat{x}) > A(k_g, \hat{x}) > g(x_1, \ldots, x_n) = f(x_1, \ldots, x_n, 0)$.

Induction step. Assume that $A(k - 2, \hat{x} + y) > f(x_1, \ldots, x_n, y)$ for some arbitrary choice of y. Since $A(k - 2, \hat{x} + y)$ is necessarily greater than $\hat{x} + y$ (and therefore greater than either $\hat{x}$ or y), this means that

$$A(k - 2, \hat{x} + y) > \max(x_1, \ldots, x_n, y, f(x_1, \ldots, x_n, y)).$$

The definition of A, monotonicity, and the fact that $k - 3 \geq k_h$ now imply that

$$
\begin{aligned}
A(k - 2, \hat{x} + y^+) &= A(k - 3, A(k - 2, \hat{x} + y)) \\
&> A(k_h, \max(x_1, \ldots, x_n, y, f(x_1, \ldots, x_n, y))) \\
&> h(x_1, \ldots, x_n, y, f(x_1, \ldots, x_n, y)) \\
&= f(x_1, \ldots, x_n, y^+).
\end{aligned}
$$

Thus $A(k - 2, \hat{x} + y) > f(x_1, \ldots, x_n, y)$ for all y. From this it follows via Lemma 5-4 that

$$
\begin{aligned}
A(k, \max(x_1, \ldots, x_n, y)) &= A(k, \max(\hat{x}, y)) \\
&> A(k - 2, 2\max(\hat{x}, y)) \\
&\geq A(k - 2, \hat{x} + y) \\
&> f(x_1, \ldots, x_n, y)
\end{aligned}
$$

for all $x_1, \ldots, x_n$, and y. $\square$

Lemmas 5-5 and 5-6 together yield:

Theorem 5-1. *For each n-variable primitive recursive function f, there exists a natural number k such that* $A(k, \max(x_1, \ldots, x_n)) > f(x_1, \ldots, x_n)$, *for all* $x_1, \ldots, x_n$.

Proof. By induction on the number of compositions and primitive recursions needed to define f. As before, we use $\hat{x}$ to denote $\max(x_1, \ldots, x_n)$.

Basis. If the derivation of f requires no compositions or primitive recursions, three cases are possible.

1. If f is the constant function whose value is c, choose $k = c$. Corollary 5-3.1 then guarantees that $A(k, \hat{x}) = A(c, \hat{x}) > c = f(x_1, \ldots, x_n)$.

2. If f is the projection function whose value is x_i, choose $k = 0$. Then $A(k, \hat{x}) = A(0, \hat{x}) = \hat{x} + 1 > x_i = f(x_1, \ldots, x_n)$.

3. If f is the successor function, choose $k = 1$. Then $A(k, x) = A(1, x) > A(0, x) = x + 1 = f(x)$.

Induction step. Assume the statement of the theorem to be true for all functions requiring w or fewer compositions and primitive recursions. Let

f be a function requiring a total of $w + 1$ compositions and primitive recursions. Two cases are possible.

1. If f is derived from the functions $g_1, \ldots, g_m$ and h by composition, the induction hypothesis must apply to each of $g_1, \ldots, g_m$, and h. Lemma 5-5 then guarantees the existence of a number k such that $A(k, \hat{x}) > f(x_1, \ldots, x_n)$.

2. If f is derived from the functions g and h by primitive recursion, the induction hypothesis must apply to g and h. Lemma 5-6 then guarantees the existence of a number k such that $A(k, \hat{x}) > f(x_1, \ldots, x_n)$.   □

Theorem 5-1 provides a formal expression of the notion that A "grows faster" than any primitive recursive function. It is now a simple matter to establish:

Theorem 5-2.   *Ackermann's function is not primitive recursive.*

Proof.   Assume that Ackermann's function is primitive recursive. Then according to Theorem 5-1, there must exist a natural number k such that

$$A(k, \max (x, y)) > A(x, y)$$

for all x and y. Setting $x = y = k$ then yields the contradiction

$$A(k, k) > A(k, k)$$

from which we conclude that A cannot be primitive recursive.   □

Ackermann's function is one of the simplest examples of a recursively defined (and effectively computable) total function that is not primitive recursive. In the following sections we will consider ways of expanding the class of primitive recursive functions so as to include such recursively defined functions as Ackermann's function.

The Monotonicity Properties

For the sake of completeness, we now provide formal proofs of Lemmas 5-1 through 5-4 and their corollaries.

Lemma 5-1. $A(x, y) > y$.

Proof.   By induction on x.

Basis. $x = 0$. Then $A(x, y) = A(0, y) = y + 1 > y$.

Induction step.   Assume, as "induction hypothesis one," that $A(n, y) > y$ for an arbitrary choice of n. To show that $A(n^+, y) > y$, we resort to a second induction argument on the variable y.

> *Basis.* $y = 0$. Then $A(n^+, y) = A(n^+, 0) = A(n, 1)$. But by induction hypothesis one, $A(n, 1) > 1 > 0$. Thus $A(n^+, 0) > 0$.

Induction step. Assume, as "induction hypothesis two," that $A(n^+, m) > m$ for an arbitrary choice of m. Then $A(n^+, m^+) = A(n, A(n^+, m))$, by definition of A. According to induction hypothesis one, $A(n, A(n^+, m)) > A(n^+, m)$. According to induction hypothesis two, $A(n^+, m) > m$ or, equivalently, $A(n^+, m) \geq m^+$. Therefore $A(n^+, m^+) > m^+$.

Thus $A(n^+, y) > y$ for all y, and the induction step is completed.   ☐

Lemma 5-2. $A(x, y + 1) > A(x, y)$.

Proof. By induction on x.

Basis. If $x = 0$, we have $A(0, y + 1) = y + 2 > y + 1 = A(0, y)$, by definition of A.

Induction step. Assume that $A(n, y + 1) > A(n, y)$. Making use of the definition of A and Lemma 5-1, we have

$$A(n^+, y + 1) = A(n, A(n^+, y)) > A(n^+, y). \qquad ☐$$

Corollary 5-2.1. If $y_2 > y_1$, then $A(x, y_2) > A(x, y_1)$.

Proof. By induction on the variable $y_2 - y_1 - 1$, using Lemma 5-2 at each step.   ☐

Lemma 5-3. $A(x + 1, y) \geq A(x, y + 1)$.

Proof. By induction on y.

Basis. If $y = 0$, then $A(x + 1, y) = A(x + 1, 0) = A(x, 1) = A(x, y + 1)$.

Induction step. Assume that $A(x + 1, n) \geq A(x, n + 1)$ and consider the value of $A(x + 1, n^+)$. By definition, $A(x + 1, n^+) = A(x, A(x + 1, n))$. According to the induction hypothesis and Lemma 5-1, $A(x + 1, n) \geq A(x, n^+) \geq n^+ + 1$. Corollary 5-2.1 therefore guarantees that $A(x, A(x + 1, n)) \geq A(x, n^+ + 1)$, from which we conclude that $A(x + 1, n^+) \geq A(x, n^+ + 1)$.   ☐

Corollary 5-3.1. $A(x, y) > x$.

Proof. Using Lemma 5-3 in a simple induction argument, we get $A(x, y) \geq A(0, x + y)$. But $A(0, x + y) = x + y + 1 > x$.   ☐

Corollary 5-3.2. If $x_2 > x_1$, then $A(x_2, y) > A(x_1, y)$.

Proof. First note that $A(x + 1, y) \geq A(x, y + 1) > A(x, y)$, then use a straightforward induction argument.   ☐

Lemma 5-4. $A(x + 2, y) > A(x, 2y)$.

Proof. By induction on y.

Basis. If $y = 0$, then $A(x + 2, y) = A(x + 2, 0) > A(x, 0) = A(x, 2y)$.

Induction step. Assume that $A(x + 2, n) > A(x, 2n)$, and consider the value of $A(x + 2, n^+)$. By definition $A(x + 2, n^+) = A(x + 1, A(x + 2, n))$, and by hypothesis $A(x + 2, n) > A(x, 2n)$. Corollary 5-2.1 therefore ensures that

$$A(x + 2, n^+) > A(x + 1, A(x, 2n)).$$

Applying Lemma 5-3 to $A(x + 1, A(x, 2n))$ yields

$$A(x + 2, n^+) > A(x, A(x, 2n) + 1).$$

According to Lemma 5-1, $A(x, 2n) \geq 2n + 1$ or, equivalently, $A(x, 2n) + 1 \geq 2n + 2 = 2(n^+)$. Applying Corollary 5-2.1 once more now yields

$$A(x + 2, n^+) > A(x, 2n^+). \qquad \square$$

5.2 μ-RECURSIVE FUNCTIONS

Although the class of primitive recursive functions contains a great many functions of practical interest, it does not include all the Turing-computable or effectively computable functions. As we have just seen, it does not even include all the effectively computable total functions. It is therefore natural to ask how the class of primitive recursive functions can be extended so as to admit a larger number of effectively computable functions.

One approach, which we pursue in this section, is to remove the upper bound from the minimalization operator. The resulting composition rule leads us to define a new class of effectively computable functions called the μ-recursive functions. With the help of an important technique known as arithmetization, it is possible to show that Ackermann's function is a μ-recursive function. The same technique can also be used to show that every Turing-computable function is μ-recursive and consequently that the class of μ-recursive functions is identical to the class of Turing-computable functions.

Definitions and Basic Properties

Suppose that g is a total function of $n + 1$ variables. Extending the notation of Chapter 4 in an obvious way, we write

$$f(x_1, \ldots, x_n) = \mu z[g(x_1, \ldots, x_n, z) = 0] \qquad (5\text{-}6)$$

to define a new function f whose values are determined as follows. If there exists a number z such that $g(x_1, \ldots, x_n, z) = 0$, the value of $f(x_1, \ldots, x_n)$ is the smallest such z; if there is no number z such that $g(x_1, \ldots, x_n, z) = 0$, the value of $f(x_1, \ldots, x_n)$ is undefined. This function f is said to be obtained from the function g by (*unbounded*) *minimalization*.

Unlike the bounded minimalization operation, unbounded minimalization will in general yield nontotal functions, even though the functions to which it is applied are required to be total. In fact, unbounded minimalization provides one of our most important means of defining nontotal functions. But it can also be useful in defining total functions, especially when it is difficult or impossible to find an appropriate limit to be used in a bounded minimalization. Of course, the use of unbounded minimalization in this way does not necessarily preserve primitive recursiveness.

Example 5-1

a) Suppose that g is the total two-variable function whose values are given by the equation $g(x, z) = |x - 3z|$. Let the function f be defined from g by unbounded minimalization, thus:

$$f(x) = \mu z[g(x, z) = 0]$$

Then f is the function whose values are given by the equation:

$$f(x) = \begin{cases} x/3 & \text{if } x \text{ is a multiple of } 3 \\ \uparrow & \text{otherwise} \end{cases}$$

b) Suppose that g is a total, one-variable function that assumes the value 0 for an infinite number of argument values. Let g' denote the two-variable function defined so that:

$$g'(x, z) = g(z) + (x \dotminus z)$$

Since g' is necessarily total, we may use it to define a new function f by unbounded minimalization, as follows:

$$f(x) = \mu z[g'(x, z) = 0]$$

A little thought shows that f is a total function whose value for argument x is the smallest z greater than or equal to x for which $g(z) = 0$. Note that f cannot be defined by bounded minimalization since we do not have enough information about g to bound the desired value of z. All we know is that for each x there must exist some $z \geq x$ for which $g(z) = 0$. $\quad\square$

We intend to define the μ-recursive functions to be those functions that can be obtained from the constant, projection, and successor functions by means of functional composition, primitive recursion, and unbounded minimalization. Before doing this, we need to understand the effect of using nontotal functions in a definition by primitive recursion.

Suppose that g and h are given partial functions of n and $n + 2$ variables, and that the new function f is specified in terms of g and h according to the usual format for a primitive recursion, namely:

$$f(x_1, \ldots, x_n, 0) = g(x_1, \ldots, x_n)$$

$$f(x_1, \ldots, x_n, y^+) = h(x_1, \ldots, x_n, y, f(x_1, \ldots, x_n, y))$$

Evidently $f(x_1, \ldots, x_n, 0)$ will be defined whenever $g(x_1, \ldots, x_n)$ is, $f(x_1, \ldots, x_n, 1)$ will be defined whenever both $f(x_1, \ldots, x_n, 0)$ and $h(x_1, \ldots, x_n, 0, f(x_1, \ldots, x_n, 0))$ are, $f(x_1, \ldots, x_n, 2)$ will be defined whenever both $f(x_1, \ldots, x_n, 1)$ and $h(x_1, \ldots, x_n, 1, f(x_1, \ldots, x_n, 1))$ are, and so on. Proceeding inductively, we see that $f(x_1, \ldots, x_n, y^+)$ will be defined iff all the values $f(x_1, \ldots, x_n, 0)$, $\ldots, f(x_1, \ldots, x_n, y)$ and $h(x_1, \ldots, x_n, y, f(x_1, \ldots, x_n, y))$ are defined. Consequently if $f(x_1, \ldots, x_n, y)$ is undefined for some value of y, it must be undefined for all succeeding values of y. However, when $f(x_1, \ldots, x_n, y)$ is defined, its value is determined by the values of g and h in the usual way.

Thus primitive recursion, like functional composition, can be applied to nontotal as well as total functions. This fact permits us to define the μ-recursive functions inductively, as follows.

1. The constant, projection, and successor functions are all μ-recursive.

2. If $g_1, \ldots, g_m$ are n-variable μ-recursive functions and h is an m-variable μ-recursive function, then the composite function $f = h \circ (g_1, \ldots, g_m)$ is also μ-recursive.

3. If g and h are n- and $(n + 2)$-variable μ-recursive functions, then the function f defined from g and h by primitive recursion is also μ-recursive.

4. If g is a total $(n + 1)$-variable μ-recursive function, then the function f defined from g by unbounded minimalization is also μ-recursive.

No functions other than those that can be defined by means of a finite number of applications of rules (1) through (4) are μ-recursive.

It should be clear that all primitive recursive functions are μ-recursive, since both composition and primitive recursion are permitted in the definition of μ-recursive functions. What remains to be determined is how the μ-recursive functions are related to such classes as the Turing-computable functions or the effectively computable functions. These relationships will occupy most of our attention for the remainder of this section.

We begin by observing that the μ-recursive functions are effectively computable. This follows from the fact that functions defined by composition, primitive recursion, or minimalization from effectively computable functions must themselves be effectively computable. If f is defined from the effectively computable functions $g_1, \ldots, g_m$ and h by composition, then $f(x_1, \ldots, x_n)$ may be determined by evaluating $g_1(x_1, \ldots, x_n), \ldots, g_m(x_1, \ldots, x_n)$ and, if each of these computations terminates, evaluating h for the resulting argument values.

If f is defined from the effectively computable functions g and h by primitive recursion, $f(x_1, \ldots, x_n, y)$ may be determined by first evaluating $g(x_1, \ldots, x_n)$ to obtain $f(x_1, \ldots, x_n, 0)$. Assuming that computation terminates, we then evaluate $h(x_1, \ldots, x_n, 0, f(x_1, \ldots, x_n, 0))$ to obtain $f(x_1, \ldots, x_n, 1)$; if that computation terminates, we then evaluate $h(x_1, \ldots, x_n, 1, f(x_1, \ldots, x_n, 1))$ to obtain $f(x_1, \ldots, x_n, 2)$, and so on. And if f is defined from the effectively computable total function g by minimalization, $f(x_1, \ldots, x_n)$ may be determined by successively evaluating $g(x_1, \ldots, x_n, 0)$, $g(x_1, \ldots, x_n, 1)$, $g(x_1, \ldots, x_n, 2)$, and so on, until such time as a value of z is found for which $g(x_1, \ldots, x_n, z) = 0$.[†]

As might be expected, it is often convenient to use predicates in the definition of functions by unbounded minimalization. Extending the practice introduced for bounded minimalization, we write

$$f(x_1, \ldots, x_n) = \mu z[P(x_1, \ldots, x_n, z)] \tag{5-7}$$

to define the function f whose value is that of the smallest z for which $P(x_1, \ldots, x_n, z)$ holds or is undefined if no such z exists. If we now define a μ-*recursive predicate* to be a predicate whose characteristic function is μ-recursive, it is a simple matter to show that the function f specified in Eq. (5-7) is μ-recursive whenever the predicate P is. (In practice, we will never need to use anything but primitive recursive predicates in definitions by minimalization.)

We eventually wish to show that the class of μ-recursive functions is identical to the class of Turing-computable functions. For the time being, we content ourselves with the easy half of the proof, namely showing that all μ-recursive functions are Turing computable. We begin with two preliminary results.

Lemma 5-7. *Let the function f be defined from the Turing-computable functions g and h by primitive recursion. Then f is also Turing computable.*

Argument. The argument is virtually the same as that given for Theorem 4-2. The only difference is that the functions g and h are now permitted to be nontotal. However, it is easily seen that the construction used in Chapter 4 works equally well for nontotal functions. □

Lemma 5-8. *Let g be any total $(n + 1)$-variable Turing-computable function, and let f be defined from g by minimization, so that:*

$$f(x_1, \ldots, x_n) = \mu z[g(x_1, \ldots, x_n, z) = 0]$$

Then f is also Turing computable.

Argument. Since g is Turing computable, there must be some Turing machine **T** that evaluates g in the format described in Lemma 3-1. Then a machine **T̂**

[†] Note that because g is a total function, there is no danger of the computation of $f(x_1, \ldots, x_n)$ being "trapped" in the evaluation of an undefined value of g. It is just this consideration that motivates the requirement that the function g be total in definitions by minimalization.

that evaluates the function f can be designed to operate as follows. When presented with the initial tape pattern

$$\underline{0} \; \bar{x}_1 \; 0 \cdots 0 \; \bar{x}_n \, 0 \cdots$$

$\hat{\mathbf{T}}$ first appends to this pattern a block consisting of a single 1. The tape now contains the pattern

$$\underline{0} \; \bar{x}_1 \; 0 \cdots 0 \; \bar{x}_n \, 0 \, \bar{z} \, 0 \cdots$$

where $z = 0$. The machine $\hat{\mathbf{T}}$ now acts like the machine $\mathbf{T}$, evaluating $g(x_1, \ldots, x_n, z)$ and producing the pattern:

$$0 \; \bar{x}_1 \; \; 0 \cdots 0 \; \bar{x}_n \, 0 \, \bar{z} \, \underline{0} \; \overline{g(x_1, \ldots, x_n, z)} \, 0 \cdots$$

This evaluation is guaranteed to terminate because g is total. The machine $\hat{\mathbf{T}}$ now examines the block $\overline{g(x_1, \ldots, x_n, z)}$. If this block represents the natural number 0, z is the desired value of $f(x_1, \ldots, x_n)$, and $\hat{\mathbf{T}}$ halts in front of the z-block. Otherwise, $\hat{\mathbf{T}}$ erases the block $g(x_1, \ldots, x_n, z)$, increases the length of the z-block by one, and evaluates g for the new value of z. This process of evaluating g, examining the result, and incrementing z is repeated until such time as z may reach a value for which $g(x_1, \ldots, x_n, z) = 0$, at which point $\hat{\mathbf{T}}$ halts with $\bar{z}$ as its output. If no such value of z exists, $\hat{\mathbf{T}}$ never halts. In this way $\hat{\mathbf{T}}$ does in fact compute $f(x_1, \ldots, x_n)$.   □

We now have everything that is needed to establish:

Theorem 5-3.   *Every μ-recursive function f is Turing computable.*

Proof. The proof is accomplished by induction on the number of compositions, primitive recursions, and minimalizations needed to define the function f. The basis of the proof is established by noting that the constant, projection, and successor functions are all Turing computable. The induction step follows easily from Theorem 3-2 and Lemmas 5-7 and 5-8.   □

The μ-Recursiveness of Ackermann's Function

We now propose to show that Ackermann's function is μ-recursive. The first part of the job is to devise a scheme whereby each step in an evaluation of Ackermann's function can be represented by a single natural number. We will then show that the numbers associated with successive steps in an evaluation can be obtained from a primitive recursive function and that the result of the evaluation can be extracted by a suitably defined μ-recursive function. We begin with the problem of representing evaluations numerically.

Recall that Ackermann's function is defined by three equations, or *rules*, as we will refer to them here:

Rule 1 $A(0, y) = y^+$

Rule 2 $A(x^+, 0) = A(x, 1)$

Rule 3 $A(x^+, y^+) = A(x, A(x^+, y))$

As the next example illustrates, successive applications of these defining rules can be used to evaluate Ackermann's function for any choice of argument values.

Example 5-2. To evaluate A(2, 1), we proceed as follows:

$$
\begin{array}{lll}
A(2,\ 1) & = A(1,\ A(2,\ 0)) & \text{Rule 3} \\
& = A(1,\ A(1,\ 1)) & \text{Rule 2} \\
& = A(1,\ A(0,\ A(1,\ 0))) & \text{Rule 3} \\
& = A(1,\ A(0,\ A(0,\ 1))) & \text{Rule 2} \\
& = A(1,\ A(0,\ 2)) & \text{Rule 1} \\
& = A(1,\ 3) & \text{Rule 1} \\
& = A(0,\ A(1,\ 2)) & \text{Rule 3} \\
& = A(0,\ A(0,\ A(1,\ 1))) & \text{Rule 3} \\
& = A(0,\ A(0,\ A(0,\ A(1,\ 0)))) & \text{Rule 3} \\
& = A(0,\ A(0,\ A(0,\ A(0,\ 1)))) & \text{Rule 2} \\
& = A(0,\ A(0,\ A(0,\ 2))) & \text{Rule 1} \\
& = A(0,\ A(0,\ 3)) & \text{Rule 1} \\
& = A(0,\ 4) & \text{Rule 1} \\
& = 5 & \text{Rule 1} \quad\square
\end{array}
$$

Three facts are readily apparent from a study of Rules 1 to 3 and Example 5-2. First, each step in the evaluation of $A(x,\ y)$ results either in a single natural number (i.e., the desired answer), or else in a nested expression of the form

$$A(w_1,\ A(w_2,\ A(w_3,\ \ldots\ A(w_{k-1},\ w_k)\ \cdots))).$$

Second, exactly one of the three rules is applicable at any given step. And third, no matter which rule applies, it must always be applied to the rightmost A in the given expression—i.e., it must always be applied at the innermost level of the nest.[†]

Perhaps less obvious than these observations is the fact that the evaluation procedure just outlined must always terminate after a finite number of steps. Let us therefore prove this fact.

Lemma 5-9. *For each choice of natural numbers x and y, the expression* $A(x,\ y)$ *can be reduced to a single natural number by means of a finite number of applications of Rules 1, 2, and 3.*

Proof. By induction on x.

Basis. Suppose $x = 0$. Then a single application of Rule 1 to $A(0,\ y)$ yields the result y^+.

[†] A more careful justification of these statements requires an inductive proof in which the induction variable ranges over the steps in an evaluation.

Induction step. Assume, as hypothesis H_1, that the evaluation of $A(m, y)$ eventually terminates for every choice of y. We now show, by induction on y, that the evaluation of $A(m^+, y)$ must eventually terminate.

> *Basis.* Suppose $y = 0$. Applying Rule 2 to $A(m^+, 0)$ yields the expression $A(m, 1)$. Then according to the induction hypothesis H_1, a finite number of additional steps yields an evaluation of $A(m, 1)$.

> *Induction step.* Assume, as hypothesis H_2, that the evaluation of $A(m^+, n)$ terminates in a finite number of steps, and consider the evaluation of $A(m^+, n^+)$. One application of Rule 3 yields the expression $A(m, A(m^+, n))$. By hypothesis H_2, a finite number of additional steps serves to evaluate $A(m^+, n)$—i.e. to reduce $A(m, A(m^+, n))$ to an expression of the form $A(m, w)$. Hypothesis H_1 then ensures that another finite number of steps suffices to evaluate $A(m, w)$. Thus $A(m^+, n^+)$ can indeed be evaluated in a finite number of steps.

This completes the (main) induction step, and hence the proof. □

Let us now devise a method of representing the steps of an evaluation numerically. Remember that each step consists of an expression of the form

$$A(w_1, A(w_2, \ldots, A(w_{k-1}, w_k) \cdots)).$$

Since the terms in such an expression are always associated to the right, most of the punctuation can be omitted without loss of information. In particular, the expression $A(w_1, A(w_2, \ldots, A(w_{k-1}, w_k) \cdots))$ can be unambiguously represented by the k-tuple

$$(w_1, \ldots, w_{k-1}, w_k).$$

Thus each step in the evaluation of Ackermann's function can be described by a tuple of natural numbers.

We next use a Gödel-numbering scheme to reduce the description of each step in an evaluation to a single natural number. In particular, we choose to represent the tuple $(w_1, \ldots, w_k)$ by the natural number

$$2^k 3^{w_1} \cdots p_k^{w_k}.$$

Note that the length of the tuple is encoded as the exponent of two, while the components of the tuple are encoded as the exponents of the succeeding primes. This provides a simple way of telling how many components to "look for" in a given Gödel number.

This scheme allows us to describe the evaluation of Ackermann's function with a sequence of natural numbers. Such a numbering scheme is referred to as an *arithmetization* of the evaluation of Ackermann's function. The details of the scheme are illustrated in the following example. Note that the steps of the evaluation are numbered, starting with zero, and that the zeroth step corresponds to the expression $A(x, y)$ whose value is to be determined.

Example 5-3. The evaluation exhibited in Example 5-2 can be described numerically as follows.

Step	Tuple	Gödel Number	
0	(2, 1)	$2^2 3^2 5^1$	= 180
1	(1, 2, 0)	$2^3 3^1 5^2 7^0$	= 600
2	(1, 1, 1)	$2^3 3^1 5^1 7^1$	= 840
3	(1, 0, 1, 0)	$2^4 3^1 5^0 7^1 11^0$	= 336
4	(1, 0, 0, 1)	$2^4 3^1 5^0 7^0 11^1$	= 528
5	(1, 0, 2)	$2^3 3^1 5^0 7^2$	= 1176
6	(1, 3)	$2^2 3^1 5^3$	= 1500
7	(0, 1, 2)	$2^3 3^0 5^1 7^2$	= 1960
8	(0, 0, 1, 1)	$2^4 3^0 5^0 7^1 11^1$	= 1232
9	(0, 0, 0, 1, 0)	$2^5 3^0 5^0 7^0 11^1 13^0$	= 352
10	(0, 0, 0, 0, 1)	$2^5 3^0 5^0 7^0 11^0 13^1$	= 416
11	(0, 0, 0, 2)	$2^4 3^0 5^0 7^0 11^2$	= 1936
12	(0, 0, 3)	$2^3 3^0 5^0 7^3$	= 2744
13	(0, 4)	$2^2 3^0 5^4$	= 2500
14	(5)	$2^1 3^5$	= 486 □

We now define the three-variable function ψ so that $\psi(x, y, n)$ is the Gödel number of the nth step in the evaluation of $A(x, y)$. For instance, we see from Example 5-3 that $\psi(2, 1, 8) = 1232$. In order to make ψ a total function, we agree to "extend" each evaluation of Ackermann's function as follows. If the actual evaluation of $A(x, y)$ terminates at step n_0 and this step is assigned the Gödel number z, we simply set $\psi(x, y, n) = z$ for all $n \geq n_0$. Thus $\psi(2, 1, 15) = \psi(2, 1, 16) = \psi(2, 1, 17) = \cdots = 486$. The resulting function ψ will be referred to as the *trace function* for the evaluation of A.

Our next job is to show that ψ is a primitive recursive function. For this purpose it is convenient to be able to view every natural number z as being a Gödel number of some tuple. The simplest approach is to ignore any "extra" primes in the decomposition of z. Thus the natural number

$$2^k 3^{w_1} 5^{w_2} \cdots p_k^{w_k} p_{k+1}^{w_{k+1}} \cdots p_m^{w_m}$$

will be treated as if it were the Gödel number of the k-tuple $(w_1, \ldots, w_k)$.

With this convention in mind, we now assign an auxiliary predicate and an auxiliary function to each of the rules defining Ackermann's function. For each of the values $i = 1$, $i = 2$, and $i = 3$, let Q_i denote the predicate such that:

$Q_i(z) \Leftrightarrow z$ is a Gödel number of a tuple to which Rule i applies

And let h_i denote a function such that, if $Q_i(z)$ holds, then:

$h_i(z)$ is the Gödel number of the tuple that results when Rule i is applied to the tuple represented by the number z.

If $Q_i(z)$ does not hold—i.e., if z does not represent a tuple to which Rule i applies, the value assumed by $h_i(z)$ is unimportant. (We will in fact choose these "default values" so as to ensure the primitive recursiveness of h_i.)

The predicates Q_1, Q_2, Q_3 and the functions h_1, h_2, h_3 provide an easy way of specifying the trace function ψ. Since $\psi(x, y, 0)$ is just the Gödel number of the tuple (x, y), we have:

$$\psi(x, y, 0) = 2^2 3^x 5^y \tag{5-8a}$$

And since $\psi(x, y, n^+)$ can be obtained from $\psi(x, y, n)$ by determining which rule applies to the tuple represented by $\psi(x, y, n)$ and then computing the effect of applying that rule, we can write:

$$\psi(x, y, n^+) = \begin{cases} h_1(\psi(x, y, n)) & \text{if} \quad Q_1(\psi(x, y, n)) \\ h_2(\psi(x, y, n)) & \text{if} \quad Q_2(\psi(x, y, n)) \\ h_3(\psi(x, y, n)) & \text{if} \quad Q_3(\psi(x, y, n)) \\ \psi(x, y, n) & \text{otherwise} \end{cases} \tag{5-8b}$$

Thus to show that ψ is primitive recursive, it is only necessary to show that the predicates Q_1, Q_2, Q_3 and the functions h_1, h_2, h_3 are primitive recursive.

To this end, we introduce one more auxiliary function L, whose value for argument z is the length of the tuple represented by z. Since

$$L(z) = E(0, z)$$

the function L is certainly primitive recursive.

Now, the predicate $Q_1(z)$ is to hold iff z represents a tuple containing at least two components, of which the next-to-last is zero. Thus we can write

$$Q_1(z) \iff (L(z) > 1) \wedge (E(L(z) \dot- 1, z) = 0).$$

Similar reasoning yields the following definitions of Q_2 and Q_3, as the reader may verify.

$$Q_2(z) \iff (L(z) > 1) \wedge (E(L(z) \dot- 1, z) \neq 0) \wedge (E(L(z), z) = 0)$$

$$Q_3(z) \iff (L(z) > 1) \wedge (E(L(z) \dot- 1, z) \neq 0) \wedge (E(L(z), z) \neq 0)$$

Thus the predicates Q_1, Q_2, and Q_3 are all primitive recursive.

The functions h_1, h_2, and h_3 are only slightly more difficult to deal with. First consider the case in which z represents a tuple to which Rule 1 applies. That is, suppose that the prime decomposition of z has the form:

$$z = 2^k 3^{w_1} \cdots p_{k-2}^{w_{k-2}} p_{k-1}^0 p_k^{w_k} \cdots p_m^{w_m}$$

Inspection of Rule 1 reveals that in this case the appropriate value of $h_1(z)$ is:

$$2^{k-1} 3^{w_1} \cdots p_{k-2}^{w_{k-2}} p_{k-1}^{w_k + 1}$$

We are therefore led to define h_1 as follows:

$$h_1(z) = 2^{L(z) \div 1} \left[\prod_{i=1}^{L(z) \div 2} \mathrm{pn}(i)^{E(i,\,z)} \right] \mathrm{pn}(L(z) \div 1)^{E(L(z),\,z)+1}$$

The choice of this definition ensures that h_1 is primitive recursive and that $h_1(z)$ assumes the required value whenever $Q_1(z)$ holds. (Remember that it does not matter what value $h_1(z)$ assumes when $Q_1(z)$ does not hold.)

Similar observations concerning Rules 2 and 3 lead to the following definitions of the functions h_2 and h_3.

$$h_2(z) = \left[\prod_{i=0}^{L(z) \div 2} \mathrm{pn}(i)^{E(i,\,z)} \right] \mathrm{pn}(L(z) \div 1)^{E(L(z) \div 1,\,z) \div 1} \, \mathrm{pn}(L(z))$$

$$h_3(z) = 2^{L(z)+1} \left[\prod_{i=1}^{L(z) \div 2} \mathrm{pn}(i)^{E(i,\,z)} \right] [\mathrm{pn}(L(z) \div 1)^{E(L(z) \div 1,\,z) \div 1}]$$
$$\times \, [\mathrm{pn}(L(z))^{E(L(z) \div 1,\,z)}][\mathrm{pn}(L(z) + 1)^{E(L(z),\,z) \div 1}]$$

Thus the desired functions h_1, h_2, and h_3, like the predicates Q_1, Q_2, and Q_3, are primitive recursive. And this in turn implies that the trace function ψ specified in Eqs. (5-8) is primitive recursive.

We are now ready to show that Ackermann's function is μ-recursive. For this purpose we define the two-variable function η so that $\eta(x, y)$ is the number of steps needed to evaluate $A(x, y)$. Evidently $\eta(x, y)$ is the smallest value of n for which $\psi(x, y, n)$ represents a single natural number. Thus:

$$\eta(x, y) = \mu \dot{n}[L(\psi(x, y, n)) = 1]$$

Since L and ψ are primitive recursive, it follows that η is a μ-recursive function. Next note that the outcome of the evaluation of $A(x, y)$ is just the value of the natural number represented by $\psi(x, y, \eta(x, y))$. We may therefore write

$$A(x, y) = E(1, \psi(x, y, \eta(x, y))).$$

Since E and ψ are primitive recursive and η is μ-recursive, we have finally established:

Theorem 5-4. *Ackermann's function is μ-recursive.*

It is interesting to note that the proof of Theorem 5-4 provides a new explanation of why Ackermann's function is not primitive recursive. Looking back over the argument, we see that Ackermann's function could be shown to be primitive recursive if only the function η could be shown to be primitive recursive, rather than just μ-recursive. And the only thing that prevents η from being primitive recursive is the lack of a (primitive-recursive) bound for the minimalization used to define η. Thus we may say that Ackermann's function is not primitive recursive because there is no primitive recursive bound on the number of steps needed to evaluate $A(x, y)$.

The μ-Recursiveness of the Turing-Computable Functions

It has already been noted that every μ-recursive function is Turing computable. We now turn to the problem of showing that every Turing-computable function is μ-recursive. In order to keep the discussion as simple as possible, we will consider only one-variable functions. The extension to functions of two or more variables is straightforward and will be left to the reader.

Let f be any one-variable Turing-computable function, and let **T** be a Turing machine that computes f in standard form. We assume that **T**'s tape symbols are denoted 0, 1, 2, ..., with 0 representing a blank, and that its states are denoted $q_0, q_1, q_2, \ldots$, with q_0 being the designated starting state. We will also suppose that **T** operates on a tape that is singly infinite to the right and will think of the tape squares as being numbered from left to right, starting with zero.

The method used to show that f is μ-recursive closely parallels that used to show that Ackermann's function is μ-recursive. We begin by arithmetizing the computations that **T** performs. That is, we devise a Gödel-numbering scheme for representing the various steps in **T**'s computations. We then define a primitive recursive trace function that yields the Gödel numbers of successive steps in any desired computation. Finally, we show that the outcome of any computation can be obtained from the trace function by means of an appropriately defined μ-recursive function and therefore that the given function f is μ-recursive.

First the arithmetization. Note that three things are needed to describe the configuration in which a machine finds itself at any given step of a computation: the current internal state, the location of the reading head, and the current tape pattern. Each of these items will be represented by a single natural number, referred to respectively as the *state number*, *position number*, and *tape number* associated with the given configuration. These numbers are determined as follows. The state number of a configuration is just the subscript

	0	1	2
q_0	$0\ R\ q_1$		
q_1	$1\ L\ q_2$	$2\ R\ q_1$	
q_2		$1\ L\ q_2$	$1\ R\ q_3$
q_3	$1\ L\ q_3$	$1\ R\ q_3$	

Figure 5-3

of the current internal state. The position number is the number of the currently scanned square. The tape number is the Gödel number of the pattern of symbols currently appearing on the tape. In other words, if the first $k + 1$ tape squares contain the symbols $s_0, s_1, \ldots, s_k$ and the rest of the tape is blank, the tape number of the configuration is $2^{s_0}3^{s_1} \cdots p_k^{s_k}$.

Example 5-4. Consider the Turing machine described in Fig. 5-3. The first eight steps in the computation that this machine performs on the initial tape pattern $\underline{0}\ 1\ 1\ 0\ 0\ 0\ \cdots$ are listed below, along with the corresponding state, position, and tape numbers. Note that step zero represents the machine's initial configuration.

Step	Tape pattern	State number	Position number	Tape number		
0	$\underline{0}$ 1 1 0 0 0	0	0	$2^0\ 3^1\ 5^1$	=	15
1	0 $\underline{1}$ 1 0 0 0	1	1	$2^0\ 3^1\ 5^1$	=	15
2	0 2 $\underline{1}$ 0 0 0	1	2	$2^0\ 3^2\ 5^1$	=	45
3	0 2 2 $\underline{0}$ 0 0	1	3	$2^0\ 3^2\ 5^2$	=	225
4	0 2 $\underline{2}$ 1 0 0	2	2	$2^0\ 3^2\ 5^2\ 7^1$	= 1575
5	0 2 1 $\underline{1}$ 0 0	3	3	$2^0\ 3^2\ 5^1\ 7^1$	=	315
6	0 2 1 1 $\underline{0}$ 0	3	4	$2^0\ 3^2\ 5^1\ 7^1$	=	315
7	0 2 1 $\underline{1}$ 1 0	2	3	$2^0\ 3^2\ 5^1\ 7^1\ 11^1$	= 3465

Thus each step in a Turing-machine computation can be unambiguously described by an ordered triple of natural numbers. Such a triple can in turn be reduced to a single natural number by means of a further Gödel- or Cantor-numbering scheme. In particular, if a certain computation step is characterized by state number q, position number p, and tape number t, we shall assign the Cantor number $\pi^3(q, p, t)$ to that step, and refer to it as the *configuration number* of that step. For example, the configuration number of step 2 in the computation of Example 5-4 is $\pi^3(1, 2, 45) = 1385$. In this way an entire computation can be represented by a sequence of configuration numbers.

Let us now return to the given one-variable function f and the machine $\mathbf{T}$ that computes f in standard form. We define the two-variable function $\psi_{\mathbf{T}}$ so that $\psi_{\mathbf{T}}(x, n)$ is the configuration number of the nth step in the computation that $\mathbf{T}$ performs for the argument x. As in the case of Ackermann's function, we think of extending $\mathbf{T}$'s computations indefinitely. In particular, if $\mathbf{T}$ actually halts at step n_0 for input x, we define $\psi_{\mathbf{T}}(x, n)$ to have the same value as $\psi_{\mathbf{T}}(x, n_0)$ for all $n \geq n_0$. The resulting function $\psi_{\mathbf{T}}$ will be referred to as the *trace function* for the machine $\mathbf{T}$.

Our next job is to prove that the function ψ_T is primitive recursive. We do this by showing that ψ_T can be defined by primitive recursion from suitable primitive recursive auxiliary functions. Consider first the value $\psi_T(x, 0)$. Since T computes the function f in standard form, $\psi_T(x, 0)$ must be the configuration number corresponding to the situation in which T is in state q_0 and is scanning square zero of the tape pattern $0\ \bar{x}\ 0 \cdots$. This number is given by the one-variable function Enc, where

$$\text{Enc } (x) = \pi^3\left(0,\ 0,\ \prod_{i=1}^{x+1} \text{pn } (i)\right).$$

This function is obviously primitive recursive.

In order to express $\psi_T(x, n^+)$ in terms of $\psi_T(x, n)$, we introduce the auxiliary functions h_1, h_2, and h_3. If z is the Gödel number of a possible configuration of T, then $h_1(z)$, $h_2(z)$, and $h_3(z)$ are to have the following values

$h_1(z)$	the new state number resulting from the move that T makes when in the configuration whose number is z
$h_2(z)$	the new position number resulting from the move that T makes when in the configuration whose number is z
$h_3(z)$	the new tape number resulting from the move that T makes when in the configuration whose number is z

If z does not represent a possible configuration of T, the particular values assumed by h_1, h_2, and h_3 are not important.

Note that the trace function ψ_T is readily obtained from Enc, h_1, h_2, and h_3 by primitive recursion, since:

$$\psi_T(x, 0) = \text{Enc } (x)$$
$$\psi_T(x, n^+) = \pi^3(h_1(\psi_T(x, n)),\ h_2(\psi_T(x, n)),\ h_3(\psi_T(x, n)))$$

Thus to show that ψ_T is primitive recursive it is only necessary to show that h_1, h_2, and h_3 are primitive recursive.

To this end, we introduce the auxiliary two-variable functions γ_q, γ_s, and γ_d whose values are specified as follows. If a is the number of a state of T and b is the number of a symbol used by T, then:

$\gamma_q(a, b)$	is the number of the new state that T enters after scanning symbol b in state a
$\gamma_s(a, b)$	is the number of the new symbol that T writes after scanning symbol b in state a
$\gamma_d(a, b)$	is 0, 1, or 2 according as T moves left, halts, or moves right after scanning symbol b in state a

Whenever a does not represent a state of **T** or b does not represent a symbol used by **T**, the values $\gamma_q(a, b)$, $\gamma_s(a, b)$, and $\gamma_d(a, b)$ are defined to be 0.

The detailed definitions of γ_q, γ_s, and γ_d obviously depend on the structure of **T**. But each of these functions assumes a nonzero value for only a finite number of combinations of argument values. Thus no matter what the structure of **T** is, Theorem 4-5 guarantees that each of the functions γ_q, γ_s, and γ_d is primitive recursive.

Example 5-5. In the case of the machine of Fig. 5-3, the values of the functions γ_q, γ_s, and γ_d may be specified as follows:

$$\gamma_q(a, b) = \begin{cases} 3 & \text{if } (a = 2 \wedge b = 2) \vee (a = 3 \wedge b = 1) \vee (a = 3 \wedge b = 2) \\ 2 & \text{if } (a = 1 \wedge b = 0) \vee (a = 2 \wedge b = 0) \vee (a = 2 \wedge b = 1) \\ & \quad \vee (a = 3 \wedge b = 0) \\ 1 & \text{if } (a = 0 \wedge b = 0) \vee (a = 1 \wedge b = 1) \vee (a = 1 \wedge b = 2) \\ 0 & \text{otherwise} \end{cases}$$

$$\gamma_s(a, b) = \begin{cases} 2 & \text{if } (a = 1 \wedge b = 1) \vee (a = 0 \wedge b = 2) \vee (a = 1 \wedge b = 2) \\ & \quad \vee (a = 3 \wedge b = 2) \\ 1 & \text{if } (a = 1 \wedge b = 0) \vee (a = 3 \wedge b = 0) \vee (a = 0 \wedge b = 1) \\ & \quad \vee (a = 2 \wedge b = 1) \vee (a = 3 \wedge b = 1) \vee (a = 2 \wedge b = 2) \\ 0 & \text{otherwise} \end{cases}$$

$$\gamma_d(a, b) = \begin{cases} 2 & \text{if } (a = 0 \wedge b = 0) \vee (a = 1 \wedge b = 1) \vee (a = 2 \wedge b = 2) \\ & \quad \vee (a = 3 \wedge b = 1) \\ 1 & \text{if } (a = 0 \wedge b = 1) \vee (a = 0 \wedge b = 2) \vee (a = 1 \wedge b = 2) \\ & \quad \vee (a = 2 \wedge b = 0) \vee (a = 3 \wedge b = 2) \\ 0 & \text{otherwise} \end{cases}$$

Note that whenever the machine is in a halting situation, its new state must match its old state, its new symbol must match its old symbol, and the value of γ_d must be 1. $\square$

One more definition completes our repertoire of auxiliary functions. Let $\delta(z)$ denote the number of the scanned symbol in the machine configuration represented by the number z. Since

$$\delta(z) = \mathrm{E}(\sigma_2^3(z), \sigma_3^3(z)),$$

it is clear that δ is also a primitive recursive function.

We are now prepared to establish the primitive recursiveness of h_1, h_2, and h_3. The new state number $h_1(z)$ can be obtained directly from the numbers of the state and scanned symbol associated with configuration z:

$$h_1(z) = \gamma_q(\sigma_1^3(z), \delta(z))$$

The new position number $h_2(z)$ can be obtained by adding the value of the directional function γ_d to the old position number and then subtracting 1:

$$h_2(z) = (\sigma_2^3(z) + \gamma_d(\sigma_1^3(z), \delta(z))) \div 1$$

Finally, the new tape number $h_3(z)$ can be obtained by dividing the old tape number by the prime power corresponding to the old scanned symbol, and multiplying by the prime power corresponding to the new symbol:

$$h_3(z) = \text{quo } (\text{pn}(\sigma_2^3(z))^{\delta(z)}, \sigma_3^3(z)) \cdot \text{pn}(\sigma_2^3(z))^{\gamma_s(\sigma_1^3(z), \delta(z))}$$

Thus the functions h_1, h_2, and h_3 are all primitive recursive and, as a consequence, so is the trace function ψ_T.

Paralleling the treatment of Ackermann's function, we now define a function η_T that determines when the evaluation of $f(x)$ is completed. Specifically, we define $\eta_T(x)$ to be the number of steps required by $\mathbf{T}$ to evaluate $f(x)$ whenever the evaluation of $f(x)$ terminates and to be undefined otherwise. Now, the configuration number $\psi_T(x, n)$ represents a situation in which $\mathbf{T}$ has halted iff $\psi_T(x, n) = \psi_T(x, n^+)$. Therefore

$$\eta_T(x) = \mu n \left[\psi_T(x, n) = \psi_T(x, n^+) \right]$$

and η_T is a μ-recursive function.

The outcome of $\mathbf{T}$'s computation is just the natural number represented by $\mathbf{T}$'s final tape pattern. This natural number can be extracted from the final configuration number z by the primitive recursive function Dec, where

$$\text{Dec } (z) = \tilde{\mu}y \left[\sigma_3^3(z) = \prod_{i=1}^{y+1} \text{pn } (i) \right].$$

Thus we can now write

$$f(x) = \text{Dec } (\psi_T(x, \eta_T(x))).$$

Since η_T is μ-recursive, and ψ_T and Dec are primitive recursive, it follows that f is μ-recursive.

The argument presented above can easily be extended to apply to Turing-computable functions of more than one variable, thereby establishing:

Theorem 5-5. *Every Turing-computable function is μ-recursive.*

Theorems 5-3 and 5-5 together yield:

Corollary 5-5.1. *The set of μ-recursive functions is identical to the set of Turing-computable functions.*

Corollary 5-5.1 implies that the μ-recursive functions can be indexed so as to form a standard family. There are many ways of doing this, of which the simplest is to use the indexing scheme already established for Turing computable functions. In other words, we assign the index i to the μ-recursive function computed by Turing machine $\mathbf{T}_i$. In terms of such an indexing scheme, the class of μ-recursive functions has all the characteristics of a standard family. Thus the domain, totality, and equivalence functions for the μ-recursive functions are not themselves μ-recursive, the class of μ-recursive functions obeys the Recursion Theorem, and so on.

Canonical Forms

The argument used to prove Theorem 5-5 also allows us to establish certain canonical forms for the definition of μ-recursive functions. As a first example, we have:

Theorem 5-6. *Let f be any μ-recursive function of one variable. Then there exists a two-variable primitive recursive function ρ_f and a two-place primitive recursive predicate θ_f such that*

$$f(x) = \rho_f(x, \mu n[\theta_f(x, n)]).$$

Proof. Since f is μ-recursive, it is also Turing computable. Let $\mathbf{T}$ be a Turing machine that evaluates f in standard form. Then according to the argument used to establish Theorem 5-5, there exists a primitive recursive function $\psi_\mathbf{T}$ such that

$$f(x) = \text{Dec } (\psi_\mathbf{T}(x, \mu n[\psi_\mathbf{T}(x, n) = \psi_\mathbf{T}(x, n^+)])).$$

Now define ρ_f to be the composite function $\text{Dec} \circ (\psi_\mathbf{T})$, so that

$$\rho_f(x, y) = \text{Dec } (\psi_\mathbf{T}(x, y))$$

and define θ_f to be the predicate such that

$$\theta_f(x, n) \iff \psi_\mathbf{T}(x, n) = \psi_\mathbf{T}(x, n^+).$$

Since Dec and $\psi_\mathbf{T}$ are primitive recursive, so are ρ_f and θ_f. Moreover,

$$f(x) = \rho_f(x, \mu n[\theta_f(x, n)])$$

as required.   □

The extension of Theorem 5-6 to functions of more than one variable is straightforward. As an obvious consequence, we have:

Corollary 5-6.1. In defining a μ-recursive function, it is never necessary to apply unbounded minimalization more than once, and that application need only be made to a primitive recursive function.

It must be remembered that the function ρ_f and the predicate θ_f mentioned in Theorem 5-6 depend on the choice of the given μ-recursive function f. We now show that the dependence of ρ_f on f can be removed. Specifically, we propose to show that every one-variable μ-recursive function f can be expressed in terms of a *fixed* primitive recursive function ρ and a suitably chosen predicate $\hat{\theta}_f$. In fact, by applying a simple encoding trick, we can eliminate the variable x as an argument of the function ρ.

The key idea is to assign the Cantor number $\pi^2(\psi_T(x, n), n)$ to the nth step in the computation that the machine **T** performs for input x. Then instead of using the predicate θ_f to select a value of n for which $\psi_T(x, n)$ represents a configuration in which **T** has halted, we may use a new predicate $\hat{\theta}_f$ to select a Cantor number $\pi^2(\psi_T(x, n), n)$ in which $\psi_T(x, n)$ represents a configuration in which **T** has halted. Since this Cantor number, call it z, contains complete information about **T**'s final tape pattern, the function ρ can be designed to extract the value of $f(x)$ directly from z. Thus ρ does not need to "know" what function f is being evaluated or what the argument of f is.

Theorem 5-7. There exists a fixed, one-variable primitive recursive function ρ with the following property. For each one-variable μ-recursive function f there is a two-place primitive recursive predicate $\hat{\theta}_f$ such that

$$f(x) = \rho(\mu z[\hat{\theta}_f(x, z)])$$

for all x.

Proof. Let ρ be the primitive recursive function such that

$$\rho(z) = \text{Dec}\ (\sigma_1^2(z)).$$

Now suppose that f is a given one-variable μ-recursive function, that **T** is a Turing machine that evaluates f in standard form, and that ψ_T is the trace function for **T**. Choose the predicate $\hat{\theta}_f$ so that

$$\hat{\theta}_f(x, z) \quad \Leftrightarrow \quad (\psi_T(x, \sigma_2^2(z)) = \sigma_1^2(z)) \wedge (\psi_T(x, \sigma_2^2(z) + 1) = \sigma_1^2(z)).$$

This predicate is certainly primitive recursive. If z is viewed as the Cantor number of an ordered pair (w, n), we see that $\hat{\theta}_f(x, z)$ holds iff $w = \psi_T(x, n) = \psi_T(x, n + 1)$. Thus $\hat{\theta}_f(x, z)$ holds only if the first component of z represents a configuration in which **T** has halted after being supplied with the input x. It follows from these observations that if $f(x)$ is defined, the quantity $\mu z[\hat{\theta}_f(x, z)]$ will also be defined and will be a Cantor number whose first component represents a configuration in which **T** has halted after being supplied with x. To obtain the value of $f(x)$ from this Cantor number, it is only necessary to extract the first component of the Cantor number and apply the function Dec to it. Thus

$$f(x) = \text{Dec }(\sigma_1^2(\mu z[\hat{\theta}_f(x, z)]))$$

or

$$f(x) = \rho(\mu z[\hat{\theta}_f(x, z)])$$

as required. $\qquad\square$

Extending Theorem 5-7 to functions of more than one variable yields:

Theorem 5-8. *Let f be any μ-recursive function of n variables. Then there exists an $(n + 1)$-place primitive recursive predicate $\hat{\theta}_f$ such that*

$$f(x_1, \ldots, x_n) = \rho(\mu z[\hat{\theta}_f(x_1, \ldots, x_n, z)])$$

where ρ is the fixed primitive recursive function defined in Theorem 5-7.

We may now make use of the equivalence between the μ-recursive functions and the Turing-computable functions to eliminate the dependence of the predicate $\hat{\theta}_f$ on the choice of the function f.

Theorem 5-9. *(Kleene's Normal Form Theorem) For each positive integer n there exists a fixed $(n + 2)$-place primitive recursive predicate τ with the following property. If f is any n-variable μ-recursive function, there must exist a natural number w such that*

$$f(x_1, \ldots, x_n) = \rho(\mu z[\tau(w, x_1, \ldots, x_n, z)])$$

where ρ is the fixed primitive recursive function of Theorems 5-7 and 5-8.

Proof. Let g denote the universal function $U^{(n+1)}$ for the family of Turing-computable functions. According to Theorem 5-5, g is μ-recursive. Define τ to be the primitive recursive predicate $\hat{\theta}_g$ associated with g as specified in Theorem 5-8. Now suppose that f is any given n-variable μ-recursive function. Note that f is necessarily Turing computable and has some index w. Thus

$$f(x_1, \ldots, x_n) = U^{(n+1)}(w, x_1, \ldots, x_n) = g(w, x_1, \ldots, x_n).$$

But, according to Theorem 5-8,

$$g(w, x_1, \ldots, x_n) = \rho(\mu z[\hat{\theta}_g(w, x_1, \ldots, x_n, z)])$$

whence

$$f(x_1, \ldots, x_n) = \rho(\mu z[\tau(w, x_1, \ldots, x_n, z)])$$

as required. $\qquad\square$

Theorems 5-6, 5-7, 5-8, and 5-9 provide simple standard forms in which all μ-recursive functions can be expressed. They also reveal something of the role of primitive recursive functions and predicates in the definition of μ-recursive functions. In particular we see that primitive recursive functions and predicates suffice for all but one critical step in the definition of a μ-recursive function. That step, for which unbounded minimalization may be required, amounts to determining when the evaluation of the function in question is completed. In fact, we can now generalize an observation made in connection with Ackermann's function: The total μ-recursive functions that are not primitive recursive are just those for which the number of steps needed to evaluate them cannot be bounded by primitive recursive functions.

5.3 GENERAL RECURSIVE FUNCTIONS

Recursive definitions of one form or another play an important part in the study of computability, for they usually provide effective ways of evaluating the functions they define. In this section we formalize the notion of a recursive definition and use that formalization to define the class of *general recursive functions*. This new class is easily shown to contain all the μ-recursive functions, and the great generality of its definition suggests that it may contain other functions as well. However, this turns out not to be the case, as we shall see.

Introduction

We have by now examined a variety of recursive schemes for defining new functions from old functions: primitive recursion, course-of-values recursion, the recursion used to define Ackermann's function, and so on. All these schemes have two features in common: they explicitly define new function values in terms of previously determined function values, and they provide effective procedures for evaluating the functions that they define. Building on these observations, let us now try to arrive at a useful formalization of the notion of a "recursive definition." We begin by examining in detail the way in which some of the familiar recursion schemes provide effective evaluation procedures.

Consider first the evaluation of Ackermann's function. Recall that this function is defined by the equations:

$$A(0, y) = y^+ \tag{5-9a}$$

$$A(x^+, 0) = A(x, 1) \tag{5-9b}$$

$$A(x^+, y^+) = A(x, A(x^+, y)) \tag{5-9c}$$

Suppose that we wish to evaluate $A(2, 0)$. The first step is obviously to apply Eq. (5-9b) to the expression $A(2, 0)$. To be precise, this "application" of

Eq. (5-9b) consists in *substituting* 1 for x in Eq. (5-9b) and writing 2 for 1^+, so as to obtain the equation

$$A(2, 0) = A(1, 1). \tag{5-9d}$$

The next step is to apply Eq. (5-9c) to $A(1, 1)$ or, more precisely, to substitute 0 for both x and y in Eq. (5-9c) and write 1 for 0^+, thereby obtaining the equation

$$A(1, 1) = A(0, A(1, 0)). \tag{5-9e}$$

Consolidating what has been done so far, we now *replace* the term $A(1, 1)$ on the right-hand side of Eq. (5-9d) by its equivalent as specified in Eq. (5-9e), thereby obtaining

$$A(2, 0) = A(0, A(1, 0)). \tag{5-9f}$$

The remainder of the evaluation proceeds in a similar way. Substituting 0 for x in Eq. (5-9b) gives

$$A(1, 0) = A(0, 1). \tag{5-9g}$$

Replacing the term $A(1, 0)$ in Eq. (5-9f) by its equivalent as given in Eq. (5-9g) yields

$$A(2, 0) = A(0, A(0, 1)). \tag{5-9h}$$

Substituting 1 for y in Eq. (5-9a) and then replacing the term $A(0, 1)$ in Eq. (5-9h) produces the equations:

$$A(0, 1) = 2 \tag{5-9i}$$

and

$$A(2, 0) = A(0, 2). \tag{5-9j}$$

One more substitution and one more replacement complete the evaluation with the equations:

$$A(0, 2) = 3 \tag{5-9k}$$

and

$$A(2, 0) = 3. \tag{5-9ℓ}$$

Note that aside from rewriting the successor of a given natural number, only two basic operations were needed to evaluate $A(2, 0)$:

1. substituting a natural number for a variable in an equation;
2. replacing a term in an equation by its equivalent as specified by another equation.

Indeed, these two operations permit the evaluation of Ackermann's function for any choice of argument values.

As a second example, consider the evaluation of the two-variable multiplication function m. Recall that multiplication is defined by primitive recursion from the addition function a, which in turn is defined by primitive recursion from the successor function. Thus the function m is completely defined by the following equations.

$$m(x, 0) = 0 \tag{5-10a}$$

$$m(x, y^+) = a(m(x, y), x) \tag{5-10b}$$

$$a(x, 0) = x \tag{5-10c}$$

$$a(x, y^+) = a(x, y)^+ \tag{5-10d}$$

In terms of these four defining equations, the evaluation of m proceeds in much the same way as the evaluation of Ackermann's function. In particular, the same two operations of substitution and replacement are sufficient for carrying out the evaluation.

Suppose, for instance, that we wish to evaluate $m(2, 2)$. Then the following sequence of substitutions and replacements is appropriate.

$m(2, 2) = a(m(2, 1), 2)$	subs. in (5-10b)	(5-10e)
$m(2, 1) = a(m(2, 0), 2)$	subs. in (5-10b)	(5-10f)
$m(2, 2) = a(a(m(2, 0), 2), 2)$	repl. in (5-10e) from (5-10f)	(5-10g)
$m(2, 0) = 0$	subs. in (5-10a)	(5-10h)
$m(2, 2) = a(a(0, 2), 2)$	repl. in (5-10g) from (5-10h)	(5-10i)
$a(0, 2) = a(0, 1)^+$	subs. in (5-10d)	(5-10j)
$a(0, 1) = a(0, 0)^+$	subs. in (5-10d)	(5-10k)
$a(0, 0) = 0$	subs. in (5-10c)	(5-10ℓ)
$a(0, 1) = 1$	repl. in (5-10k) from (5-10ℓ)	(5-10m)
$a(0, 2) = 2$	repl. in (5-10j) from (5-10m)	(5-10n)
$m(2, 2) = a(2, 2)$	repl. in (5-10i) from (5-10n)	(5-10o)
$a(2, 2) = a(2, 1)^+$	subs. in (5-10d)	(5-10p)
$a(2, 1) = a(2, 0)^+$	subs. in (5-10d)	(5-10q)
$a(2, 0) = 2$	subs. in (5-10c)	(5-10r)
$a(2, 1) = 3$	repl. in (5-10q) from (5-10r)	(5-10s)
$a(2, 2) = 4$	repl. in (5-10p) from (5-10s)	(5-10t)
$m(2, 2) = 4$	repl. in (5-10o) from (5-10t)	(5-10u)

Note that the replacements used to obtain Eqs. (5-10m), (5-10n), (5-10s), and (5-10t), like most of the substitutions, involve the rewriting of the successor of a number.

It is not hard to see that all functions that are defined by primitive recursion and composition can be evaluated in the manner just illustrated. So can functions defined by other recursion schemes. Consider, for example, the Fibonacci function f specified by Eqs. (4-42). In this case the defining equations

$$f(0) = 1$$

$$f(1) = 1$$

$$f(n^{++}) = a(f(n^{+}), f(n))$$

$$a(x, 0) = x$$

$$a(x, y^{+}) = a(x, y)^{+}$$

provide a framework within which f can be evaluated by substitutions and replacements, as the reader should convince him or herself. Functions defined by more general course-of-values recursions, or by multiple recursions, or by even more complicated types of recursions also yield to this method of evaluation.

These observations suggest that the possibility of evaluation by substitution and replacement is an essential feature of recursively defined functions. We now propose to take this feature as the sole defining characteristic of a new class of functions, called the general recursive functions. Thus we will consider a function f to be a *general recursive function* iff there exists a finite set of defining equations from which f can be correctly evaluated by successive substitutions and replacements. Note that two basic requirements are embodied in the notion of "correct evaluation." First, whenever the value of f is defined, it must be possible to obtain that value, and no other, by some sequence of substitutions and replacements. And second, whenever the value of f is not defined, it must not be possible to obtain a value for f by means of any sequence of substitutions and replacements.

Before trying to give a more precise definition of the general recursive functions, it is appropriate to consider some informal examples.

Example 5-6

a) Consider the two-variable function f_1, where

$$f_1(x, y) = \begin{cases} 1 & \text{if} \quad x \le y \\ 0 & \text{if} \quad x > y. \end{cases}$$

This function is general recursive, since it can be evaluated by substitutions and replacements from the equations

$$f_1(0, y) = 1$$
$$f_1(x^+, 0) = 0$$
$$f_1(x^+, y^+) = f_1(x, y)$$

as the reader may verify.

    b) Consider the two-variable function f_2, where

$$f_2(x, y) = \begin{cases} y & \text{if} & x \leq y \\ \uparrow & \text{if} & x > y. \end{cases}$$

This function is also general recursive, since it can be evaluated by substitutions and replacements from the equations:

$$f_2(x, x) = x$$
$$f_2(x, y^+) = f_2(x, y)^+$$

Note carefully that if $x > y$, there is no possible sequence of substitutions and replacements that will yield a value for $f_2(x, y)$.   □

    Two remarks may help clarify the nature of a general recursive function. First, note that the equations needed to define a given function will often define other functions at the same time. For instance, the four equations (5-10a) through (5-10d) used to define the multiplication function include a pair of equations that define the addition function. Thus a set of equations must in general be viewed as defining a set of functions simultaneously, one function for each different "function letter" mentioned in the equations. In practice, however, we are usually interested in one function at a time, and so will usually focus our attention on one of the functions defined by a given set of equations.

    Second, suppose that a certain set of functions is defined by a certain set of equations. Then those functions must satisfy the given equations in the sense that every substitution of natural numbers for variables in one of the equations must yield a true statement about the functions. Thus, for example, substitution of 3 for x and 7 for y in Eq. (5-10b) yields the statement $m(3, 8) = a(m(3, 7), 3)$, which is a true statement about multiplication and addition. On the other hand, not every set of functions that satisfy a given set of equations is actually defined by those equations, since it may not be possible to derive all the values of such a set of functions by means of substitutions and replacements.

Example 5-7. Consider the following set of equations, in which only one function is mentioned:

$$f(0, 0) = 0$$
$$f(x^+, y) = f(x, y^+)$$

There are many functions that satisfy these equations. Among them are the function f such that $f(x, y) = 0$ for all x and y, the function f such that $f(x, y) = x + y$ for all x and y, the function f such that $f(x, y) = (x + y)^2$ for all x and y, and so on. Yet the only equation of the form $f(u_1, u_2) = u$ (where u_1, u_2, and u are natural numbers) that can be derived from the given equations by substitutions and replacements is $f(0, 0) = 0$. Thus among all the functions that satisfy these equations, the equations actually define the function f whose value is 0 for $x = y = 0$ and is undefined for all other choices of x and y.   □

It must also be remembered that some sets of equations do not define functions at all. For in order for a set of equations to define a function f, it must be possible to obtain at most one function value from the given equations for each choice of argument values.

Example 5-8. Consider the following equations:

$$f(x, 0) = x^+ \tag{5-11a}$$

$$f(x, y^+) = f(x^+, y) \tag{5-11b}$$

$$f(x^+, y^+) = f(x, f(x, y)) \tag{5-11c}$$

These equations do *not* define a function f, since conflicting values can be derived from them. For example, each of the equations $f(1, 1) = 3$ and $f(1, 1) = 2$ can be obtained from the given equations by an appropriate sequence of substitutions and replacements, as shown below.

Derivation 1

$f(1, 1) = f(2, 0)$	subst. in (5-11b)	(5-11d)
$f(2, 0) = 3$	subst. in (5-11a)	(5-11e)
$f(1, 1) = 3$	repl. in (5-11d) from (5-11e)	(5-11f)

Derivation 2

$f(1, 1) = f(0, f(0, 0))$	subst. in (5-11c)	(5-11g)
$f(0, 0) = 1$	subst. in (5-11a)	(5-11h)
$f(1, 1) = f(0, 1)$	repl. in (5-11g) from (5-11h)	(5-11i)
$f(0, 1) = f(1, 0)$	subst. in (5-11b)	(5-11j)
$f(1, 0) = 2$	subst. in (5-11a)	(5-11k)
$f(0, 1) = 2$	repl. in (5-11j) from (5-11k)	(5-11ℓ)
$f(1, 1) = 2$	repl. in (5-11i) from (5-11ℓ)	(5-11m)

□

Formal Definitions

So far our discussion of general recursive functions has been informal and, as a consequence, somewhat imprecise. Let us now reformulate the notion of a general recursive function in more precise terms, so that we will be able to establish some of the basic properties of such functions. To do this, we must specify exactly what form defining equations are to take, how natural numbers are to be represented, how substitution and replacement are to be carried out, and so on. In making this specification, it will prove convenient to distinguish between the symbols used to represent functions in equations and the names of the functions defined by those equations.

The equations that provide the formal definition of general recursive functions are expressions made up of variables, functions letters, the special symbols 0 and S, and certain punctuation marks. Intuitively, variables play their usual role, function letters stand for functions, 0 stands for the natural number zero, and S stands for the successor function. Formally, a *variable* is defined to by any one of the symbols x_1, x_2, x_3, etc. A *function letter* is defined to be any one of the symbols $f_0^1, f_1^1, f_2^1, \ldots; f_0^2, f_1^2, f_2^2, \ldots; f_0^3, f_1^3, f_2^3, \ldots$; and so on. The symbol f_i^n will be referred to as the ith n-variable function letter.[†]

The basic building blocks of equations are *terms*, which are defined inductively as follows.

1. Every variable is a term.
2. The symbol 0 is a term.
3. If t is a term, so is the expression $S(t)$.
4. If $t_1, \ldots, t_n$ are terms, so is the expression $f_i^n(t_1, \ldots, t_n)$ for each $i \geq 0$.

Only those expressions that can be built up through a finite number of applications of these four rules are to be considered terms.

The terms that represent natural numbers will play an important part in the subsequent discussion. They are called *numerals* and are defined inductively as follows.

1. The term 0 is a numeral.
2. If n is a numeral, so is the term $S(n)$.

Thus the numerals are the terms 0, $S(0)$, $S(S(0))$, $S(S(S(0)))$, ... For convenience, these terms will usually be abbreviated $\bar{0}, \bar{1}, \bar{2}, \bar{3}, \ldots$, respectively.

Example 5-9

a) Each of the following expressions is a term:

$$\bar{17}; \quad f_1^3(x_1, x_2, x_3); \quad f_4^3(\bar{3}, \bar{17}, \bar{5}); \quad f_0^2(x_1, \bar{3}); \quad f_6^2(\bar{3}, f_1^1(\bar{2}));$$
$$f_1^3(S(x_2), x_1, f_5^1(\bar{2})); \quad f_1^2(f_1^2(x_1, S(\bar{6})), \bar{5}).$$

[†] In practice, we will often use symbols such as x, y, z, etc. to represent variables and symbols such as f, g, h, etc. to represent function letters. This usage is to be viewed as a form of abbreviation for the official variables and function letters defined here.

Remember that the expressions $\overline{17}$, $\overline{3}$, $\overline{5}$, etc. are really abbreviations for certain terms. Thus $f_0^2(x_1, \overline{3})$, for example, stands for the term $f_0^2(x_1, S(S(S(\overline{0}))))$.

b) None of the following expressions is a term:

$$f_3^2(x_1, x_2, x_3); \qquad S(x_1, \overline{7}); \qquad f_1^1(x_1) + \overline{3}; \qquad f_1^2(f_1^1(x)), x_2).$$

In each case the reader should be sure to understand why the given expression does not exhibit the form required of a term.   □

We now define a (*formal*) *equation* to be an expression of the form $t_1 = t_2$, where t_1 and t_2 are terms. Thus $S(S(x_1)) = x_2$, $f_0^2(\overline{3}, \overline{5}) = \overline{4}$, and $S(f_1^2(x_1, \overline{0})) = f_3^1(S(x_1))$ are examples of equations. In such expressions the symbol $=$ should be thought of as an abstract symbol, not as a representation of the identity relation. In particular, the expression $t_1 = t_2$ must not be interpreted as saying that t_1 and t_2 are the same terms. (If we wish to indicate that the terms t_1 and t_2 are identical, we must resort to some such notation as $t_1 \equiv t_2$.)

In order to specify how substitution and replacement are to be carried out on formal equations, we must first introduce the process of *term substitution*. Suppose that t_1, t_2, and t_3 are any three terms, not necessarily distinct. By the *substitution* of t_3 for t_2 in t_1 we mean the process of replacing *every* occurrence of the term t_2 within t_1 by an occurrence of the term t_3. The term that results from this replacement will be denoted $t_1(t_3/t_2)$. Note that if there are no occurrences of t_2 in t_1, then $t_1(t_3/t_2)$ is identical to t_1.

Example 5-10. Let t_1 through t_8 denote the following terms.

$$
\begin{aligned}
&t_1\!: \ f_1^1(x_1) &\qquad &t_5\!: \ f_2^2(f_1^1(\overline{3}), x_1)\\
&t_2\!: \ x_1 &\qquad &t_6\!: \ f_3^2(x_1, \overline{2})\\
&t_3\!: \ \overline{2} &\qquad &t_7\!: \ f_4^3(x_1, f_1^1(\overline{3}), f_2^2(f_1^1(\overline{3}), x_1))\\
&t_4\!: \ f_1^1(\overline{3}) &\qquad &t_8\!: \ S(x_1)
\end{aligned}
$$

Then the substitutions specified at the left below yield the terms given at the right.

$$
\begin{aligned}
&t_1(t_3/t_2)\!: &\quad &f_1^1(\overline{2})\\
&t_4(t_3/t_2)\!: &\quad &f_1^1(\overline{3})\\
&t_5(t_6/t_4)\!: &\quad &f_2^2(f_3^2(x_1, \overline{2}), x_1)\\
&t_7(t_8/t_2)\!: &\quad &f_4^3(S(x_1), f_1^1(\overline{3}), f_2^2(f_1^1(\overline{3}), S(x_1)))\\
&t_7(t_3/t_4)\!: &\quad &f_4^3(x_1, \overline{2}, f_2^2(\overline{2}, x_1))
\end{aligned}
$$
   □

The concept of term substitution allows us to formalize the substitution and replacement operations used to evaluate general recursive functions. These operations will be presented in the form of rules for deriving new equations from given equations. We begin with substitution, which is straightforward.

The Substitution Rule (SR). If x_i is a variable, n is a numeral, and t_1 and t_2 are terms, then the equation $t_1(n/x_i) = t_2(n/x_i)$ can be derived from the equation $t_1 = t_2$.

Simply put, this rule permits the substitution of the numeral n for each occurrence of the variable x_i in the equation $t_1 = t_2$.

Example 5-11. Applying the substitution rule to the equation
$$f_2^2(S(x_1),\, S(x_2)) = f_2^2(x_1,\, f_2^2(S(x_1),\, x_2))$$
with x_2 as x_i and $\overline{7}$ as n yields the equation
$$f_2^2(S(x_1),\, S(\overline{7})) = f_2^2(x_1,\, f_2^2(S(x_1),\, \overline{7}))$$
which can also be written as
$$f_2^2(S(x_1),\, \overline{8}) = f_2^2(x_1,\, f_2^2(S(x_1),\, \overline{7})).\qquad\square$$

Replacement is harder to deal with, since we must provide for the substitution of one term for some, but not necessarily all, occurrences of another term. In particular, if $t_1 = t_2$ and $t_3 = t_4$ are given equations, we must provide for the substitution of t_4 for an arbitrary subset of the occurrences of t_3 in the equation $t_1 = t_2$. This partial substitution may be described by means of an auxiliary equation $w_1 = w_2$, obtained from $t_1 = t_2$ by writing a new variable x_v in place of those occurrences of t_3 that are to be replaced by t_4. For then the given equation $t_1 = t_2$ can be represented as the result of substituting t_3 for every occurrence of x_v in $w_1 = w_2$, while the desired new equation can be represented as the result of substituting t_4 for every occurrence of x_v in $w_1 = w_2$.

This artifice allows us to formalize the replacement operation as follows.

The Replacement Rule (RR). Let $t_1 = t_2$ and $t_3 = t_4$ be given equations. Suppose there exists an equation $w_1 = w_2$ and a variable x_v (not appearing in $t_1 = t_2$ or $t_3 = t_4$) such that $t_1 = w_1(t_3/x_v)$ and $t_2 = w_2(t_3/x_v)$. Then from the given equations $t_1 = t_2$ and $t_3 = t_4$ can be derived the new equation $w_1(t_4/x_v) = w_2(t_4/x_v)$.

The role of the auxiliary equation $w_1 = w_2$ in the application of the replacement rule may be better understood with the aid of an example.

Example 5-12. The replacement rule can be applied to the equations
$$f_1^2(f_1^1(\overline{0}),\, x_1) = f_2^3(f_1^1(\overline{0}),\, x_2,\, f_1^1(\overline{0}))$$
and
$$f_1^1(\overline{0}) = \overline{4}$$
to obtain the equation
$$f_1^2(f_1^1(\overline{0}),\, x_1) = f_2^3(\overline{4},\, x_2,\, f_1^1(\overline{0})).$$

In this application the terms t_1, t_2, t_3, and t_4 are respectively $f_1^2(f_1^1(\bar{0}), x_1)$, $f_2^3(f_1^1(\bar{0}), x_2, f_1^1(\bar{0}))$, $f_1^1(\bar{0})$, and $\bar{4}$. An appropriate auxiliary equation $w_1 = w_2$ might be

$$f_1^2(f_1^1(\bar{0}), x_1) = f_2^3(x_3, x_2, f_1^1(\bar{0})).$$

Here x_3, which does not appear in either of the given equations $t_1 = t_2$ or $t_3 = t_4$, serves as the variable x_v. Note that the given equation $f_1^2(f_1^1(\bar{0}), x_1) = f_2^3(f_1^1(\bar{0}), x_2, f_1^1(\bar{0}))$ is obtained from $w_1 = w_2$ by substituting $f_1^1(\bar{0})$ for x_3, while the new equation $f_1^2(f_1^1(\bar{0}), x_1) = f_2^3(\bar{4}, x_2, f_1^1(\bar{0}))$ is obtained from $w_1 = w_2$ by substituting $\bar{4}$ for x_3.   □

The substitution and replacement rules allow us to formalize the process of deriving new equations from given equations. Suppose that $\mathcal{E}$ is a finite set of formal equations. We say that an equation e is *derivable from* $\mathcal{E}$ if there exists a finite sequence of equations $e_1, e_2, \ldots, e_m$ such that:

1. e_m is the given equation e;
2. for each equation e_k in the sequence, either:
   a) e_k is a member of $\mathcal{E}$; or
   b) e_k can be obtained from the equation e_i by the substitution rule, for some $i < k$; or
   c) e_k can be obtained from the equations e_i and e_j by the replacement rule, for some $i < k$ and $j < k$.

Such a sequence of equations $e_1, \ldots, e_m$ (where e_m is e) is called a (*formal*) *derivation* of the equation e from the set $\mathcal{E}$. We often express the fact that e is derivable from $\mathcal{E}$ by writing $\mathcal{E} \vdash e$.

Example 5-13. Let $\mathcal{E}$ be the set consisting of the three equations

$$f_1^1(\bar{0}) = \bar{0}$$

$$f_1^1(S(x_1)) = S(S(f_1^1(x_1)))$$

$$f_2^1(x_2) = f_1^1(f_1^1(x_2)).$$

Then the following sequence of equations is a formal derivation of the equation $f_2^1(\bar{2}) = f_1^1(\bar{4})$. A justification for each member of the sequence is provided in brackets at the right.

e_1: $f_2^1(x_2) = f_1^1(f_1^1(x_2))$ [member of $\mathcal{E}$]

e_2: $f_2^1(\bar{2}) = f_1^1(f_1^1(\bar{2}))$ [from e_1 by SR]

e_3: $f_1^1(S(x_1)) = S(S(f_1^1(x_1)))$ [member of $\mathcal{E}$]

e_4: $f_1^1(\bar{2}) = S(S(f_1^1(\bar{1})))$ [from e_3 by SR]

e_5: $f_2^1(\bar{2}) = f_1^1(S(S(f_1^1(\bar{1}))))$ [from e_2 and e_4 by RR]

e_6: $f_1^1(\bar{1}) = S(S(f_1^1(\bar{0})))$ [from e_3 by SR]

e_7: $f_1^1(\bar{0}) = \bar{0}$ [member of $\mathcal{E}$]

e_8: $f_1^1(\bar{1}) = \bar{2}$ [from e_6 and e_7 by RR]

e_9: $f_2^1(\bar{2}) = f_1^1(\bar{4})$ [from e_5 and e_8 by RR]   □

The concept of a formal derivation makes possible a precise definition of the class of general recursive functions. Let f be any (partial) n-variable number-theoretic function. Then f is a *general recursive function* iff there exists a finite set of formal equations $\mathcal{E}$ and a function letter f_k^n such that, for every choice of natural numbers $u_1, \ldots, u_n$, and u,

$$\mathcal{E} \vdash f_k^n(\bar{u}_1, \ldots, \bar{u}_n) = \bar{u} \qquad \text{iff} \qquad f(u_1, \ldots, u_n) = u.$$

When such a set of equations $\mathcal{E}$ exists, we say that $\mathcal{E}$ *defines* the function f with f_k^n as *principal function letter*.

As a simple example, we note that the set of equations $\mathcal{E}$ of Example 5-13 defines two functions. The choice of f_1^1 as principal function letter yields the function f_1 such that $f_1(x) = 2x$, while the choice of f_2^1 as principal function letter yields the function f_2 such that $f_2(x) = 4x$. Thus both of these functions are general recursive functions.

The General Recursiveness of the μ-Recursive Functions

The formal definition of the general recursive functions provides the foundation needed to investigate the relationship between the general recursive functions and the μ-recursive functions. Our first job is to show that every μ-recursive function is also general recursive. This is done by noting that the constant, projection, and successor functions are general recursive, and then showing that general recursiveness is preserved by composition, primitive recursion, and minimalization. The first step is straightforward.

Lemma 5-10. Each of the constant, projection, and successor functions is a general recursive function.

Proof. The constant function $C_k^{(n)}$ is defined by the single equation

$$f_0^n(x_1, \ldots, x_n) = \bar{k}$$

and so is general recursive. The projection function $P_i^{(n)}$ is defined by the equation

$$f_0^n(x_1, \ldots, x_n) = x_i$$

and so is general recursive. The successor function S is defined by the single equation

$$f_0^1(x_1) = S(x_1)$$

and so is general recursive.   □

In order to prove that general recursiveness is preserved by composition, primitive recursion, and minimalization, we need a preliminary result concerning formal derivations. Consider any such derivation whose last equation is of the form $f_k^n(\bar{u}_1, \ldots, \bar{u}_n) = \bar{u}$ where $\bar{u}_1, \ldots, \bar{u}_n$, and $\bar{u}$ are numerals. Suppose that somewhere within the derivation an equation e has been obtained by means of the replacement rule, and that e contains one or more occurrences of a variable x_i. Now, if the equation e is in fact essential to the derivation, some subsequent application of the substitution rule must substitute a numeral for x_i. Note that this substitution could just as well be made in the equations from which e was obtained. It follows by induction that the given derivation can be replaced by an equivalent derivation in which no application of the replacement rule is followed by an application of the substitution rule, and therefore no application of the replacement rule involves an equation containing a variable.

This observation provides a canonical form for derivations of the values of general recursive functions.

Lemma 5-11. *Let the n-variable general recursive function f be defined by a certain set of equations $\mathcal{E}$, with f_k^n as principal function letter. Then if $f(u_1, \ldots, u_n) = u$, there must exist a derivation of the equation $f_k^n(\bar{u}_1, \ldots, \bar{u}_n) = \bar{u}$ in which no application of the replacement rule involves an equation containing a variable.*

Derivations of the type specified in Lemma 5-11 will be referred to as *standard form derivations.*

We now use Lemma 5-11 to show that general recursiveness is preserved by functional composition. To make the argument easier to follow, we will freely use symbols such as x, y, and z to stand for variables and symbols such as f, g, and h to stand for function letters. Once the basic argument is understood, it can easily be recast in terms of the standard symbols for variables and function letters.

Lemma 5-12. *Let $g_1, \ldots, g_m$ be n-variable general recursive functions, and let h be an m-variable general recursive function. Then the composite function $f = h \circ (g_1, \ldots, g_m)$ is also a general recursive function.*

Proof. Since $g_1, \ldots, g_m$ and h are general recursive functions, each one must be defined by some set of equations. Let these sets be denoted $\mathcal{G}_1, \ldots, \mathcal{G}_m$ and $\mathcal{H}$, respectively. Assume that the function letters used in the various equations have been chosen so that no one letter appears in two different defining sets. Assume further that the principal function letters associated with the definitions of the functions $g_1, \ldots, g_m$ and h are respectively $\mathcal{g}_1, \ldots, \mathcal{g}_m$ and $\mathcal{h}$. Now define $\mathcal{F}$ to be the set consisting of all the equations of $\mathcal{G}_1, \ldots, \mathcal{G}_m$ and $\mathcal{H}$, together with the one new equation

$$\mathcal{f}(x_1, \ldots, x_n) = \mathcal{h}(\mathcal{g}_1(x_1, \ldots, x_n), \ldots, \mathcal{g}_m(x_1, \ldots, x_n)) \qquad (5\text{-}12)$$

where $\mathcal{f}$ is a function letter that does not appear in any of the sets $\mathcal{G}_1, \ldots, \mathcal{G}_m$ or $\mathcal{H}$. We propose to show that the set of equations $\mathcal{F}$ defines the composite function $f = h \circ (g_1, \ldots, g_m)$ with $\mathcal{f}$ as principal function letter. To do this, we must show that an equation of the form $\mathcal{f}(\bar{u}_1, \ldots, \bar{u}_n) = \bar{u}$ can be derived from $\mathcal{F}$ iff $f(u_1, \ldots, u_n) = u$. The argument breaks down into two parts.

1.   Suppose that $f(u_1, \ldots, u_n) = u$ for some choice of natural numbers $u_1, \ldots, u_n$, and u. Then there must exist natural numbers $w_1, \ldots, w_m$ such that $g_1(u_1, \ldots, u_n) = w_1, \ldots, g_m(u_1, \ldots, u_n) = w_m$, and $h(w_1, \ldots, w_m) = u$. Since the functions $g_1, \ldots, g_m$ and h are defined by the sets of equations $\mathcal{G}_1, \ldots, \mathcal{G}_m$ and $\mathcal{H}$ with $\mathcal{g}_1, \ldots, \mathcal{g}_m$ and $\mathcal{h}$ as principal function letters, this means that the equation $\mathcal{g}_i(\bar{u}_1, \ldots, \bar{u}_n) = \bar{w}_i$ is derivable from $\mathcal{G}_i$, for $1 \le i \le m$, and that $\mathcal{h}(\bar{w}_1, \ldots, \bar{w}_m) = \bar{u}$ is derivable from $\mathcal{H}$. Since $\mathcal{F}$ includes all the equations of $\mathcal{G}_1, \ldots, \mathcal{G}_m$ and $\mathcal{H}$, these same equations

$$\mathcal{g}_1(\bar{u}_1, \ldots, \bar{u}_n) = \bar{w}_1$$
$$\vdots$$
$$\mathcal{g}_m(\bar{u}_1, \ldots, \bar{u}_n) = \bar{w}_m$$

and

$$\mathcal{h}(\bar{w}_1, \ldots, \bar{w}_m) = \bar{u}$$

must be derivable from $\mathcal{F}$. And since $\mathcal{F}$ includes the equation (5-12), the equation

$$\mathcal{f}(\bar{u}_1, \ldots, \bar{u}_n) = \mathcal{h}(\mathcal{g}_1(\bar{u}_1, \ldots, \bar{u}_n), \ldots, \mathcal{g}_m(\bar{u}_1, \ldots, \bar{u}_n))$$

is also derivable from $\mathcal{F}$. At this point it is obvious that $m + 1$ applications of the replacement rule will yield the equation

$$\mathcal{f}(\bar{u}_1, \ldots, \bar{u}_n) = \bar{u}.$$

Thus the fact that $f(u_1, \ldots, u_n) = u$ implies that the equation $\mathcal{f}(\bar{u}_1, \ldots, \bar{u}_n) = \bar{u}$ is derivable from $\mathcal{F}$.

2.   Suppose that the equation $\mathcal{f}(\bar{u}_1, \ldots, \bar{u}_n) = \bar{u}$ is derivable from $\mathcal{F}$, hence that there exists a standard form derivation of $\mathcal{f}(\bar{u}_1, \ldots, \bar{u}_n) = \bar{u}$ from $\mathcal{F}$. Note that the only member of $\mathcal{F}$ that contains the symbol $\mathcal{f}$ is the equation (5-12). Thus in order to construct a standard form derivation of $\mathcal{f}(\bar{u}_1, \ldots, \bar{u}_n) = \bar{u}$, it is necessary to make substitutions in (5-12) so as to obtain the equation

$$\mathcal{f}(\bar{u}_1, \ldots, \bar{u}_n) = \mathcal{h}(\mathcal{g}_1(\bar{u}_1, \ldots, \bar{u}_n), \ldots, \mathcal{g}_m(\bar{u}_1, \ldots, \bar{u}_n)). \qquad (5\text{-}13)$$

Since the symbol h does not appear together with $g_1, \ldots, g_m$ in any defining equation except (5-12), the only way of eventually obtaining $f(\bar{u}_1, \ldots, \bar{u}_n) = \bar{u}$ is by making replacements in (5-13) from equations of the form $g_1(\bar{u}_1, \ldots, \bar{u}_n) = \bar{w}_1, \ldots, g_m(\bar{u}_1, \ldots, \bar{u}_n) = \bar{w}_m$. Such replacements will yield an equation of the form

$$f(\bar{u}_1, \ldots, \bar{u}_n) = h(\bar{w}_1, \ldots, \bar{w}_m).$$

At this point the only way of reaching the desired end result is to make a replacement from an equation of the form $h(\bar{w}_1, \ldots, \bar{w}_m) = \bar{u}$. Thus if $f(\bar{u}_1, \ldots, \bar{u}_n) = \bar{u}$ is to be derivable from $\mathcal{F}$ it is necessary that equations of the form

$$g_1(\bar{u}_1, \ldots, \bar{u}_n) = \bar{w}_1$$
$$\vdots$$
$$g_m(\bar{u}_1, \ldots, \bar{u}_n) = \bar{w}_m$$

and

$$h(\bar{w}_1, \ldots, \bar{w}_m) = \bar{u}$$

be derivable from $\mathcal{F}$. Now, if $g_i(\bar{u}_1, \ldots, \bar{u}_n) = \bar{w}_i$ is derivable from $\mathcal{F}$, it must also be derivable from $\mathcal{G}_i$, since the symbol g_i does not appear in the left-hand side of any defining equation other than those of $\mathcal{G}_i$. And since the function g_i is defined by $\mathcal{G}_i$, the fact that $g_i(\bar{u}_1, \ldots, \bar{u}_n) = \bar{w}_i$ is derivable from $\mathcal{G}_i$ means that $g_i(u_1, \ldots, u_n) = w_i$. Similar reasoning shows that $h(w_1, \ldots, w_m) = u$, from which it follows that $f(u_1, \ldots, u_n) = u$. Thus the fact that $f(\bar{u}_1, \ldots, \bar{u}_n) = \bar{u}$ is derivable from $\mathcal{F}$ implies that $f(u_1, \ldots, u_n)$ equals u.

To summarize, we have shown that the equation $f(\bar{u}_1, \ldots, \bar{u}_n) = \bar{u}$ is derivable from $\mathcal{F}$ iff $f(u_1, \ldots, u_n) = u$. This means that the equations $\mathcal{F}$ define the function f and that as a consequence f is a general recursive function. $\square$

We turn next to the case of primitive recursion. Since this case uses much of the same reasoning presented in the proof of Lemma 5-12, we give only an outline of the argument.

Lemma 5-13. *Let g and h be n- and $(n + 2)$-variable general recursive functions, respectively. Then the $(n + 1)$-variable function f defined from g and h by primitive recursion is also general recursive.*

Argument. Suppose that g and h are defined by the sets of equations $\mathcal{G}$ and $\mathcal{H}$, respectively. Assume that these sets have no function letters in common, and that the principal function letters associated with g and h are g and h, respectively. Define $\mathcal{F}$ to be the set consisting of all the equations of $\mathcal{G}$ and $\mathcal{H}$ plus the two new equations

$$f(x_1, \ldots, x_n, \bar{0}) = g(x_1, \ldots, x_n)$$

and

$$f(x_1, \ldots, x_n, S(y)) = h(x_1, \ldots, x_n, y, f(x_1, \ldots, x_n, y))$$

where f is a function letter that does not appear in $\mathcal{G}$ or $\mathcal{H}$. Then the set $\mathcal{F}$ defines the function f with f as principal function letter. Establishing this fact rigorously requires an induction argument on the variable y, both parts of which rely on reasoning similar to that used in the proof of Lemma 5-12. The details are left to the reader.   □

Showing that general recursiveness is preserved by minimalization requires some preliminary work. Suppose that g is any total, $(n+1)$-variable general recursive function. Define the new $(n+1)$-variable function $\hat{g}$ so that

$$\hat{g}(x_1, \ldots, x_n, 0) = 1$$

$$\hat{g}(x_1, \ldots, x_n, y^+) = \text{sg}(g(x_1, \ldots, x_n, y) \cdot \hat{g}(x_1, \ldots, x_n, y)).$$

Since $\hat{g}$ can be obtained from g by means of primitive recursion and appropriate compositions, it follows from Lemmas 5-12 and 5-13 that $\hat{g}$ is general recursive. Moreover, it is easy to see that $\hat{g}$ is total and that

$$\hat{g}(x_1, \ldots, x_n, y) = \begin{cases} 0 & \text{if } g(x_1, \ldots, x_n, z) = 0 \text{ for some } z \text{ and} \\ & y \text{ exceeds the least such } z \\ 1 & \text{otherwise} \end{cases}$$

The association of such a function $\hat{g}$ with every total general recursive function g leads directly to our next result.

Lemma 5-14. *Let g be any total, $(n+1)$-variable general recursive function. Then the n-variable function f, where*

$$f(x_1, \ldots, x_n) = \mu z[g(x_1, \ldots, x_n, z) = 0]$$

is also general recursive.

Argument. Since g is total and general recursive, the associated function $\hat{g}$ is also general recursive. Suppose that $\hat{g}$ is defined by the set of equations $\hat{\mathcal{G}}$, with $\hat{g}$ as principal function letter. Let $\mathcal{F}$ be the set containing all the equations of $\hat{\mathcal{G}}$ plus the two new equations

$$f(x_1, \ldots, x_n) = h(\hat{g}(x_1, \ldots, x_n, y), \hat{g}(x_1, \ldots, x_n, S(y)), y) \quad (5\text{-}14)$$

$$h(\bar{1}, \bar{0}, y) = y \quad (5\text{-}15)$$

where f and h are function letters not appearing in $\hat{\mathcal{G}}$. Then the set $\mathcal{F}$ defines the function f with f as principal function letter.

To see this, first suppose that $f(u_1, \ldots, u_n) = u$ for some choice of natural numbers $u_1, \ldots, u_n$ and u. According to the definitions of f and $\hat{g}$, this means that $\hat{g}(u_1, \ldots, u_n, u) = 1$ and $\hat{g}(u_1, \ldots, u_n, u+1) = 0$. Thus the equations $\hat{g}(\bar{u}_1, \ldots, \bar{u}_n, \bar{u}) = \bar{1}$ and $\hat{g}(\bar{u}_1, \ldots, \bar{u}_n, S(\bar{u})) = \bar{0}$ are both derivable from $\hat{\mathcal{G}}$, hence from $\mathcal{F}$. The obvious substitutions and replacements in Eqs. (5-14) and (5-15) then yield the equation $f(\bar{u}_1, \ldots, \bar{u}_1) = \bar{u}$. Conversely, suppose that the equation $f(\bar{u}_1, \ldots, \bar{u}_n) = \bar{u}$ is derivable from $\mathcal{F}$. Using reasoning similar to that of Lemma 5-12, we find that this can only be the case because the equations $\hat{g}(\bar{u}_1, \ldots, \bar{u}_n, \bar{u}) = \bar{1}$ and $\hat{g}(\bar{u}_1, \ldots, \bar{u}_n, S(\bar{u})) = \bar{0}$ are derivable from $\mathcal{F}$, and ultimately from $\mathcal{G}$. But this implies that $\hat{g}(u_1, \ldots, u_n, u) = 1$ and $\hat{g}(u_1, \ldots, u_n, u + 1) = 0$, which means that $f(u_1, \ldots, u_n) = u$. Thus $f(\bar{u}_1, \ldots, \bar{u}_n) = \bar{u}$ is derivable from $\mathcal{F}$ iff $f(u_1, \ldots, u_n) = u$.   ☐

We are now ready to establish the first link between the general recursive functions and the μ-recursive functions.

Theorem 5-10. *Every μ-recursive function is a general recursive function.*

Proof. The proof is accomplished by induction on the number of applications of composition, primitive recursion, and minimalization needed to define the μ-recursive function in question. The basis is established by Lemma 5-10, while the induction step follows from Lemmas 5-12, 5-13, and 5-14.   ☐

Thus the class of general recursive functions is at least as large as the class of μ-recursive functions.

The μ-Recursiveness of the General Recursive Functions

We now turn to the question of whether there are any general recursive functions that are not μ-recursive. In other words, are there any general recursive functions that cannot be obtained from the constant, projection, and successor functions by suitably chosen compositions, primitive recursions, and minimalizations? In spite of the comprehensiveness of the definition of general recursive functions, the answer turns out to be no. It is this important fact that we now set out to prove.

The approach that we will take is similar to that used to establish the μ-recursiveness of the Turing-computable functions. We first arithmetize the process of evaluating general recursive functions. That is, we assign natural numbers to all the terms, equations, and sequences of equations that might make up a derivation. Then for each n-variable general recursive function f we define a primitive recursive predicate that holds for the natural numbers $u_1, \ldots, u_n$ and z iff z is the number of a derivation of an equation of the form $f(\bar{u}_1, \ldots, \bar{u}_n) = \bar{u}$, where f is the principal function letter associated with f. Application of the minimalization operator to this predicate then yields the number of the "first" such derivation. Once this number has been determined, the corresponding function value $f(u_1, \ldots, u_n)$ can be extracted from it by an appropriate primitive recursive function.

The assignment of numbers to terms will be done inductively. In this way the structure of the numbering scheme can be made to reflect the structure of the terms being represented. Similarly, the numbers assigned to equations will be made to reflect the way that equations are formed from terms. The number associated with the expression s (whether s is a term or an equation) will be denoted $\langle s \rangle$ and referred to as the *Gödel number* of s, even though we will not be using a standard Gödel-numbering scheme.

We begin by assigning odd Gödel numbers to the formal variables, so that

$$\langle x_i \rangle = 2i - 1.$$

The remaining terms are assigned even Gödel numbers according to the following scheme. First, we set

$$\langle 0 \rangle = 2.$$

Then for each term t we assign to the term $S(t)$ the number

$$\langle S(t) \rangle = 2 \cdot 3^{\langle t \rangle}.$$

And for each function letter f_i^n and each choice of terms $t_1, \ldots, t_n$ we assign to the term $f_i^n(t_1, \ldots, t_n)$ the number

$$\langle f_i^n(t_1, \ldots, t_n) \rangle = 2^{i+2} \, 3^{\langle t_1 \rangle} 5^{\langle t_2 \rangle} \cdots p_n^{\langle t_n \rangle}.$$

In this way each term gets assigned exactly one natural number.

The Gödel number of an equation must indicate the terms from which that equation is formed. We accordingly assign to the equation $t_1 = t_2$ the number

$$\langle t_1 = t_2 \rangle = \pi^2(\langle t_1 \rangle, \langle t_2 \rangle).$$

Finally, the Gödel number assigned to the sequence of equations $e_0, e_1, \ldots, e_m$ will be chosen in the obvious way:

$$\langle e_0, e_1, \ldots, e_m \rangle = 2^{\langle e_0 \rangle} 3^{\langle e_1 \rangle} \cdots p_m^{\langle e_m \rangle}.$$

Note that there is no intrinsic way of determining whether a particular natural number is meant to represent a term, or an equation, or a sequence of equations. Instead, we must rely on the context to distinguish among these three possibilities.

Example 5-14

a) The Gödel numbers of the variables $x_1, x_2, x_3, x_4, x_5, \ldots$, are respectively $1, 3, 5, 7, 9, \ldots$.

b) The Gödel numbers of the first four numerals are:

$$\langle \bar{0} \rangle = 2 \qquad\qquad \langle \bar{2} \rangle = 2 \cdot 3^{18}$$
$$\langle \bar{1} \rangle = 2 \cdot 3^2 = 18 \qquad \langle \bar{3} \rangle = 2 \cdot 3^{2 \cdot 3^{18}}$$

c) The Gödel numbers of the terms $S(x_1)$, $S(x_2)$, and $S(S(x_1))$ are respectively $2 \cdot 3^1 = 6$, $2 \cdot 3^3 = 54$, and $2 \cdot 3^6 = 1458$.

d) The following are examples of Gödel numbers of terms containing function letters.

$$\langle f_0^1(\bar{0}) \rangle = 2^2 \cdot 3^2 = 36 \qquad \langle f_0^1(f_0^1(\bar{0})) \rangle = 2^2 \cdot 3^{36}$$

$$\langle f_0^1(\bar{1}) \rangle = 2^2 \cdot 3^{18} \qquad\qquad \langle f_1^2(\bar{0}, \bar{1}) \rangle = 2^3 \cdot 3^2 \cdot 5^{18}$$

$$\langle f_3^1(x_1) \rangle = 2^5 \cdot 3^1 = 96 \qquad \langle f_0^2(x_1, \bar{2}) \rangle = 2^2 \cdot 3^1 \cdot 5^{2 \cdot 3^{18}}$$

e) The Gödel number of the equation $f_0^1(x_1) = S(x_1)$ is $\pi^2(12, 6) = 183.\square$

We now consider several functions and predicates that prove useful in the manipulation of Gödel numbers. We begin by introducing functions that convert numerals into their Gödel numbers and vice versa. These functions are denoted gn and val, and are defined as follows.

$$\text{gn } (n) = \langle \bar{n} \rangle$$

$$\text{val } (z) = \begin{cases} n & \text{if} \quad z = \langle \bar{n} \rangle \\ z + 1 & \text{otherwise} \end{cases}$$

We next provide predicates that determine whether a given natural number is the Gödel number of a variable, numeral, term, or equation. Specifically, we define the predicates Var, Num, Term, and Eqn so that:

$\text{Var}(z) \Leftrightarrow z$ is the Gödel number of some variable

$\text{Num}(z) \Leftrightarrow z$ is the Gödel number of some numeral

$\text{Term}(z) \Leftrightarrow z$ is the Gödel number of some term

$\text{Eqn}(z) \Leftrightarrow z$ is the Gödel number of some equation

It is a more or less routine matter to show that the functions gn and val and the predicates Var, Num, Term, and Eqn are all primitive recursive. The detailed derivations are given at the end of this section.

The ability to detect whether a Gödel number represents a particular kind of expression makes it possible to "decode" the Gödel number of a complicated equation and determine how that equation is formed. And this in turn makes it possible to determine whether one equation could have been obtained from another by the substitution rule or whether one equation could have been obtained from two others by the replacement rule. For this purpose we introduce the predicates Subs and Rep, which are defined so that:

$\text{Subs}(z_1, z_2) \quad \Leftrightarrow \quad z_1$ and z_2 are the Gödel numbers of equations, call them e_1 and e_2, respectively, and e_2 is derivable from e_1 by the substitution rule

$\text{Rep}(z_1, z_2, z_3) \quad \Leftrightarrow \quad z_1, z_2,$ and z_3 are the Gödel numbers of equations, call them e_1, e_2, and e_3, respectively, and e_3 is derivable from e_1 and e_2 by using the replacement rule to make a replacement in e_1 from e_2

It follows from the primitive recursiveness of gn, val, Var, Num, Term, and Eqn that the predicates Subs and Rep are also primitive recursive. Again the detailed derivations are given at the end of this section.

The predicates Subs and Rep provide the basic tools needed to show that every general recursive function is μ-recursive. We consider here only the case of one-variable functions. Once the argument for this special case is understood, it can easily be generalized to the case of n-variable functions.

Suppose, then, that f is a one-variable general recursive function, and that f is defined by the (finite) set of equations $\mathcal{E}$ using f_0^1 as principal function letter. We first define the predicate $DS_{\mathcal{E}}$ so that

$$DS_{\mathcal{E}}(z) \quad \Leftrightarrow \quad z \text{ is the Gödel number of an equation in the defining set } \mathcal{E}.$$

Since $\mathcal{E}$ is a finite set, $DS_{\mathcal{E}}$ holds for only a finite number of natural numbers and so is necessarily a primitive recursive predicate.

Next we introduce a predicate that determines whether a given number is the Gödel number of a formal derivation of a value of the function f. Specifically, we define the predicate $Der_{\mathcal{E}}$ so that

$Der_{\mathcal{E}}(u, z) \quad \Leftrightarrow \quad z$ is the Gödel number of a sequence of equations that constitutes a formal derivation (from $\mathcal{E}$) of an equation of the form $f_0^1(\bar{u}) = \bar{n}$, where $\bar{n}$ is a numeral.

Note that $Der_{\mathcal{E}}(u, z)$ is to hold iff: (i) each exponent in the prime decomposition of z represents an equation that either belong to $\mathcal{E}$, or is derivable from an earlier equation by substitution, or is derivable from two earlier equations by replacement; and (ii) the last exponent in the prime decomposition of z represents an equation of the form $f_0^1(\bar{u}) = \bar{n}$. This observation leads to the following specification of $Der_{\mathcal{E}}$.

$$Der_{\mathcal{E}}(u, z) \quad \Leftrightarrow \quad (\forall k)^{|z|} \left[DS_{\mathcal{E}}(E(k, z)) \vee (\exists i)^{k \doteq 1} (Subs(E(i, z), E(k, z))) \right.$$

$$\left. \vee (\exists i)^{k \doteq i} (\exists j)^{k \doteq i} (Rep(E(i, z), E(j, z), E(k, z))) \right]$$

$$\wedge [\sigma_1^2(E(|z|, z)) = 4 \cdot 3^{gn(u)} \wedge Num(\sigma_2^2(E(|z|, z)))]$$

The reader is encouraged to convince him or herself that this specification accurately reflects the definition of $Der_{\mathcal{E}}$ and that, as a consequence, $Der_{\mathcal{E}}$ is primitive recursive.

Remember that an equation of the form $f_0^1(\bar{u}) = \bar{n}$ can be derived from $\mathcal{E}$ iff the function f is defined for argument u. Thus for any given choice of u, the predicate $Der_{\mathcal{E}}(u, z)$ holds for some natural number z iff $f(u)$ is defined. Now consider the function $\eta_{\mathcal{E}}$, where

$$\eta_{\mathcal{E}}(u) = \mu z[Der_{\mathcal{E}}(u, z)].$$

This function is obviously μ-recursive. Moreover, the value of $\eta_\varepsilon(u)$ will be the Gödel number of a derivation of an equation of the form $f_0^1(\bar{u}) = \bar{n}$ whenever $f(u)$ is defined, and will be undefined whenever $f(u)$ is undefined. In effect, η_ε produces the "first" Gödel number of a derivation of the value of $f(u)$.

Once this critical Gödel number z has been obtained, the value of f can be determined by "decoding" the right-hand side of the last equation in the derivation represented by z. In particular, we can write

$$f(u) = \text{val}\,(\sigma_2^2(\text{E}(|\eta_\varepsilon(u)|, \eta_\varepsilon(u)))).$$

Since η_ε is μ-recursive and val is primitive recursive, we conclude that the function f must in fact be μ-recursive.

The preceding argument can be extended in a straightforward way to accommodate functions of more than one variable. As a result, we have:

Theorem 5-11. *Every general recursive function is also μ-recursive.*

Theorems 5-10 and 5-11 together yield:

Corollary 5-11.1. *The set of general recursive functions is identical to the set of μ-recursive functions.*

In spite of its apparent breadth, then, the definition of the general recursive functions does not yield any functions that are not μ-recursive. Thus we find that three rather differently defined classes of effectively computable functions— namely the Turing-computable, μ-recursive, and general recursive functions— all turn out to be equivalent. The significance of this equivalence will be discussed in the next chapter. In the meantime, it is appropriate to note the critical role played by arithmetization in establishing the equivalence, since it is this technique that has enabled us to show that all Turing-computable functions and all general recursive functions are μ-recursive.

Derivation of the Auxiliary Functions and Predicates

We now establish the primitive recursiveness of the various functions and predicates needed to obtain the predicate Der_ε.

The functions gn and val are easy to deal with. Recall that $\text{gn}(n)$ is defined to be the Gödel number of the numeral $\bar{n}$, while $\text{val}(z)$ is defined to have the value n if z is the Gödel number of $\bar{n}$ and to have the value $z + 1$ otherwise. Thus

$$\text{gn}\,(0) = 2$$
$$\text{gn}\,(n^+) = 2 \cdot 3^{\text{gn}\,(n)}$$

and

$$\text{val}\,(z) = \mu^z n[\text{gn}(n) = z]$$

from which it follows that both gn and val are primitive recursive.

Next consider the four predicates Var, Num, Term, and Eqn, which are defined as follows:

Var$(z) \Leftrightarrow z$ is the Gödel number of a variable

Num$(z) \Leftrightarrow z$ is the Gödel number of a numeral

Term$(z) \Leftrightarrow z$ is the Gödel number of a term

Eqn$(z) \Leftrightarrow z$ is the Gödel number of an equation

The primitive recursiveness of Var and Num follows easily from the properties of the Gödel numbering scheme, since:

$$\text{Var}(z) \Leftrightarrow \text{Od}(z)$$

and

$$\text{Num}(z) \Leftrightarrow (\overset{z}{\exists} x)[\text{gn}(x) = z]$$

The primitive recursiveness of Term can be established by showing that its characteristic function χ is primitive recursive. Note that $\chi(z)$ is to have the value 1 only if: (i) z is the number of a variable; or (ii) z is the number of the numeral $\bar{0}$; or (iii) z has the form $2 \cdot 3^a$, where a is the number of some term; or (iv) z has the form $2^i 3^{a_1} 5^{a_2} \cdots p_n^{a_n}$, where $i \geq 2$ and each of $a_1, \ldots, a_n$ is the number of some term. Thus χ can be described by the following (informal) course-of-values recursion.

$$\chi(0) = 0$$

$$\chi(z^+) = \begin{cases} 1 & \text{if} \quad (\text{Var } (z^+)) \vee (z^+ = 2) \\ & \qquad \vee (|z^+| = 1 \wedge \text{E}(0, z^+) = 1 \wedge \chi(\text{E}(1, z^+)) = 1) \\ & \qquad \vee \left((\text{E}(0, z^+) > 1) \wedge \left(\overset{|z^+|}{\underset{i=1}{\forall i}} \right) [\chi(\text{E}(i, z^+)) = 1] \right) \\ 0 & \text{otherwise} \end{cases}$$

Finally, since

$$\text{Eqn } (z) \quad \Leftrightarrow \quad \text{Term } (\sigma_1^2(z)) \wedge \text{Term } (\sigma_2^2(z))$$

the predicate Eqn must also be primitive recursive.

In order to show that the predicates Subs and Rep are primitive recursive, we introduce the new three-variable function tsub. The role of this function is to compute the effect of substituting one term for another in a third. Specifically, if u, v, and w are the Gödel numbers of the terms t_u, t_v, and t_w, then tsub(u, v, w) is to be the Gödel number of the term $t_w(t_u/t_v)$. If u, v, and w do not all represent Gödel numbers of terms, the value of tsub(u, v, w) is unimportant; we more or less arbitrarily assign the value w. Thus:

$$\text{tsub } (u, v, w) = \begin{cases} \langle t_w(t_u/t_v) \rangle & \text{if} \quad u = \langle t_u \rangle, \; v = \langle t_v \rangle, \; w = \langle t_w \rangle \\ w & \text{otherwise} \end{cases}$$

Note that whenever t_v and t_w are the same term, the result of substituting t_u for t_v is just the term t_u. Consider, then, the case in which t_v and t_w are distinct terms. If t_w is a variable or the numeral 0, the result of substituting t_u for t_v is t_w. If t_w is a more complicated term, the effect of substituting t_u for t_v can be obtained by separately substituting t_u for t_v in each of the terms from which t_w is formed. These observations suggest the following course-of-values specification of tsub:

$$\mathrm{tsub}(u, v, 0) = 0$$

$$\mathrm{tsub}(u, v, w^+) = \begin{cases} u & \text{if} \quad v = w^+ \\ w & \text{if} \quad v \neq w^+ \wedge (\mathrm{Var}(w^+) \vee w^+ = 2) \\ h(u, v, w) & \text{if} \quad v \neq w^+ \wedge \mathrm{Term}(w^+) \wedge \mathrm{Var}(w^+) \wedge w^+ \neq 2 \\ w^+ & \text{otherwise} \end{cases}$$

where

$$h(u, v, w) = 2^{E(0,\, w^+)} \prod_{i=1}^{|w^+|} \mathrm{pn}(i)^{\mathrm{tsub}(u,\, v,\, E(i,\, w^+))}$$

The reader is encouraged to convince him or herself that this specification accurately represents the function tsub, and consequently that tsub is primitive recursive.

We are now ready to deal with Subs and Rep. Recall that $\mathrm{Subs}(z_1, z_2)$ is to hold iff z_1 and z_2 are Gödel numbers of equations e_1 and e_2 such that e_2 is obtained by substituting a certain numeral for each appearance of a certain variable in e_1. Thus we can write:

$$\mathrm{Subs}(z_1, z_2) \quad \Leftrightarrow \quad \mathrm{Eqn}(z_1) \wedge \mathrm{Eqn}(z_2) \wedge \overset{z_2}{(\exists n)}\overset{z_1}{(\exists v)}[\mathrm{Num}(n) \wedge \mathrm{Var}(v)$$
$$\wedge\; (\sigma_1^2(z_2) = \mathrm{tsub}(n, v, \sigma_1^2(z_1))) \wedge (\sigma_2^2(z_2) = \mathrm{tsub}(n, v, \sigma_2^2(z_1)))]$$

(Note that the Gödel number of a variable appearing in the equation e_1 cannot exceed the Gödel number of e_1, nor can the Gödel number of a numeral appearing in e_2 exceed the Gödel number of e_2.) The predicate Subs is therefore primitive recursive.

Recall that $\mathrm{Rep}\,(z_1, z_2, z_3)$ is to hold iff z_1, z_2, and z_3 are the Gödel numbers of equations e_1, e_2, and e_3 such that e_3 is obtained by replacement in e_1 from e_2. According to the replacement rule, this means that $\mathrm{Rep}\,(z_1, z_2, z_3)$ is to hold iff there exists an equation e such that (i) e_1 is obtained by replacing every occurrence of some variable in e by the left-hand side of e_2, and (ii) e_3 is obtained by replacing every occurrence of that same variable by the right-hand side of e_2. Note that the variable in question can be any variable that does not appear in e_1 or e_3. Since every odd number is the Gödel number of some variable, it follows that $v(z_1, z_2, z_3) = 2(z_1 + z_3) + 1$

represents an appropriate variable. These observations suggest the following specification of the predicate Rep.

$$\text{Rep} (z_1, z_2, z_3) \Leftrightarrow \text{Eqn} (z_1) \wedge \text{Eqn} (z_2) \wedge \text{Eqn} (z_3) \wedge (z_1 \neq z_3)$$

$$\wedge (\exists z)[\text{Eqn} (z) \wedge (\sigma_1^2(z_1) = \text{tsub} (\sigma_1^2(z_2), v(z_1, z_2, z_3), \sigma_1^2(z)))$$

$$\wedge (\sigma_2^2(z_1) = \text{tsub} (\sigma_1^2(z_2), v(z_1, z_2, z_3), \sigma_2^2(z)))$$

$$\wedge (\sigma_1^2(z_3) = \text{tsub} (\sigma_2^2(z_2), v(z_1, z_2, z_3), \sigma_1^2(z)))$$

$$\wedge (\sigma_2^2(z_3) = \text{tsub} (\sigma_2^2(z_2), v(z_1, z_2, z_3), \sigma_2^2(z)))]$$

To establish the primitive recursiveness of Rep, it will suffice to show that the quantification used in the preceding expression can be bounded by a primitive recursive function of z_1, z_2, and z_3. To this end, note that if e_3 is derivable from e_1 and e_2 by replacement, an appropriate auxiliary equation e can be obtained by substituting the variable represented by $v(z_1, z_2, z_3)$ for certain of the appearances of the left-hand side of e_2 in e_1. Now, the Gödel number of the resulting equation is certainly no larger than that of the equation that would result from substituting $v(z_1, z_2, z_3)$ for *every* appearance of the left-hand side of e_2 in e_1. Thus the value

$$w(z_1, z_2, z_3) = \pi^2 (\text{tsub} (v(z_1, z_2, z_3), \sigma_1^2(z_2), \sigma_1^2(z_1)),$$

$$\text{tsub} (v(z_1, z_2, z_3), \sigma_1^2(z_2), \sigma_2^2(z_1)))$$

provides an upper bound on the Gödel number z corresponding to the desired equation e. Since w obviously represents a primitive recursive function, it follows that Rep is primitive recursive.

5.4 SUMMARY

Many recursive definition schemes have been shown to yield primitive recursive functions when the functions to which they are applied are primitive recursive. But not all definition schemes have this property. Ackermann's function provides an example of a total, effectively computable function that can be defined recursively from primitive recursive functions but that is not itself primitive recursive. The recursive specification of Akermann's function differs from those studied earlier in that it is "nested"; i.e., it defines new function values in terms of values of the function for arguments which are themselves function values. As a consequence, Ackermann's function is not primitive recursive. In fact, it is possible to show that the value of Ackermann's function grows faster than that of any primitive recursive function.

The class of primitive recursive functions can be enlarged in several ways. One is to introduce a new composition rule, obtained by eliminating the bound on the minimalization operator. Use of the resulting unbounded minimalization operator simplifies the definition of many total functions and makes it possible to define nontotal functions from total functions. The functions that can be obtained from the constant, projection, and successor functions by means of functional composition, primitive recursion, and unbounded minimalization are called μ-recursive functions. Thus the μ-recursive functions include all the primitive recursive functions. They also include many other functions, among them Ackermann's function. However, it is readily shown that every μ-recursive function is necessarily Turing computable.

Arithmetization provides one of the most important methods of establishing the μ-recursiveness of an effectively computable function. Several steps are typically involved in arithmetizing the computation of a function f. First a scheme is devised for using natural numbers to represent the various steps in an evaluation of f. Then a "trace function" is defined whose values give, as a function of n, the number representing the nth step in the desired evaluation. Next the minimalization operator is used to identify the step at which the evaluation terminates. Finally, the value of the function f is obtained by "decoding" the number representing the last step of the evaluation. Assuming that the trace function and the necessary decoding function can be shown to be primitive recursive—as is invariably the case—it follows that the given function f is μ-recursive.

This approach can be used to show not only that Ackermann's function is μ-recursive, but that in fact every Turing-computable function is μ-recursive. Thus the class of μ-recursive functions is identical to the class of Turing-computable functions. Arithmetization of the evaluation of Turing-computable functions also reveals certain canonical forms in which every μ-recursive function can be expressed. In particular, we find that only one application of unbounded minimalization is necessary in the specification of a μ-recursive function, and that that application can always be made to a primitive recursive function or predicate. We also find that the only thing that can prevent a μ-recursive function from being primitive recursive is the lack of a primitive recursive bound on the number of steps needed to evaluate the function.

A more inclusive way of enlarging the class of primitive recursive functions is to permit any recursive scheme whatever to be used in the definition of a function. In formalizing this approach, we accept as a valid defining scheme any set of equations from which unique function values can be derived using two basic operations: substituting a number for a variable, and replacing a term by an equivalent term. The functions that can be defined in this way are called general recursive functions. The class of general recursive functions is readily shown to include all the μ-recursive functions, and therefore all the Turing-computable functions. And by arithmetizing the process of deriving the values of a general recursive function, it is possible to prove that every general recursive function is necessarily μ-recursive.

In conclusion, we find that, although the primitive recursive functions include a great many useful functions, they do not include all the effectively computable functions. The Turing-computable, μ-recursive, and general recursive functions represent three possible extensions of the class of primitive recursive functions. The fact that these three extensions turn out to be identical provides the basis for the formal definition of computability discussed in the next chapter.

5.5 PROBLEMS

Section 5.1

5.1.1   Consider the one-variable function A_*, where $A_*(x) = A(x, x)$.

a) Use Theorem 5-1 to prove that A_* is not primitive recursive.

b) Let f be any specified primitive recursive function of one variable. Prove that $A_*(x) > f(x)$ for all sufficiently large values of x.

5.1.2   In this problem we investigate the number of steps needed to evaluate Ackermann's function. Specifically, we consider the two-variable function B, where $B(x, y)$ is the number of applications of Eqs. (5-4) needed to evaluate $A(x, y)$.

a) Give a recursive definition of B.

b) Establish each of the following inequalities:

    i) $B(x^+, y) > y$ for all x and y

    ii) $B(x, y) > x$ for all x and y

    iii) $B(x^+, y^+) > A(x, y)$ for all x and y

c) Prove that B is not primitive recursive.

5.1.3   The one-variable primitive recursive function f is derived from certain constant, projection, and successor functions by a total of c functional compositions and r primitive recursions. The largest constant function value appearing in the derivation is k. Specify a natural number m such that $A(m, x) > f(x)$ for all x. Can the value of m be decreased if we require only that $A(m, x) > f(x)$ for sufficiently large x?

5.1.4   Let the one-variable functions $h_0, h_1, \ldots, h_i, \ldots$ be defined so that $h_i(x) = A(x, i)$. Thus $h_i(0), h_i(1), \ldots, h_i(n), \ldots$ are the values appearing in the ith column of Fig. 5-2. Prove that none of the functions $h_0, h_1, \ldots, h_i, \ldots$ are primitive recursive. *Hint:* Assume the contrary and make use of Lemma 5-4.

Section 5.2

5.2.1 In each of the following parts, make use of unbounded minimalization to establish the μ-recursiveness of the specified function f. In parts (b) through (d), assume that g denotes a total μ-recursive function.

a) $f(x) = \begin{cases} \sqrt{x} & \text{if } x \text{ is a perfect square} \\ \uparrow & \text{otherwise} \end{cases}$

b) $f(x) = \begin{cases} g^{-1}(x) & \text{if } x \in \text{ran } g \\ \uparrow & \text{otherwise} \end{cases}$ (Assume that g is one-to-one)

c) $f(x) = $ the $(x + 1)$st smallest argument value z for which $g(z) = 0$ provided there exist at least $x + 1$ such argument values, and undefined otherwise

d) $f(x) = \begin{cases} 0 & \text{if there exists a sequence of } x \text{ or more consecutive} \\ & \text{argument values for which } g \text{ has the value } 0 \\ \uparrow & \text{otherwise} \end{cases}$

5.2.2 Let f be any three-variable primitive recursive function, and let g be defined from f by minimalization, thus:

$$g(x, y) = \mu z[f(x, y, z) = 0]$$

a) Explain why the following definition does not establish the μ-recursiveness of the function h:

$$h(x) = \mu y[g(x, y) = 0]$$

b) Assume that the preceding definition means that $h(x)$ is to be the least value of y for which $g(x, y) = 0$ if such a value exists, and is to be undefined otherwise. Show that h is in fact μ-recursive.

c) Show that the function h' is also μ-recursive, where

$$h'(x) = \mu y[g(x, y) = k]$$

and k is a constant.

5.2.3 This problem shows that the μ-recursive functions can be defined without the use of primitive recursion as long as certain additional base functions are allowed.

a) Suppose that the $(n + 1)$-variable function f is obtained from the n-variable function g and the $(n + 2)$-variable function h by primitive recursion. Define the new function Ψ so that $\Psi(x_1, \ldots, x_n, y)$ is the Gödel number of the sequence $f(x_1, \ldots, x_n, 0), \ldots, f(x_1, \ldots, x_n, y)$ if the values in question are all defined, and is undefined otherwise. Show that Ψ can be derived from g and h using only unbounded minimalization, composition, logical connectives, the elementary comparison predicates, the extraction function E, and various of the constant, projection, and successor functions.

    b) Show that every μ-recursive function can be obtained from the following base functions by means of composition and minimalization alone: the constant, projection, and successor functions; addition and proper subtraction; the two sign functions; and the extraction function E.

5.2.4   The two-variable function f is defined recursively as follows.

$$f(0, y) = y^+$$
$$f(x^+, 0) = f(x, 1)$$
$$f(x^+, y^+) = f(x, f(x, f(x^+, y)))$$

Use an approach similar to that taken with Ackermann's function to show that f is μ-recursive.

5.2.5   A certain one-tape, singly infinite Turing machine **T** computes the two-variable function f according to the following conventions. At the beginning of the evaluation of $f(x_1, x_2)$, the machine is presented with a tape containing the pattern $\underline{0}\ \bar{x}_1\ 0\ \bar{x}_2\ 0\ \cdots$ and is made to scan the first square, as indicated. The only way that **T** can halt is by running off the end of its tape. If and when that happens, the value of $f(x_1, x_2)$ is taken to be the total number of 1's appearing on the tape. If **T** does not run off its tape, the value of $f(x_1, x_2)$ is undefined.

    Use a modification of the argument preceding Theorem 5-5 to show that f is μ-recursive. Indicate which of the auxiliary functions described in the text can be used as is and which must be changed. Give new derivations for the functions that must be changed. *Suggestion:* Let the first tape square be assigned the number 1.

5.2.6   Show that the one-variable function f computed by the following algorithm is μ-recursive.

    To evaluate $f(x)$:

1.   Set y equal to x.
2.   If y is even, replace it by $y/2$; if y is odd but not 1, replace it by $3y + 1$.
3.   Repeat step (2) until such time as the current value of y becomes 1, at which point stop and take $f(x)$ to be the sum of all the values of y produced in the course of the computation, including the original value x and the final value 1.

This algorithm may be illustrated by the following sample computation: For $x = 3$, the successive values of y are 3, 10, 5, 16, 8, 4, 2, and 1, whence $f(3) = 49$.

5.2.7   Let g be a specified one-variable primitive recursive function. Consider the one-variable function f computed by the following algorithm.

To evaluate $f(x)$:

1.   Establish a list L that initially contains just the argument value x.
2.   Select the number that appears most often in the current list (in case of ties, pick the largest of the numbers involved), apply the function g to it, and append the resulting value to the list.
3.   Repeat step (2) until such time as the number 0 is produced, at which point take the length of the list (including the 0) to be the value of $f(x)$.

Give a derivation establishing that f is μ-recursive. *Suggestion:* Let the list L be represented by the number $p_0^{e_0} p_1^{e_1} \cdots p_k^{e_k}$, where e_i is the number of times that i appears in the list.

5.2.8   Let P be a given two-place primitive recursive predicate, and consider the set S defined as follows: A given natural number n is to be a member of S iff there exists a finite sequence of numbers $n_0, n_1, \ldots, n_m$ (with $m > 0$) such that

    i) $n_0 = 0$
    ii) $n_m = n$
    iii) $P(n_i, n_{i+1})$ holds for all i in the range $0 \le i < m$

Use a Gödel-numbering scheme to show that the function f is μ-recursive, where

$$f(x) = \begin{cases} x & \text{if} & x \in S \\ \uparrow & \text{if} & x \notin S. \end{cases}$$

5.2.9   This problem considers two alternative definitions of unbounded minimalization in which the given function g is not required to be total.

  a) Suppose that minimalization is defined so that

$$\mu y[g(x_1, \ldots, x_n, y) = 0]$$

denotes the smallest value y_0 for which $g(x_1, \ldots, x_n, y_0) = 0$, provided $g(x_1, \ldots, x_n, u)\!\downarrow$ for all $u < y_0$, and is undefined otherwise. Show that this definition leads to the same class of μ-recursive functions as the definition used in the text.

  b) Suppose that minimalization is defined so that

$$\mu y[g(x_1, \ldots, x_n, y) = 0]$$

denotes the smallest value y_0 for which $g(x_1, \ldots, x_n, y_0) = 0$, regardless of whether $g(x_1, \ldots, x_n, u)\downarrow$ for all $u < y_0$. Does this definition lead to the same class of μ-recursive functions as that used in the text? Justify your answer. Comment on the effective computability of functions defined using this form of minimalization.

5.2.10 Analyze each of the following claims and indicate whether or not it is valid. Justify your reasoning briefly.

a) One student claims to have discovered a very interesting total one-variable Turing-computable function f that grows rapidly but is easy to compute. Specifically, it is said that f grows faster than Ackermann's function (that is, that $f(x) > A(x, x)$ for large x) but that $f(x)$ can be evaluated in fewer than $p(x)$ steps, where p is a certain primitive recursive function.

b) Another student claims to have found an equally interesting total one-variable Turing computable function g. This function assumes only the values 0 and 1, but is said to be so hard to compute that there is no total μ-recursive function r such that the number of Turing machine steps needed to evaluate $g(x)$ is at most $r(x)$.

5.2.11 Show that not every one-variable μ-recursive function f can be expressed in the form

$$f(x) = \mu y[g(x, y) = 0]$$

where g is a total μ-recursive function. (Thus the minimalization referred to in Corollary 5-6.1 must in general be followed by a functional composition.) *Hint:* Consider the function f such that:

$$f(x) = \begin{cases} x & \text{if} \quad \varphi_x(x)\downarrow \\ \uparrow & \text{otherwise} \end{cases}$$

where φ_x denotes the xth μ-recursive function of one variable.

5.2.12 Let the μ-recursive functions be indexed by the indices of the Turing machines that compute them, as suggested in the text. Show that under this indexing scheme, the S-M-N functions are all primitive recursive. *Hint:* Consider the number of steps required to evaluate a given S-M-N function by a Turing machine.

Section 5.3

5.3.1 Show how to obtain each of the values specified at the left of the following display from the defining equations given at the right by means of appropriate substitutions and replacements. To save steps, substitutions of

values for two or more variables in the same equation may be made simultaneously. In part (c) it is not necessary to derive values of the addition function,
which is denoted by the letter a; equations of the form $a(m, n) = m + n$
may be introduced as needed.

a) $f(3, 5) = 1$ 1. $f(0, y) = 1$
                      2. $f(x^+, 0) = 0$
                      3. $f(x^+, y^+) = f(x, y)$

b) $f(4, 2) = 2$ 1. $f(0, y) = y$ 4. $a(x, 0) = x$
                      2. $f(x, y) = f(y, x)$ 5. $a(x, y^+) = a(x, y)^+$
                      3. $f(x, a(x, y)) = f(x, y)$

c) $f(5, 13) = 2$ 1. $f(x, y) = h(g(x, r, n), n, l(a(r, k), x), y)$
                      2. $h(u, n, 1, u) = n$
                      3. $l(v, v^+) = 1$
                      4. $g(x, r, 0) = r$
                      5. $g(x, r, n^+) = a(g(x, r, n), x)$

5.3.2 Establish the general recursiveness of each of the functions described
below by supplying an appropriate set of defining equations. In parts (d)
through (h) you need not include equations for the addition function a.

a) $f_1(x) = sg(x)$ e) $f_5(x, y) = \pi^2(x, y)$
b) $f_2(x, y) = x \div y$ f) $f_6(x) = \sigma_1^2(x)$
c) $f_3(x, y) = \max(x, y)$ g) $f_7(x, y) = quo(x, y)$
d) $f_4(x, y, u, v) = \begin{cases} u & \text{if } x = y \\ v & \text{if } x \ne y \end{cases}$ h) $f_8(x, y) = rem(x, y)$

5.3.3 For each of the sets of equations given below, determine whether that set
defines a function f. If it does, describe the function; if it does not, exhibit a
conflicting pair of derivations.

a) 1. $f(x, y) = h(x, y)$ 3. $h(x, x) = 0$
    2. $f(x, y) = h(y, x)$ 4. $h(x^+, y) = h(x, y)^+$

b) 1. $f(x, x) = x$
    2. $f(x, y^+) = f(x, y)^+$
    3. $f(x^+, y) = f(x, y)^+$

c) 1. $f(a(x, 2)) = a(g(x), 3)$ 4. $g(1) = 0$
    2. $f(a(x, x)) = g(a(x, x))^+$ 5. $a(x, 0) = x$
    3. $g(x) = f(x)^+$ 6. $a(x, y^+) = a(x, y)^+$

5.3.4 For each of the sets of equations given below, describe the general recursive function f defined by those equations.

a) 1. $f(g(y)) = y$
   2. $f(x^+) = h(a(x^+, k^+), g(f(x)^+), f(x))$
   3. $h(w, w, z) = z$
   4. $g(0) = 1$
   5. $g(x^+) = a(g(x), g(x))$
   6. $a(x, 0) = x$
   7. $a(x, y^+) = a(x, y)^+$

b) 1. $f(x, y) = h(a(x, z), y, x, z)$
   2. $f(x, y) = g(a(y, z^+), x, y)$
   3. $h(w, w, x, z) = f(x, z)$
   4. $g(w, w, y) = y$
   5. $a(x, 0) = x$
   6. $a(x, y^+) = a(x, y)^+$

5.3.5 Provide a set of defining equations for each of the functions specified below. Try to use as few equations as possible.

a) $q(x) = \lfloor \sqrt{x} \rfloor$ (7 equations are sufficient)

b) $p(x) = \begin{cases} 1 & \text{if} & x \text{ is prime} \\ \uparrow & \text{if} & x = 0 \text{ or } x = 1 \\ 0 & \text{otherwise} \end{cases}$ (9 equations are sufficient)

5.3.6 Consider the following set of defining equations.

1. $f(x, y) = h(m(x, x), y, x)$ 5. $a(0, x) = x$
2. $h(x, x, y) = y$ 6. $a(x, 1) = x^+$
3. $m(x, 0) = 0$ 7. $a(x, y^+) = a(x, y)^+$
4. $m(x, y^+) = a(m(x, y), x)$

a) Explain why the sequence of equations given below is not a legitimate derivation of the equation $f(2, 1) = h(m(2, 2), 1, x)$.

8. $a(0, 1) = 1$	subs (5)
9. $m(1, 0) = 0$	subs (3)
10. $m(1, 1) = a(m(1, 0), 1)$	subs (4)
11. $a(m(1, 0), 1) = a(0, 1)$	repl (10, 9)
12. $m(1, 1) = 1$	repl (10, 11, 8)
13. $f(x, m(1, 1)) = h(m(x, x), m(1, 1), x)$	subs (1)
14. $f(x, 1) = h(m(x, x), 1, x)$	repl (13, 12)
15. $f(2, 1) = h(m(2, 2), 1, x)$	subs (14)

b) Convert the following (legitimate) derivation to standard form.

8. $m(x, 1) = a(m(x, 0), x)$ subs (4)
9. $m(x, 1) = a(0, x)$ repl (8, 3)
10. $m(x, 1) = x$ repl (9, 5)
11. $f(1, y) = h(m(1, 1), y, 1)$ subs (1)
12. $m(1, 1) = 1$ subs (10)
13. $f(1, y) = h(1, y, 1)$ repl (11, 12)
14. $h(1, 1, 1) = 1$ subs (2)
15. $f(1, 1) = h(1, 1, 1)$ subs (13)
16. $f(1, 1) = 1$ repl (15, 14)

5.3.7   Let g and h be total two-variable general recursive functions. Show that each of the functions described below is also general recursive. In each case specify an appropriate set of defining equations.

a) $f_1(x) = \begin{cases} x & \text{if} \quad x \in \text{ran } g \\ \uparrow & \text{otherwise} \end{cases}$

b) $f_2(x, y, z) = \begin{cases} g(x, y) & \text{if} \quad y \geq x \quad \text{and} \quad z = 0 \\ h(x, f_2(x, y, \lfloor z^+/2 \rfloor)) & \text{if} \quad y \geq x \quad \text{and} \quad z > 0 \\ \uparrow & \text{otherwise} \end{cases}$

c) $f_3(x, n) =$ the $(n + 1)$st smallest argument value z for which $g(x, z) = 0$ if there are at least $n + 1$ such values of z, and undefined otherwise

5.3.8   Let g be a two-variable primitive recursive function whose value is always nonzero. Let B denote the set defined inductively as follows: (i) Every odd number is a member of B; and (ii) if x and y are members of B, so is $g(x, y)$. Now consider the function f, where

$$f(n) = \begin{cases} 1 & \text{if} \quad n \in B \\ \uparrow & \text{if} \quad n \notin B \end{cases}$$

a) Use a simplified version of the argument leading to Theorem 5-11 to show that f is μ-recursive. (Note that if $n \in B$, there must exist a sequence of numbers that may be thought of as a "derivation" of n.)

b) Suppose that the function g is defined by the set of equations $\mathcal{G}$ with g as principal function letter. Show that f is general recursive by specifying a set of defining equations for it.

5.3.9   In this problem we consider a way of representing general recursive functions by natural numbers. Suppose that the equations $e_0, \ldots, e_m$ define the n-variable function f with principal function letter f_u^n, and that the Gödel numbers of $e_0, \ldots, e_m$ are respectively $z_0, \ldots, z_m$. Then we assign the natural number $i = \pi^2(u, \langle z_0, \ldots, z_m \rangle)$ to the function f. Note that each function will in this way be assigned many numbers, although not every number will be assigned to a function.

a) Show that for each $n > 0$ there exists a μ-recursive function $V^{(n+1)}$ such that

$$V^{(n+1)}(i, x_1, \ldots, x_n) = f(x_1, \ldots, x_n)$$

whenever i is a natural number assigned to the n-variable function f. (The value assumed by $V^{(n+1)}$ when i does not represent an n-variable function is completely arbitrary.)

b) Comment on the difficulty involved in using the assignment presented above as the basis for an indexing scheme for the general recursive functions.

Chapter 6

Computability and Decidability

In this chapter we examine the related concepts of a computable function, a computable or semicomputable set, and a solvable decision problem. The formal definition of a computable function, introduced in Section 6.1, is based on the properties of Turing-computable and μ-recursive functions developed in the preceding chapters. As we shall see, the class of computable functions turns out to be the smallest class of number-theoretic functions that satisfies the five basic properties discussed in Chapter 1. The concept of a computable function will be used in Sections 6.2 and 6.3 to develop appropriate notions of "computability" for sets, relations, and decision problems.

6.1 COMPUTABLE FUNCTIONS

Chapter 1 introduced the notion of an effectively computable function, by which we mean a function that can be evaluated by a "mechanistic" procedure. We have subsequently studied several classes of functions that meet this criterion of being subject to mechanistic evaluation. Building on our understanding of these classes, we now propose a formal counterpart of the intuitive notion of effective computability. After motivating the formal definition, we will investigate its relationship to the five basic properties discussed in Chapter 1.

Recursive Functions and the Church-Turing Thesis

Recall that the concept of Turing computability is based on a set of elementary operations—namely reading, writing, and shifting—that admit a simple, mechanistic realization. Consequently all the computations performed by Turing machines satisfy the informal requirement of "mechanizability" and all the number-theoretic functions evaluated by Turing machines satisfy the intuitive notion of being "effectively computable." The class of Turing-computable

287

functions turns out to be a very large one and would seem to include most, if not all, of the effectively computable functions.

In a similar way, the concept of μ-recursiveness is based on a set of elementary functions—namely the constant, projection, and successor functions—that satisfy the intuitive notion of being effectively computable. Then because the operations of functional composition, primitive recursion, and minimalization can each be implemented in a simple, mechanistic way, it follows that all the μ-recursive functions are effectively computable. Indeed, as we have seen, the class of μ-recursive functions is identical to the class of Turing-computable functions.

Although Turing computability and μ-recursiveness are the most commonly discussed ways of modeling the notion of effective computability, they are not the only ones. We have already examined the possibility of defining and evaluating functions through the use of arbitrary sets of equations. But, as we have seen, the resulting class of general recursive functions turns out to be identical to the class of Turing-computable or μ-recursive functions.

Another approach involves the concept of a "normal algorithm" as proposed by A. A. Markov. Here the objects to be operated upon are strings of symbols and the basic operations consist in replacing one string by another in accordance with specified "rewriting rules." By adopting appropriate conventions for representing numbers, normal algorithms can be used to evaluate number-theoretic functions. And because the allowable rewriting rules are all inherently "mechanizable," it follows that the associated functions are all effectively computable.

Markov's rewriting rules permit moderately complicated string transformations to be accomplished in single steps. It is nevertheless possible to arithmetize the computations carried out by normal algorithms and thereby to establish that all the associated number-theoretic functions are μ-recursive. Conversely, it is possible to show that the computations of any given Turing machine can be "simulated" through the use of suitably chosen rewriting rules. Thus the class of number-theoretic functions that can be evaluated using Markov's scheme is identical to the class of Turing-computable or μ-recursive functions.

Similar arguments can be used to show that all the other proposed general-purpose computing models also yield the class of Turing-computable or μ-recursive functions. The fact that a number of different approaches lead to the same set of functions suggests the fundamental importance of that set. Its members are referred to as the *recursive functions* or, synonymously, as the *computable functions*. Thus the recursive functions include all the primitive recursive, Turing-computable, μ-recursive, and general recursive functions. The predicates whose characteristic functions are recursive are in turn called *recursive predicates* or *computable predicates*.

The recursive functions are, as we have noted, all effectively computable.

It is now natural to ask whether the class of recursive functions includes every effectively computable number-theoretic function. All available evidence suggests that it does. Let us look briefly at this evidence.

First, there is the great generality of the recursive functions. It is difficult to think of a function that is effectively computable in the intuitive sense but that is not included among the recursive functions. In particular, because of the simplicity and generality of the basic Turing machine operations, it is difficult to think of an effective computation that cannot be carried out by a suitably designed Turing machine.

Second, there is the difficulty of enlarging the class of recursive functions. As we have seen, there are a number of extensions or generalizations of the Turing-machine model that preserve the property of effective computability: the use of extra reading heads, extra tapes, multidimensional tapes, etc. However, none of these extensions or generalizations enlarges the class of functions computable by Turing machines. Similarly, there are a number of ways of generalizing the formulation of the μ-recursive functions so as to permit the use of a broad range of recursive definition schemes. But even the most comprehensive of these extensions—namely the one leading to the class of general recursive functions—fails to yield any new functions. Thus it seems to be very difficult, if not impossible, to enlarge the class of recursive functions without sacrificing effective computability.

Third, and perhaps most compelling, there is the evidence provided by the arithmetization process. A computation carried out by an effective algorithm must certainly consist of a well-defined sequence of simple, mechanistic operations. In practice, each of the states that such a computation might assume between consecutive operations can be represented by a string of symbols, hence ultimately by a single natural number. Then as long as the effect of each basic operation can be described by a total recursive function, and as long as the conditions in which the various operations are to be applied (or in which the computation is to terminate) can be described by recursive predicates, it is possible to show that the function evaluated by the algorithm in question is recursive. Because recursive functions and predicates can represent very complicated operations and conditions, it seems inevitable that every effective algorithm must evaluate a recursive function.

The arguments just outlined led the logicians Church and Turing, among others, to hypothesize that the effectively computable functions are precisely the recursive functions. This hypothesis will be referred to as:

The Church-Turing Thesis. *Every effectively computable number-theoretic function is recursive (computable) and vice versa.*

The identification of effective computability and recursiveness proposed by Church and Turing has been generally adopted by other mathematicians; we shall do likewise.

It is important to recognize that the Church-Turing Thesis is not a theorem. That is, it is not a statement that is subject to formal proof. In order to "prove" the Church-Turing Thesis, we would first have to give a formal definition of effective computability, and that would require the translation of a rather vague intuitive notion into precise terms. Actually, we may view the Church-Turing Thesis itself as an attempt at such a definition, for in adopting the thesis we accept the appropriateness of equating the informal notion of effective computability with the formal notion of recursiveness.

The Church-Turing Thesis, at least as presented above, refers only to number-theoretic functions. In order to deal with other kinds of functions, we must somehow relate them to number-theoretic functions. A little thought shows that there are many nonnumber-theoretic functions whose evaluations are closely associated with the evaluation of number-theoretic functions. Among these are the functions that operate on strings of symbols. Since strings of symbols can be represented by natural numbers, every such string-manipulating function can be represented by a corresponding number-theoretic function. Indeed, any function from one countable set to another can be so represented through the use of a suitable natural-number coding scheme. And in cases of practical interest, it is normally possible to choose a coding scheme that can be implemented in what is obviously an effective manner. The familiar Gödel numbering provides a good example.

Example 6-1. Consider the function F that maps each string on the alphabet $\{a, b, c\}$ into the reverse of that string, so that $F(bcab) = bacb$, $F(aacbc) = cbcaa$, and so on. Let us agree to represent the individual symbols a, b, and c by the numbers $\langle a \rangle = 1$, $\langle b \rangle = 2$, and $\langle c \rangle = 3$. Then each string $s_0 s_1 \cdots s_n$ on the alphabet $\{a, b, c\}$ can be represented by the Gödel number $\langle s_0 s_1 \cdots s_n \rangle = 2^{\langle s_0 \rangle} 3^{\langle s_1 \rangle} \cdots p_n^{\langle s_n \rangle}$. Using this representation, we find that the number-theoretic counterpart of F is a function f such that $f(\langle s_0 s_1 \cdots s_n \rangle) = \langle F(s_0 s_1 \cdots s_n) \rangle = \langle s_n \cdots s_1 s_0 \rangle$. Of course, this specification is not complete, since not every natural number is the Gödel number of a string. But there are many ways of completing the definition of f, among the simplest being to require that $f(z) = 0$ whenever z is not the Gödel number of a string.   □

Representation schemes such as that of Example 6-1 allow us to deal with a broad class of nonnumber-theoretic functions. Suppose that F is such a function, that some means has been chosen for representing the arguments and values of F numerically, and that as a result F may be recast as a number-theoretic function f. Then as long as there exist effective procedures for converting back and forth between the original arguments and values of F and their associated numerical representations, it follows that F will be effectively computable iff f is effectively computable. According to the Church-Turing thesis, this means that F will be effectively computable iff f is recursive.

Of course, we have not yet said what procedures for translating arguments and values of F into numbers (and vice versa) are to be considered "effective." It certainly appears reasonable to suppose that the effective translation schemes include all those that can be carried out by suitably designed Turing machines. Here we assume that the arguments and values of F are basically patterns of symbols and that what the desired machines must do is convert these patterns into unary representations of natural numbers and vice versa. Since experience suggests that such translation schemes exist for all countable domains, we are led to extend the Church-Turing thesis and identify the effectively computable functions with those that can be translated, by means of suitable Turing machines, into recursive number-theoretic functions.

The adoption of this more general version of the Church-Turing Thesis tends to blur the distinction between number-theoretic and nonnumber-theoretic functions. It is common, in fact, to refer to all effectively computable functions as "recursive" or "computable" functions. We will tend to reserve the word *recursive* for number-theoretic functions and to use *computable* as a general term for any effectively computable function, whether number-theoretic or not.

Because of the relationship between the recursive functions and the effectively computable functions, the class of recursive functions plays an important role in the study of effective processes in general. The existence of effective procedures for generating given sets, for determining whether an object belongs to a given set, or for determining whether a given relationship holds among specified objects all depend on the recursiveness of appropriate functions. Thus it is important to understand the basic properties of the recursive functions and to be able to determine which functions are recursive and which are not.

The cleanest way of establishing that a function is recursive is to supply a μ-recursive derivation for it—i.e., to show how the function can be obtained from known recursive functions by means of composition, primitive recursion, and minimalization. This process should by now have become fairly routine. From this point on, we will rely less on the formal approach and begin to make use of the Church-Turing Thesis. Thus we will often use an informal description of how a certain function might be evaluated as a justification that that function is recursive. The next example illustrates this approach.

Example 6-2. Let g be a total one-variable recursive function such that $g(n + 1) > g(n)$ for all n. Now consider the function f, where

$$f(x) = \begin{cases} 1 & \text{if } x \in \text{ran } g \\ 0 & \text{if } x \notin \text{ran } g. \end{cases}$$

Note that the following procedure allows us to determine the value of $f(x)$ for any natural number x.

> Evaluate $g(0)$, $g(1)$, $g(2)$, ... in sequence until finding the first argument value k such that $g(k) \geq x$. (Since g is strictly monotonic, such a value must eventually be reached.) If $g(k) = x$, then x certainly belongs to the range of g, so set $f(x) = 1$. If $g(k) > x$, the monotonicity of g guarantees that no larger argument value will yield x as a function value. In this case x does not belong to the range of g, so set $f(x) = 0$.

Since g is both total and recursive, the procedure just outlined is an effective one. The Church-Turing Thesis then implies that f is a recursive function. □

A similar approach will be taken with nonnumber-theoretic functions. As previously noted, such a function is to be considered computable if it can be translated, via appropriate Turing machines, into a recursive number-theoretic function. We will not always bother to present this translation explicitly or to establish the recursiveness of the resulting number-theoretic function. Very often we will simply describe an effective procedure for evaluating the given function and rely on the extended version of the Church-Turing Thesis as justification for concluding that the function is computable.

Example 6-3. Again consider the function F that maps each finite string on the alphabet $\{a, b, c\}$ into the reverse of that string. To establish the computability of F, we must specify an appropriate scheme for representing strings by natural numbers, show that the conversion from strings to numbers and vice versa can be accomplished by suitable Turing machines, and show that the associated number-theoretic function f is recursive. But since the choice of a representation scheme, the design of the necessary Turing machines, and the recursive derivation of f are all quite straightforward, we would not normally present them. Instead, we simply note that there exists an effective procedure for evaluating $F(s_0 s_1 \cdots s_n)$: begin with the last symbol of $s_0 s_1 \cdots s_n$, append the immediately preceding symbol, then the next preceding symbol, and so on until s_0 is reached. Given the existence of this procedure, the extended Church-Turing Thesis implies the computability of F.   □

We now turn to the characteristics of the recursive functions as a family. Since the recursive functions form a denumerably infinite set, they can be indexed by the natural numbers. It is common practice to use the symbol φ to denote a recursive function and to write $\varphi_i^{(n)}$ to denote the ith n-variable recursive function according to some agreed upon indexing scheme. Perhaps the simplest such scheme, and the one that we now officially adopt, is to use the indices already assigned to the Turing-computable functions. Thus from now on we will write $\varphi_i^{(n)}$ to denote the n-variable recursive function that is evaluated by the *i*th Turing machine. As usual, we will omit the superscipt (n) when the context permits.

It follows immediately from this choice of an indexing scheme that all five of the basic properties discussed in Chapter 1 must hold for the family of recursive functions. As a result, all the consequences of these properties must also hold. Thus the diagonal domain function d, where

$$d(x) = \begin{cases} 1 & \text{if} \quad \varphi_x(x)\downarrow \\ 0 & \text{if} \quad \varphi_x(x)\uparrow \end{cases}$$

is not recursive. Similarly, such functions as the general domain functions, the totality function, and the equivalence function are not recursive. On the other hand the Recursion Theorem and all its consequences must hold for the recursive functions.

Of course, Properties 1, 2, and 4 must hold regardless of the indexing scheme chosen; only Properties 3 and 5 depend on the details of the index assignment. It is therefore natural to ask whether there are other indexing schemes for which Properties 3 and 5 also hold. The fact is that there are many such schemes. We refer to these schemes as *acceptable indexings* and to the scheme described here as the *standard indexing*. It is interesting to note that all the acceptable indexings are closely related to the standard indexing.

To explore this relationship, we will, as usual, use $\varphi_i^{(n)}$ to denote the *n*-variable recursive function that is assigned the index i in the standard indexing. Let a second indexing scheme be proposed, and let $\theta_i^{(n)}$ be used to denote the *n*-variable function that is assigned the index i in this new scheme. In general, of course, $\varphi_i^{(n)}$ and $\theta_i^{(n)}$ will not be the same function. Now suppose that it is possible to translate back and forth between the two schemes in an effective way. Suppose, in particular, that there exist recursive functions $f_1, f_2, f_3, \dots$ and $t_1, t_2, t_3, \dots$ such that

$$\theta_{t_n(i)}^{(n)} = \varphi_i^{(n)} \qquad \text{and} \qquad \varphi_{f_n(i)}^{(n)} = \theta_i^{(n)} \tag{6-1}$$

for all $n \geq 1$ and all $i \geq 0$. To make sure that the roles of $f_1, f_2, \dots$ and $t_1, t_2, \dots$ are understood, we note that $t_n(i)$ is an index in the new scheme of the *n*-variable function whose index in the standard scheme is i, while $f_n(i)$ is an index in the standard scheme of the *n*-variable function whose index in the new scheme is i.

It is now easy to show that the new indexing scheme is necessarily an acceptable indexing. As usual, let $U^{(n+1)}$ denote the $(n+1)$-variable universal function for the standard indexing. Then the function $V^{(n+1)}$, where

$$V^{(n+1)}(i, x_1, \dots, x_n) = U^{(n+1)}(f_n(i), x_1, \dots, x_n)$$

is the $(n+1)$-variable universal function for the new indexing, since

$$V^{(n+1)}(i, x_1, \dots, x_n) = \varphi_{f_n(i)}^{(n)}(x_1, \dots, x_n) = \theta_i^{(n)}(x_1, \dots, x_n).$$

Similarly, let $s_{n,m}$ denote an S-M-N function for the standard indexing. Then the function $\sigma_{n,m}$, where

$$\sigma_{n,m}(i, x_1, \ldots, x_m) = t_n(s_{n,m}(f_{n+m}(i), x_1, \ldots, x_m))$$

serves as a corresponding S-M-N function for the new indexing, since

$$\theta^{(n)}_{\sigma_{n,m}(i, x_1, \ldots, x_m)}(y_1, \ldots, y_n) = \theta^{(n)}_{t_n(s_{n,m}(f_{n+m}(i), x_1, \ldots, x_m))}(y_1, \ldots, y_n)$$

$$= \varphi^{(n)}_{s_{n,m}(f_{n+m}(i), x_1, \ldots, x_m)}(y_1, \ldots, y_n)$$

$$= \varphi^{(n+m)}_{f_{n+m}(i)}(x_1, \ldots, x_m, y_1, \ldots, y_n)$$

$$= \theta^{(n+m)}_i(x_1, \ldots, x_m, y_1, \ldots, y_n).$$

Because the functions $V^{(n+1)}$ and $\sigma_{n,m}$ are obviously recursive, it follows that the new indexing scheme satisfies Properties 3 and 5.

Conversely, it can be shown that if a given nonstandard indexing scheme is acceptable, there must exist recursive functions $f_1, f_2, \ldots$ and $t_1, t_2, \ldots$ that translate back and forth between the indices of the given scheme and those of the standard scheme. Suppose that, as above, $U^{(n+1)}$ and $s_{n,m}$ denote universal and S-M-N functions for the standard indexing and that $V^{(n+1)}$ and $\sigma_{n,m}$ denote the corresponding functions for a certain acceptable indexing scheme. If we define t_n so that $t_n(i) = \sigma_{n,1}(u_{n+1}, i)$, where u_{n+1} is an index of $U^{(n+1)}$ in the *new* scheme, we see that

$$\theta^{(n)}_{t_n(i)}(x_1, \ldots, x_n) = \theta^{(n)}_{\sigma_{n,1}(u_{n+1}, i)}(x_1, \ldots, x_n) = \theta^{(n+1)}_{u_{n+1}}(i, x_1, \ldots, x_n)$$

$$= U^{(n+1)}(i, x_1, \ldots, x_n) = \varphi^{(n)}_i(x_1, \ldots, x_n).$$

Thus the recursive function t_n computes indices in the new scheme from given indices in the standard scheme. An analogous argument shows that the reverse translation is accomplished by defining f_n so that $f_n(i) = s_{n,1}(v_{n+1}, i)$, where v_{n+1} is an index of $V^{(n+1)}$ in the *standard* scheme.

We conclude, then, that an indexing scheme is an acceptable one iff there exist recursive functions that compute indices in the given scheme from indices in the standard scheme and vice versa. Thus all the acceptable indexings are closely related to the standard one derived from Turing machine descriptions. Since in most applications the only thing that matters is that Properties 3 and 5 hold, it is common practice to ignore the details of how indices are assigned and simply assume that $\varphi^{(n)}_i$ denotes the ith n-variable function according to some acceptable indexing scheme.

There are, on the other hand, situations in which it is convenient to make use of some special feature of the standard indexing scheme. Of particular importance in this regard is the fact that in the standard scheme there is a well-defined step-by-step evaluation procedure associated with each function $\varphi^{(n)}_i$, namely the computation performed by the Turing machine T_i. Relying on the fact that it is always possible to translate back and forth between acceptable indexings, we shall feel free to attribute step-by-step evaluation procedures to every acceptable indexing, without concerning ourselves with the details of those procedures.

Thus for every acceptable indexing scheme, and for each positive integer n, we may define the predicate $\hat{H}^{(n+2)}$ and the function $\hat{U}^{(n+2)}$ so that:

$$\hat{H}^{(n+2)}(i, x_1, \ldots, x_n, w) \Leftrightarrow \text{the evaluation of } \varphi_i(x_1, \ldots, x_n) \text{ terminates}$$
$$\text{with a defined value within } w \text{ steps}$$

and

$$\hat{U}^{(n+2)}(i, x_1, \ldots, x_n, w) = \begin{cases} \varphi_i(x_1, \ldots, x_n) + 1 & \begin{array}{l}\text{if the evaluation of} \\ \varphi_i(x_1, \ldots, x_n) \text{ terminates with a} \\ \text{defined value within } w \text{ steps}\end{array} \\ 0 & \text{otherwise.} \end{cases}$$

It follows from the preceding remarks, as the reader may verify, that these predicates and functions are recursive. This fact is often useful in the translation of informal algorithms into precise derivations.

Example 6-4. Assuming that φ_i denotes the ith one-variable recursive function according to some acceptable indexing, let the function f be defined so that

$$f(x, y) = \begin{cases} 1 & \text{if } \varphi_x(x)\!\downarrow \text{ or } \varphi_y(y)\!\downarrow \\ \uparrow & \text{otherwise.} \end{cases}$$

This function may be evaluated by the following (informal) algorithm. Begin by carrying out the first step in the evaluation of $\varphi_x(x)$ and the first step in the evaluation of $\varphi_y(y)$. If either evaluation is found to terminate in one step with a defined value, set $f(x, y) = 1$. Otherwise carry out the first two steps in the evaluations of $\varphi_x(x)$ and $\varphi_y(y)$, then the first three steps, and so on, until such time as one evaluation or the other may yield a defined value, at which point set $f(x, y) = 1$. Since this procedure provides an effective way of evaluating f, we conclude that f is recursive. To convert the informal procedure into a formal derivation, we note that the predicate $\hat{H}^{(3)}$ can be used to detect the situation in which one of the evaluations has terminated. Specifically, we observe that $\hat{H}^{(3)}(x, x, n) \vee \hat{H}^{(3)}(y, y, n)$ will hold for given numbers x, y, and n iff the evaluation of $\varphi_x(x)$ or the evaluation of $\varphi_y(y)$ terminates with a defined value within n steps. As a consequence, we can write

$$f(x, y) = C_1(\mu n[\hat{H}^{(3)}(x, x, n) \vee \hat{H}^{(3)}(y, y, n)])$$

from which it follows that f is recursive.   □

The Role of the Basic Properties

We now set out to show that the class of recursive functions is the smallest class that satisfies the five properties of Chapter 1. Specifically, we propose to show that every class of functions satisfying the five basic properties must include all the recursive functions or, equivalently, all the μ-recursive functions. To this end, let $\mathcal{F}$ be a given class of functions for which Properties 1 through 5 hold, and let the n-variable members of $\mathcal{F}$ be denoted $F_0^{(n)}, F_1^{(n)}, F_2^{(n)}, \ldots$ in the usual way. Since $\mathcal{F}$ contains the constant, projection, and successor functions and is closed under composition, we need only show that it is also closed under primitive recursion and minimalization.

First consider the case of primitive recursion. In particular, consider the special case in which g and h are respectively one- and three-variable members of $\mathcal{F}$ and f is the two-variable function defined from g and h by primitive recursion, so that:

$$f(x, 0) = g(x)$$
$$f(x, y^+) = h(x, y, f(x, y)) \tag{6-2}$$

As a first step toward showing that f is necessarily a member of $\mathcal{F}$, we note that the values of f can be expressed in terms of repeated composition of the function h with itself and with g, thus:

$$f(x, 0) = g(x)$$
$$f(x, 1) = h(x, 0, g(x))$$
$$f(x, 2) = h(x, 1, h(x, 0, g(x))) \tag{6-3}$$
$$f(x, 3) = h(x, 2, h(x, 1, h(x, 0, g(x))))$$
$$\vdots$$
$$f(x, y) = h(x, y - 1, h(x, y - 2, \cdots h(x, 0, g(x)) \cdots))$$

The next, and crucial, step is to observe that with the addition of two auxiliary variables, the values of f can in effect be defined "backwards." Suppose we could establish the existence in $\mathcal{F}$ of a four-variable function θ such that:

$$\theta(x, y, y, w) = w$$
$$\theta(x, y, y - 1, w) = h(x, y - 1, w)$$
$$\theta(x, y, y - 2, w) = h(x, y - 1, h(x, y - 2, w)) \tag{6-4}$$
$$\theta(x, y, y - 3, w) = h(x, y - 1, h(x, y - 2, h(x, y - 3, w)))$$
$$\vdots$$
$$\theta(x, y, 0, w) = h(x, y - 1, h(x, y - 2, \cdots h(x, 0, w) \cdots))$$

By comparison with Eqs. (6-3), we see that the value of $f(x, y)$ is given by $\theta(x, y, 0, g(x))$. Thus if θ is a member of $\mathcal{F}$, it follows that f must also be a member of $\mathcal{F}$.

To establish that $\mathcal{F}$ must in fact contain a function θ having the properties required by Eqs. (6-4), we need a more compact representation of those properties. Note that according to Eqs. (6-4), the value of $\theta(x, y, y - (k + 1), w)$ can be obtained from that of $\theta(x, y, y - k, w)$ by making an appropriate substitution for w. Specifically,

$$\theta(x, y, y - (k + 1), w) = \theta(x, y, y - k, h(x, y - (k + 1), w)) \quad \text{for all } k < y.$$

Since $y - k$ is the successor of $y - (k + 1)$, we may now rephrase the constraints of Eqs. (6-4) in the form:

$$.\theta(x, y, z, w) = \begin{cases} w & \text{if } z = y \\ \theta(x, y, z^+, h(x, z, w)) & \text{if } z < y \end{cases} \tag{6-5}$$

Thus it suffices to show that $\mathcal{F}$ must contain a function θ that satisfies Eq. (6-5).

This is readily done with the aid of the Recursion Theorem. Since $\mathcal{F}$ satisfies the five basic properties, it can be shown to contain a total index-computing function γ such that, for each index u:

$$F^{(4)}_{\gamma(u)}(x, y, z, w) = \begin{cases} w & \text{if } z = y \\ F^{(4)}_{u}(x, y, z^+, h(x, z, w)) & \text{if } z \neq y \end{cases}$$

The Recursion Theorem, which necessarily holds for $\mathcal{F}$, then guarantees the existence of a natural number m such that $F^{(4)}_{\gamma(m)} = F^{(4)}_{m}$. Thus

$$F^{(4)}_{m}(x, y, z, w) = F^{(4)}_{\gamma(m)}(x, y, z, w) = \begin{cases} w & \text{if } z = y \\ F^{(4)}_{m}(x, y, z^+, h(x, z, w)) & \text{if } z \neq y \end{cases}$$

and $F^{(4)}_{m}$ is the desired function θ.

The approach just described is easily extended to arbitrary definitions by primitive recursion.

Theorem 6-1. *Let $\mathcal{F}$ be a class of functions satisfying Properties 1 through 5, and let g and h be respectively n- and $(n + 2)$-variable members of $\mathcal{F}$. Then the $(n + 1)$-variable function f defined from g and h by primitive recursion must also be a member of $\mathcal{F}$.*

Proof. It is a routine matter to show that $\mathcal{F}$ must contain a total function γ such that[†]

$$F^{(n+3)}_{\gamma(u)}(x_1, \ldots, x_n, y, z, w)$$

$$= \begin{cases} w & \text{if } z = y \\ F^{(n+3)}_{u}(x_1, \ldots, x_n, y, z^+, h(x_1, \ldots, x_n, z, w)) & \text{if } z \neq y. \end{cases}$$

[†] Care must be taken to ensure that $F^{(n+3)}_{\gamma(u)}(x_1, \ldots, x_n, y, z, w)$ assumes the value w whenever $z = y$, even in those cases where $F^{(n+3)}_{u}(x_1, \ldots, x_n, y, z^+, h(x_1, \ldots, x_n, z, w))$ is not defined. To this end, we first define a function β such that

$$F^{(n+3)}_{\beta(u)}(x_1, \ldots, x_n, y, z, w) = F^{(n+3)}_{u}(x_1, \ldots, x_n, y, z^+, h(x_1, \ldots, x_n, z, w))$$

for all u. The desired index-computing function γ can then be obtained by noting that $F^{(n+3)}_{\gamma(u)}(x_1, \ldots, x_n, y, z, w) = U^{(n+4)}(\Lambda(z, y, p, \beta(u)), x_1, \ldots, x_n, y, z, w)$, where p is an index of the projection function $P^{(n+3)}_{n+3}$.

Since $\mathcal{F}$ satisfies Properties 1 through 5, the Recursion Theorem holds. Thus for some natural number m we have

$$F_m^{(n+3)}(x_1, \ldots, x_n, y, z, w)$$

$$= \begin{cases} w & \text{if } z = y \\ F_m^{(n+3)}(x_1, \ldots, x_n, y, z^+, h(x_1, \ldots, x_n, z, w)) & \text{if } z \neq y. \end{cases} \quad (6\text{-}6)$$

We must now show that the desired function f can be obtained from $F_m^{(n+3)}$. Upon substituting the value k for z and the value $f(x_1, \ldots, x_n, k)$ for w in Eq. (6-6), we see that

$$F_m^{(n+3)}(x_1, \ldots, x_n, y, k, f(x_1, \ldots, x_n, k))$$
$$= F_m^{(n+3)}(x_1, \ldots, x_n, y, k^+, h(x_1, \ldots, x_n, k, f(x_1, \ldots, x_n, k)))$$
$$= F_m^{(n+3)}(x_1, \ldots, x_n, y, k^+, f(x_1, \ldots, x_n, k^+))$$

for all $k < y$. A simple induction argument now suffices to prove that

$$F_m^{(n+3)}(x_1, \ldots, x_n, y, 0, f(x_1, \ldots, x_n, 0))$$
$$= F_m^{(n+3)}(x_1, \ldots, x_n, y, y, f(x_1, \ldots, x_n, y)).$$

But according to Eq. (6-6), $F_m^{(n+3)}(x_1, \ldots, x_n, y, y, f(x_1, \ldots, x_n, y)) = f(x_1, \ldots, x_n, y)$. Therefore:

$$f(x_1, \ldots, x_n, y) = F_m^{(n+3)}(x_1, \ldots, x_n, y, y, f(x_1, \ldots, x_n, y))$$
$$= F_m^{(n+3)}(x_1, \ldots, x_n, y, 0, f(x_1, \ldots, x_n, 0))$$
$$= F_m^{(n+3)}(x_1, \ldots, x_n, y, 0, g(x_1, \ldots, x_n))$$

Since $F_m^{(n+3)}$ and g are members of $\mathcal{F}$, and since $\mathcal{F}$ is closed under composition, it follows that the function f must also be a member of $\mathcal{F}$. $\quad\square$

We now turn to definitions by (unbounded) minimalization. As an introductory example, consider the case in which g is a total two-variable member of $\mathcal{F}$ and f is defined from g by minimalization, thus:

$$f(x) = \mu z[g(x, z) = 0].$$

Still assuming that $\mathcal{F}$ satisfies Properties 1 through 5, we wish to show that f necessarily belongs to $\mathcal{F}$.

As with primitive recursion, the key step is choosing an appropriate auxiliary function. Assume for the moment that we can establish the existence in $\mathcal{F}$ of a two-variable function M such that

$$M(x, z) = \begin{cases} 0 & \text{if } g(x, z) = 0 \\ M(x, z^+) + 1 & \text{if } g(x, z) \neq 0. \end{cases} \quad (6\text{-}7)$$

Note that if $\mu z[g(x, z) = 0]$ is defined and equal to r, then $g(x, 0), \ldots, g(x, r-1)$ must all be nonzero. According to Eq. (6-7), this means that $M(x, r) = 0$,

$M(x, r - 1) = 1$, $M(x, r - 2) = 2$, ..., and $M(x, 0) = r$. If $\mu z[g(x, z) = 0]$ is undefined, $g(x, z)$ must be nonzero for all values of z. Thus $M(x, z)$ must equal $M(x, z + 1) + 1$ for all z. This means that $M(x, 0)$ must be undefined, for if $M(x, 0)$ were defined and equal to k, it would follow that $M(x, k) = 0$ and $M(x, k) = M(x, k + 1) + 1$, an obvious impossibility.

These observations imply that the value of the function f is given by

$$f(x) = M(x, 0). \tag{6-8}$$

Thus all that remains is to show that $\mathcal{F}$ necessarily contains a function M satisfying Eq. (6-7). As in the case of primitive recursion, this is readily done with the aid of the Recursion Theorem. Generalizing the argument to the case of functions with arbitrarily many variables, we have:

Theorem 6-2. *Let $\mathcal{F}$ be a class of functions satisfying Properties 1 through 5, and let g be a total, $(n + 1)$-variable member of $\mathcal{F}$. Then the n-variable function f defined from g by minimalization must also be a member of $\mathcal{F}$.*

Proof. It is a routine matter to show that $\mathcal{F}$ must contain a total function γ such that

$$F^{(n+1)}_{\gamma(u)}(x_1, \ldots, x_n, z) = \begin{cases} 0 & \text{if } g(x_1, \ldots, x_n, z) = 0 \\ F^{(n+1)}_u(x_1, \ldots, x_n, z^+) + 1 & \text{if } g(x_1, \ldots, x_n, z) \neq 0. \end{cases}$$

The Recursion Theorem then ensures the existence of an index m such that

$$F^{(n+1)}_m(x_1, \ldots, x_n, z) = \begin{cases} 0 & \text{if } g(x_1, \ldots, x_n, z) = 0 \\ F^{(n+1)}_m(x_1, \ldots, x_n, z^+) + 1 & \text{if } g(x_1, \ldots, x_n, z) \neq 0. \end{cases}$$

Noting that $\mu z[g(x_1, \ldots, x_n, z) = 0]$ is defined and equal to r iff $F^{(n+1)}_m(x_1, \ldots, x_n, 0) = r$, we see that the desired function f can be expressed in the form

$$f(x_1, \ldots, x_n) = F^{(n+1)}_m(x_1, \ldots, x_n, 0).$$

Since $\mathcal{F}$ satisfies the five basic properties, it follows that f must belong to $\mathcal{F}$. $\square$

Theorems 6-1 and 6-2 lead immediately to the following important result.

Theorem 6-3. *Any class of functions that satisfies Properties 1 through 5 must contain all the recursive functions.*

Proof. The formal proof proceeds by induction on the number of composition rules needed to define a given recursive function φ. The basis is established by the fact that Property 1 is satisfied. The induction step is established by the fact that the class in question must be closed under functional composition (Property 2), primitive recursion (Theorem 6-1), and minimalization (Theorem 6-2). $\square$

Theorem 6-3 tells us that the class of recursive functions is in fact the *smallest* class of functions that satisfies Properties 1 through 5. These five properties may therefore be viewed as providing an inductive definition of the class of recursive functions. Remembering that the properties were chosen because they represent essential features of practical programming languages, we conclude that those features completely characterize the set of functions that can be evaluated in an algorithmic manner.

Relative Computability

Very often the ability to determine the values of one function implies the ability to determine the values of another function. We say that the function f is *effectively computable* in terms of, or *relative to*, the function f_0 if the ability to compute arbitrary values of f_0 would provide an effective procedure for evaluating f. As a simple example, consider the diagonal domain function d and the general domain function $D^{(2)}$ for the class of recursive functions. Since $d(x) = D^{(2)}(x, x)$, it is clear that if we had a way of determining arbitrary values of $D^{(2)}$ we could evaluate d. Thus the function d is effectively computable relative to $D^{(2)}$.

In the remainder of this section we will introduce some of the ways in which this notion of relative computability can be made precise. We begin with what is probably the most natural formulation—that in which a function f is to be evaluated in terms of a designated total function f_0. Then we will consider the less obvious but often more convenient formulation in which a function is to be evaluated in terms of a designated set A. In later sections we will make use of these concepts to investigate the computability of various sets and relations.

Relative computability is most easily formalized in the context of μ-recursive functions. Suppose that f_0 is a designated total number-theoretic function. Then a second function f is said to be *μ-recursive relative to f_0* iff f can be derived from f_0 and the constant, projection, and successor functions by means of composition, primitive recursion, and minimalization. In other words, f is μ-recursive relative to f_0 when the assumption that f_0 is μ-recursive allows us to show that f is μ-recursive. Note that if f_0 is in fact μ-recursive, the functions that are μ-recursive relative to it are precisely the μ-recursive functions. And if f is in fact μ-recursive, it is necessarily μ-recursive relative to *any* total number-theoretic function f_0.

Example 6-5. Suppose that f_0 is the (total) two-variable domain function $D^{(2)}$ for the class of recursive functions. Now consider the two-variable function f defined so that

$$f(x, n) = \begin{cases} \text{the } (n + 1)\text{st argument value for which } \varphi_x^{(1)} \text{ is defined, or} \\ \uparrow \quad \text{if } \varphi_x^{(1)} \text{ is not defined for at least } n + 1 \text{ argument values.} \end{cases}$$

Note that f can be derived from $D^{(2)}$ through the use of primitive recursion, minimalization, and appropriate primitive recursive functions and predicates, since:

$$f(x, 0) = \mu y[D^{(2)}(x, y) = 1]$$
$$f(x, n^+) = \mu y[D^{(2)}(x, y) = 1 \wedge y > f(x, n)]$$

Thus f is μ-recursive relative to $D^{(2)}$.   □

It is also possible to formalize the notion of relative computability in the context of Turing machines. This can be done in several ways, one of which we now outline. We consider machines that operate on two tapes: a "main tape" upon which the machine operates in the usual way, and an "auxiliary tape" which the machine can read but not write upon. If the function f is to be evaluated in terms of a designated total function f_0, we assume that the auxiliary tape is inscribed with an (infinite) unary listing of all the values of f_0. The main tape is initially inscribed with the argument of f. The machine computes the value of f much as an ordinary machine would, except that it is able to "look up" values of f_0 on the auxiliary tape, copy those values onto the main tape, and use them as needed in its computation. A machine that operates in this way is said to *evaluate f relative to f_0*.

Of course, the values of f_0 must be arranged on the auxiliary tape in such a way that there is an effective procedure for locating any desired value in a finite number of steps. The following conventions (among others) will ensure that this is the case. If f_0 is a one-variable function, its values are to be listed in the obvious order: $f_0(0), f_0(1), f_0(2)$, etc. If f_0 is an n-variable function, with $n \geq 2$, its values are to be listed according to the Cantor numbers of its arguments: $f_0(\sigma_1^n(0), \ldots, \sigma_n^n(0)), f_0(\sigma_1^n(1), \ldots, \sigma_n^n(1)), f_0(\sigma_1^n(2), \ldots, \sigma_n^n(2))$, etc. In this way we ensure that the location of any value of f_0 can readily be determined from the pertinent argument values.

We now define a function f to be *Turing computable relative to* a designated total function f_0 iff there exists some two-tape machine **M** that evaluates f relative to f_0 in the manner outlined above. Of course, there is no requirement that **M** actually "use" any of the values of f_0 in its computations. Thus if f is a Turing-computable function, it is necessarily Turing computable relative to any total number-theoretic function f_0.

Example 6-6. Again let f_0 be the two-variable domain function $D^{(2)}$ for the class of recursive functions. Let the one-variable function f be defined so that

$$f(x) = \sum_{i=0}^{x} \varphi_x(i),$$

where undefined values of $\varphi_x(i)$ are counted as zero in the summation. To show that f is Turing computable relative to f_0, we note that f can be evaluated relative to f_0 by a two-tape machine **M** that behaves as follows. When provided with a listing of the values of $D^{(2)}$ on its auxiliary tape and an argument $\bar{x}$ on its main tape, **M** successively computes the numbers $m_0 = \pi(x, 0)$, $m_1 = \pi(x, 1)$, and so on up to $m_x = \pi(x, x)$. After each of these computations, **M** uses the value m_i obtained to locate the m_ith block on its auxiliary tape. According to our conventions, this block represents the value of $D^{(2)}(x, i)$. If $D^{(2)}(x, i) = 1$, so that $\varphi_x(i)\!\downarrow$, **M** acts like a universal machine to evaluate $\varphi_x^{(1)}(i)$ and adds the resulting value to a running sum. If $D^{(2)}(x, i) = 0$, so that $\varphi_x(i)\!\uparrow$, **M** simply skips to the next value of i. Since a machine operating in this manner will eventually produce the desired value $f(x)$, we conclude that f is Turing computable relative to $f_0 = D^{(2)}$. □

As might be expected, the concepts of relative Turing computability and relative μ-recursiveness are equivalent. If f_0 is a specified total function, we know that the constant, projection, and successor functions are all Turing computable relative to f_0. And f_0 is certainly Turing computable relative to itself. Moreover, it is not hard to see that relative Turing computability is preserved by functional composition, primitive recursion, and minimalization. From these observations it follows that any function f that is μ-recursive relative to f_0 must also be Turing computable relative to f_0.

Conversely, it is possible to use an arithmetization argument to show that any function f that is Turing computable relative to f_0 must also be μ-recursive relative to f_0. The arithmetization is similar to that previously applied to ordinary Turing-machine computations, except for the treatment of the auxiliary tape. Since the pattern appearing on this tape is infinitely long, it cannot all be represented in the configuration numbers assigned to the various steps in the evaluation of f. Instead, we limit the initial configuration number to be a representation of the *first* block on the auxiliary tape, together with the main tape pattern, the initial state, and the head position. Then if at any point in the subsequent computation the machine leaves the already encoded portion of its auxiliary tape, the function f_0 is used to extend the representation of the auxiliary tape by one block. Leaving the details of the process to the reader, we state our conclusion as:

Theorem 6-4. *The function f is μ-recursive relative to the total function f_0 iff it is Turing computable relative to f_0.*

In view of Theorem 6-4, it is natural to assign a single name to the concepts of relative μ-recursiveness and relative Turing computability. We say that a function f is *recursive relative to* the total function f_0 iff f is μ-recursive (or Turing computable) relative to f_0. Arguments similar to those that motivated the Church-Turing Thesis may now be advanced for:

The Relativized Church-Turing Thesis. (*Functional form*) *The function f is effectively computable relative to the total function f_0 iff f is recursive relative to f_0.*

The definition of relative recursiveness leads to several simple but useful consequences.

Theorem 6-5

   a) *Every total function f_0 is recursive relative to itself.*

   b) *If f_2 is recursive relative to the total function f_1 and f_1 is recursive relative to the total function f_0, then f_2 is recursive relative to f_0.*

Part (a) of this theorem has already been noted; part (b) follows easily from the Relativized Church-Turing Thesis.

The definition of relative μ-recursiveness also gives us:

Theorem 6-6. *If f is recursive relative to the total function f_0, and f_0 is itself a recursive function, then f must also be a recursive function.*

This theorem is most useful when rephrased in the following equivalent form.

Corollary 6-6.1. *If f is recursive relative to the total function f_0, and if f is not a recursive function, then f_0 must not be a recursive function.*

This corollary formalizes one of the most general methods of establishing nonrecursiveness. To prove that a certain total function f_0 is not recursive, it suffices to find another function f that is known not to be recursive, but that can be shown to be recursive relative to f_0. Here it is important to remember the direction of the argument: the function that is already known to be non-recursive must be recursive relative to the function that is to be proven non-recursive, not the other way around.

What we have considered so far is probably the most natural and obvious form of relative computability: that in which a desired function f is computed in terms of the values of a specified total function f_0. In spite of its naturalness, this formulation has certain disadvantages, one of the most serious being that it does not generalize nicely to the case in which f_0 is nontotal.[†] Thus in the theory of recursive functions it is common practice to use a slightly simpler formulation of relative recursiveness in which a function f is evaluated in terms of the membership or nonmembership of numbers in a specified set.

Informally, we say that a function f is effectively computable in a given set of natural numbers A if the ability to determine whether any arbitrary natural number belongs to A provides an effective method of evaluating f. Similarly, we say that f is effectively computable in a given n-ary number-theoretic relation R

[†] In particular, the most obvious Turing-machine and μ-recursive formulations do not agree.

if the ability to determine whether R holds for any arbitrary n-tuple of natural numbers provides an effective method of evaluating f. These informal notions are made precise in the obvious way: The function f is defined to be *recursive in* a given set of natural numbers A (or in a given number-theoretic relation R) iff f is recursive relative to the characteristic function of A (or the characteristic function of R).

Example 6-7. Let A be the set $\{x \mid \varphi_x(x)\downarrow\}$ and let χ_A be its characteristic function. Now consider the function f such that:

$$f(x) = \begin{cases} \varphi_x(x) & \text{if} \quad \varphi_x(x)\downarrow \\ 0 & \text{otherwise} \end{cases}$$

It is easy to see that the ability to determine whether a number belongs to A would make it possible to evaluate the function f. To put the argument in formal terms, we note that

$$f(x) = U^{(2)}(\Lambda(\chi_A(x), 1, x, w), x)$$

where w is an index of the constant function $C_0^{(1)}$. Thus f is recursive relative to the function χ_A and therefore by definition recursive in the set A.   □

The hypothesis that recursiveness in a set or relation accurately reflects the intuitive notion of effective computability in terms of membership in a set or relation provides still another version of the Church-Turing Thesis:

The Relativized Church-Turing Thesis. (*Relational form*) *The function f is effectively computable in a given set A (or relation R) iff it is recursive in A (or R).*

Recursiveness in a set or relation may at first appear to be a more restricted concept than recursiveness relative to a function. But as the next theorem shows, the two concepts are essentially equivalent. Thus no real loss of generality is entailed in adopting the set or relational version.

Theorem 6-7. *A function f is recursive relative to the total n-variable function f_0 iff f is recursive in f_0 as an $(n + 1)$-ary relation.*

Proof. For convenience we use R_0 to denote f_0 as a relation and χ_0 to denote the characteristic function of R_0. Since f_0 is total, and since by definition

$$\chi_0(x_1, \ldots, x_n, z) = \begin{cases} 1 & \text{if} \quad f_0(x_1, \ldots, x_n) = z \\ 0 & \text{if} \quad f_0(x_1, \ldots, x_n) \neq z, \end{cases}$$

it follows that χ_0 is recursive relative to f_0. Theorem 6-5(b) now implies that if a given function f is recursive in R_0—and so by definition recursive relative to χ_0—it must also be recursive relative to f_0. Conversely, since

$$f_0(x_1, \ldots, x_n) = \mu z[\chi_0(x_1, \ldots, x_n, z) = 1]$$

we see that f_0 is recursive relative to χ_0. Thus if the function f is recursive relative to f_0, it must also be recursive relative to χ_0, and so by definition recursive in R_0.   □

The concept of recursiveness in a set is fundamental to the theory of recursive functions. As we have seen, it incorporates the notion of recursiveness relative to a given total function. In addition, it is readily extended, through the use of characteristic functions, to include the possibility of one *set* being recursive in another. But before exploring such matters it is appropriate to examine the basic notions of computability as they pertain to sets.

6.2 COMPUTABLE SETS

In this section we consider the application of computational ideas to number-theoretic sets and relations. We begin by defining two important classes of sets of natural numbers: the recursive sets and the recursively enumerable sets. We then investigate the properties of these two classes and establish the basic relationships between them. An understanding of these properties and relationships will prove helpful in our investigation of decision problems in Section 6.3.

Basic Definitions

There are two fundamental computational processes that might be associated with a specified set A of natural numbers. One is the process of determining, for any given natural number n, whether or not n is a member of A. The other is the process of generating, one by one, the various members of A. As we shall see, these two processes are related, but not equivalent. But before we can describe this relationship, we must provide careful definitions of the underlying concepts.

Consider first the matter of deciding whether a number belongs to a set. Let A be a specified set of natural numbers, and let χ_A be its characteristic function, so that:

$$\chi_A(x) = \begin{cases} 1 & \text{if} \quad x \in A \\ 0 & \text{if} \quad x \notin A \end{cases}$$

Then an effective procedure for determining whether an arbitrary natural number belongs to A is really nothing more than an effective procedure for evaluating χ_A. When such a procedure exists, we say that the set A is *effectively computable*.

The relationship between effectively computable and recursive functions suggests a formal counterpart for the notion of an effectively computable set. We simply define the set A to be *recursive* iff its characteristic function is recursive. As usual, the term *computable* is sometimes used as a synonym for recursive.

Example 6-8

a) Let S be the set consisting of all the perfect squares. Note that χ_S, the characteristic function of S, can be expressed in the form:

$$\chi_S(x) = \begin{cases} 1 & \text{if} & \overset{x}{(\exists z)[z^2 = x]} \\ 0 & \text{otherwise} \end{cases}$$

Thus χ_S is a recursive function, from which it follows that S is a recursive set. In a similar way the set of primes, the set of multiples of 7, and the set of divisors of 120 can all be seen to be recursive sets.

b) Let K be the set $\{x \mid \varphi_x(x){\downarrow}\}$. Then the characteristic function of K is the function χ_K, where:

$$\chi_K(x) = \begin{cases} 1 & \text{if} & \varphi_x(x){\downarrow} \\ 0 & \text{if} & \varphi_x(x){\uparrow} \end{cases}$$

Since χ_K is just the diagonal domain function for the class of recursive functions, it follows that χ_K is not a recursive function and therefore that K is not a recursive set. ☐

Several basic consequences follow easily from the definition of a recursive set. For instance, consideration of the relevant characteristic functions enables us to show that the class of recursive sets is closed under the operations of union, intersection, and set difference.

Lemma 6-1. *If A and B are recursive sets, so are $A \cup B$, $A \cap B$, and $A - B$.*

The proof of this lemma provides a simple exercise for the reader.

Recall that any function whose value is nonzero for only a finite number of argument values is necessarily recursive. Thus every finite set of natural numbers has a recursive characteristic function and is therefore recursive. Since the set of natural numbers N is certainly recursive, it follows from Lemma 6-1 that the complement with respect to N of any finite set is also recursive.[†] Noting that the complement of a finite set is referred to as a *cofinite set*, we may summarize these observations in:

Lemma 6-2. *Every finite or cofinite set of natural numbers is recursive.*

We now turn to the matter of generating the members of a set. We say that a set is *effectively enumerable* iff there exists an effective algorithm for generating a listing of the members of that set, one after another. This listing may be either empty, or finite, or infinite. In other words, the algorithm may never produce any output at all, or it may terminate after generating a finite number

---

[†] In this section, the complement of a set A will always mean the complement with respect to N and will always be denoted $\bar{A}$.

of elements, or it may continue producing elements indefinitely. The order in which the members of the set are produced is unimportant, and repetitions are allowed. All we require is that: (i) every element produced by the algorithm belong to the given set; and (ii) no matter which member of the set we consider, the algorithm will produce that member within a finite number of steps.

Example 6-9. Let A be the set of composite numbers—i.e., the set of natural numbers greater than 1 that are not prime. Now consider the following algorithm. For each natural number k, beginning with $k = 2$ and proceeding in increasing order, successively compute the products $2k$, $3k$, ..., k^2. This algorithm generates the (infinite) sequence of numbers 4, 6, 9, 8, 12, 16, 10, 15, 20, 25, 12, 18, ... Every member of this sequence is obviously a composite number. Moreover, any specified composite number will necessarily be produced within some finite number of steps. In particular, the composite number $p_{i_1}^{e_1} p_{i_2}^{e_2} \cdots p_{i_n}^{e_n}$ will certainly be produced as a multiple of $p_{i_1}^{e_1 - 1} p_{i_2}^{e_2} \cdots p_{i_n}^{e_n}$, if not before. Thus the algorithm generates precisely the members of A, from which it follows that A is an effectively enumerable set. $\square$

It should be evident from the definition that every finite set (including the empty set) is effectively enumerable. An algorithm that generates a finite set may terminate once it has produced all the members of that set or, since repetitions are permitted, it may continue indefinitely. On the other hand an algorithm that generates an infinite set cannot terminate but must continue to produce new elements indefinitely.

The notion of effective enumerability can be formalized in several ways. One of the most natural approaches is based on a simple variant of the Turing-machine model. Suppose that a Turing machine is provided with two tapes, a "main" tape on which it operates in the usual way, and an "output" tape on which it can write but not read. The machine is to be started with both tapes blank and is to behave as follows. After some initial computation, the machine writes the unary representation of a natural number a_0 on its output tape. After further computation, it writes the representation of a second number a_1 immediately to the right of the representation of a_0. It then goes on to compute and write the representations of numbers a_2, a_3, etc. This process may continue indefinitely, or the machine may halt at any point. In any event, the result is a (possibly infinite) sequence of natural-number representations on the machine's output tape. This sequence may be viewed as a listing, perhaps with repetitions, of the members of a set. We accordingly define a set of natural numbers A to be *Turing enumerable* iff there exists some two-tape Turing machine that generates a listing of the members of A in the manner just outlined.

The Turing-machine model of effective enumerability makes it easy to visualize the process of generating the members of a set, and can be particularly useful in informal arguments. However, it does not usually provide a convenient framework in which to carry out precise proofs. For this latter purpose we must resort to a functional approach.

Suppose that there is an effective algorithm for generating the members of a certain nonempty set A. Without loss of generality we may assume that the members of A are produced as an infinite sequence $a_0, a_1, a_2, \ldots, a_n, \ldots$. For if A is finite and the given algorithm terminates after a finite number of steps, the desired infinite sequence can be obtained by executing the algorithm over and over again. Now note that the members of the infinite sequence $a_0, a_1, \ldots, a_n$, $\ldots$ can be viewed as the values of a total one-variable function f, where $f(n) = a_n$. This function is effectively computable, since to evaluate $f(n)$ we need only use the given algorithm to generate the first $n + 1$ members of the sequence $a_0, a_1, \ldots, a_n, \ldots$. Thus every effectively enumerable set must be the range of a total, effectively computable function. Conversely, if a set A is the range of a total, effectively computable one-variable function f, then A must be effectively enumerable, since the members of A can be generated by evaluating $f(0), f(1), f(2), \ldots, f(n), \ldots$ in sequence.

These observations suggest the following formalization. We define the set A to be *recursively enumerable* iff (i) A is empty, or (ii) A is the range of a total one-variable recursive function f. In the latter case, we refer to f as an *enumerating function* for A and say that f *enumerates* A. In the literature of recursive-function theory, the term *semicomputable* is sometimes used as a synonym for recursively enumerable, and the term recursively enumerable is often abbreviated to *r.e.*

Example 6-10. The total function f, where $f(x) = x^3$, enumerates the set of perfect cubes. Since f is recursive, we conclude that the set of perfect cubes is recursively enumerable. In a similar way we see that the set of multiples of seven and the set of primes are recursively enumerable. As Example 6-9 suggests, the set of composite numbers is also recursively enumerable, although in this case it is not as easy to specify an appropriate enumerating function. □

Although we will not give the details here, it is possible to show that recursive enumerability is equivalent to Turing enumerability. This equivalence, together with the obvious appropriateness of the underlying definitions, prompts us to extend the Church-Turing Thesis and accept the concept of recursive or Turing enumerability as the formal counterpart of effective enumerability. In doing so, it is convenient to adopt recursive enumerability and the functional definition that it represents as the basic formal concept. In particular, all our theorems and proofs will be phrased in terms of recursive enumerability.

We will have to learn more about the properties of recursive enumerability before we can easily show that certain interesting sets are recursively enumerable. But the definition by itself allows us to show that some sets are not recursively enumerable. The next theorem provides an important example of this possibility.

Theorem 6-8.   *The set* $T = \{x \mid \varphi_x^{(1)}$ *is total*$\}$ *is not recursively enumerable.*

Proof. Assume that T is recursively enumerable and hence is the range of some total one-variable recursive function f. Define the auxiliary function h so that

$$h(x) = U^{(2)}(f(x), x) + 1 = \varphi_{f(x)}(x) + 1.$$

Since f is recursive, h must also be recursive. Let w be any index of h. Note that because ran $f = T$, the function $\varphi_{f(x)}^{(1)}$ must be total for every choice of x, from which it follows that h must also be total. This in turn means that $w \in T$ and, consequently, that $w \in$ ran f. Suppose in particular that $w = f(u)$. Then we have

$$h(u) = \varphi_{f(u)}(u) + 1 = \varphi_w(u) + 1 = h(u) + 1.$$

From this contradiction, we conclude that the set T is not in fact recursively enumerable.   □

Recursive versus Recursively Enumerable Sets

Recursiveness and recursive enumerability play an important role in the study of computable functions and decision problems. But before we can appreciate this role, we must understand the nature of the relationship between the two concepts. In particular, we must understand why recursiveness and recursive enumerability are not equivalent.

  We begin by showing that every recursive set is also recursively enumerable. The informal argument goes as follows. If A is empty, it is certainly recursively enumerable. If A is nonempty and recursive, there exists an effective algorithm for determining whether or not an arbitrary natural number belongs to A. Successively applying this algorithm to the numbers 0, 1, 2, ... will then enable us to generate a listing of the members of A. Recast in precise functional terms, this argument yields:

Lemma 6-3.   *Every recursive set is recursively enumerable.*

Proof. Let A be a recursive set, and let χ_A be its characteristic function. If A is empty, it is recursively enumerable by definition. Assume, then, that A is nonempty, and let α be some arbitrarily chosen member of A. Define the function f so that:

$$f(n) = \begin{cases} n & \text{if} & \chi_A(n) = 1 \\ \alpha & \text{if} & \chi_A(n) = 0 \end{cases}$$

Since χ_A is total and recursive, it follows that f is also recursive. And f is certainly total. Then because $x \in$ ran f iff $x \in A$, it follows that A is recursively enumerable.   □

The converse of Lemma 6-3 is not true. However, if both a set and its complement are recursively enumerable, then that set must be recursive. This can be seen informally as follows.

Suppose that the set A and its complement $\bar{A}$ are both recursively enumerable. Then there must exist effective procedures for generating A and $\bar{A}$, and any given natural number must eventually be produced by one procedure or the other. Thus we may determine whether or not a given natural number belongs to A by executing the two generation procedures simultaneously until the number in question shows up, and then simply noting which procedure produced it. Since this approach is evidently an effective one, we conclude that A is recursive.

This argument is formalized in:

Lemma 6-4. *If the sets A and $\bar{A}$ are both recursively enumerable, then A must be recursive.*

Proof. If either A or $\bar{A}$ is empty, the set A is certainly recursive. Assume, then, that A and $\bar{A}$ are both nonempty and that both are recursively enumerable. In particular, assume that A is the range of the total recursive function f_0 and that $\bar{A}$ is the range of the total recursive function f_1. Consider the auxiliary function g, defined so that

$$g(x) = \mu z[(f_0(z) = x) \vee (f_1(z) = x)].$$

Since f_0 and f_1 are both total and recursive, it follows that the predicate $(f_0(z) = x) \vee (f_1(z) = x)$ is recursive, hence that the function g is recursive. And since any given natural number x must belong either to the range of f_0 or to the range of f_1, it follows that g must be total. (The value of $g(x)$ is the first position occupied by x in whichever of the two generating processes produces it.) We may now express the characteristic function of A in the form:

$$\chi_A(x) = \begin{cases} 1 & \text{if} & f_0(g(x)) = x \\ 0 & \text{if} & f_0(g(x)) \neq x \end{cases}$$

As a consequence, χ_A must be a recursive function and A must be a recursive set. $\square$

Lemmas 6-1, 6-3, and 6-4 together yield:

Theorem 6-9. *The set A is recursive iff both A and $\bar{A}$ are recursively enumerable.*

Proof. If A is recursive, Lemma 6-1 implies that the set $\bar{A} = N - A$ is also recursive, whence Lemma 6-3 ensures that both A and $\bar{A}$ are recursively enumerable. And if A and $\bar{A}$ are both recursively enumerable, Lemma 6-4 guarantees that A is recursive. $\square$

As we shall see, there are some sets—among them $\{x \mid \varphi_x(x)\!\downarrow\}$—that are recursively enumerable but not recursive. Thus the existence of a procedure for

generating the members of a set does not imply the existence of a procedure for
determining whether an arbitrary natural number belongs to that set. To be
sure, an enumeration procedure provides a way of identifying the numbers that
do belong to the set. But in general an enumeration procedure does not enable
us to identify the numbers that do *not* belong to the set, for in general there will
be no point in the generation process at which it can be definitely concluded
that a given number is never going to be produced. It is for this reason that the
ability to enumerate the nonmembers of a set is a prerequisite for the recursiveness of that set.

At this point it is appropriate to mention a restricted form of recursive
enumerability. We say that the infinite set A is *recursively enumerable in increasing order* iff A is the range of a total recursive function f such that
$f(n + 1) > f(n)$ for all n. In other words, A is recursively enumerable in increasing order iff there exists a total recursive function f such that $f(0)$ is the smallest
member of A, $f(1)$ is the next smallest, and so on. Obviously, any set that is
recursively enumerable in increasing order is recursively enumerable in the
unrestricted sense. The converse, however, is not true.

It is easy to see that every infinite *recursive* set must be recursively enumerable in increasing order. All that is required to obtain such an enumeration is to
examine the natural numbers one by one in sequence, test each one for membership in the given set, and list the ones that are members. Perhaps more
interesting is the fact that any set that can be enumerated in increasing order
must be recursive. For to determine whether a given natural number n belongs
to such a set, it suffices to generate the members of the set in increasing order
until a value equal to or exceeding n is produced. If n appears in this list, it
obviously belongs to the set; if n does not appear in this list, it will certainly not
be produced later in the generation process and so cannot belong to the set.[†]

The preceding informal arguments are summarized in:

Theorem 6-10. *An infinite set A is recursive iff it is recursively enumerable in
increasing order.*

Proof. a) First suppose that the infinite set A is recursive, so that its characteristic function χ_A is recursive. Let α be the smallest member of A, and define the
function f so that:

$$f(0) = \alpha$$
$$f(n + 1) = \mu x[x > f(n) \wedge \chi_A(x) = 1]$$

The function f is evidently recursive, since it is defined from the (total) recursive
function χ_A and appropriate primitive recursive predicates by means of primitive recursion and minimalization. And by definition, the values $f(0), f(1), f(2),$
... are just the members of A in increasing numerical order. Since A is infinite, f

---

[†] This reasoning has already been used in Example 6-2.

is defined for all argument values. Thus f is a total recursive function that enumerates A in increasing order.

b) Now suppose that the infinite set A is recursively enumerable in increasing order. In particular, suppose that A is the range of the total recursive function f, where $f(n + 1) > f(n)$. A simple induction argument shows that $f(n) \geq n$ and that, as a consequence, x is in the range of f iff x is one of the values $f(0), f(1), \ldots, f(x)$. The characteristic function χ_A can therefore be expressed in the form:

$$\chi_A(x) = \begin{cases} 1 & \text{if} \quad \overset{x}{(\exists n)}[f(n) = x] \\ 0 & \text{otherwise} \end{cases}$$

Since f is recursive, χ_A must also be recursive, which means that A must be a recursive set. $\square$

Thus in the case of infinite sets, recursiveness is equivalent to recursive enumerability in increasing order. This fact often helps us to show that certain sets are recursive. For example, the fact that the function f, where $f(n) = n^3$, enumerates the perfect cubes in increasing order immediately tells us that the set of perfect cubes is recursive. A similar approach applies to such sets as the set of primes, the set of factorials, and so on.

Criteria for Recursive Enumerability

Although the formal definition of recursive enumerability is simple and accurately reflects the intuitive notion of effective enumerability, it can be awkward to use. We now consider two criteria that, in effect, provide alternative definitions of recursive enumerability. Use of these alternative definitions often simplifies the job of showing that a set is recursively enumerable. And one of the new definitions also provides the conventional way of assigning "names" to the recursively enumerable sets.

If the set A is recursively enumerable, it must by definition either be empty or else be the range of a total recursive function. In either case, A is of necessity the range of some partial recursive function. We now propose to show that the converse is also true: If A is the range of a partial recursive function, then A must be recursively enumerable. The essence of the argument goes as follows.

Suppose that the nonempty set A is the range of a certain partial recursive function φ and that φ is evaluated by a certain Turing machine **T**. Then the members of A can be enumerated by carrying out the following sequence of simulations. First simulate 0 steps in the evaluation of $\varphi(0)$, then 1 step in the evaluation of $\varphi(0)$, then 0 steps in the evaluation of $\varphi(1)$, then 2 steps in the evaluation of $\varphi(0)$, and so on, following the path shown in Fig. 6-1. As this figure indicates, the ith simulation is that of the first $\sigma_2(i)$ steps in the evaluation of $\varphi(\sigma_1(i))$, where σ_1 and σ_2 are the usual Cantor decoding functions. After each simulation, determine whether the machine **T** halts and yields a defined value within the number of steps simulated. If so, add the resulting value to the list of members of A; if not, go on to the next simulation.

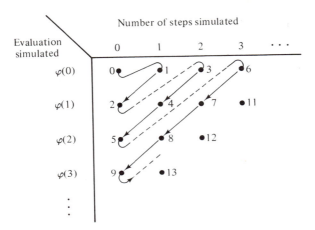

Number of steps simulated

Evaluation
simulated

$\varphi(0)$

$\varphi(1)$

$\varphi(2)$

$\varphi(3)$

Figure 6-1

Every value produced by this procedure must certainly belong to the range of the function φ and hence must be a member of A. Moreover, if the number a belongs to A, then $a = \varphi(u)$ for some number u and the evaluation of $\varphi(u)$ requires at most some finite number of steps n. Since the $\pi(u, n)$th stage in the enumeration procedure consists in the simulation of the first n steps in the evaluation of $\varphi(u)$, the number a will necessarily be added to the list at this stage. Thus the procedure outlined above in fact generates the members of the set A.

Formalizing the preceding argument and filling in the details, we obtain:

Theorem 6-11. *The set A is recursively enumerable iff A is the range of some one-variable (partial) recursive function.*

Proof. a) First suppose that A is recursively enumerable. If A is empty, then A is the range of the empty function $\Phi^{(1)}$, which is certainly recursive. Otherwise A is by definition the range of a one-variable recursive function.

b) Next suppose that A is the range of the one-variable (partial) recursive function φ. If A is empty, it is certainly recursively enumerable. Assume, then, that A is nonempty. Let **T** be a Turing machine that evaluates φ in standard form. As in Chapter 5, we use $\psi_{\mathbf{T}}$ to denote the trace function associated with the computations performed by **T**, so that $\psi_{\mathbf{T}}(x, n)$ is the configuration number of the nth step in the computation that **T** performs for the argument x. And we again use Dec to denote the function that extracts the output value associated with the halting configuration described by a given configuration number. Now consider the two new functions w and f, defined as follows:

$$w(0) = \mu m[\psi_{\mathbf{T}}(\sigma_1(m), \sigma_2(m)) = \psi_{\mathbf{T}}(\sigma_1(m), \sigma_2(m) + 1)]$$

$$w(k^+) = \mu m[\psi_{\mathbf{T}}(\sigma_1(m), \sigma_2(m)) = \psi_{\mathbf{T}}(\sigma_1(m), \sigma_2(m) + 1) \wedge m > w(k)]$$

$$f(k) = \mathrm{Dec}\,(\psi_{\mathbf{T}}(\sigma_1(w(k)), \sigma_2(w(k))))$$

Since ψ_T and Dec are known to be recursive, the functions w and f must also be recursive.

Remember that $\psi_T(\sigma_1(m), \sigma_2(m))$ is the configuration number of the $\sigma_2(m)$th step in the evaluation of $\varphi(\sigma_1(m))$, and that $\psi_T(\sigma_1(m), \sigma_2(m)) = \psi_T(\sigma_1(m), \sigma_2(m) + 1)$ iff the evaluation of $\varphi(\sigma_1(m))$ terminates within $\sigma_2(m)$ steps. Therefore $w(k)$ is the number of the kth member of the series of simulations described in Fig. 6-1 that results in a defined output, and $f(k)$ is the associated output value. It follows from these observations that the range of f is identical to the range of φ, and so is just the set A.

We now argue that the function f is necessarily total. Because A is nonempty, there must be at least one value x_0 for which $\varphi(x_0)$ is defined. Let z be the number of steps needed to evaluate $\varphi(x_0)$. Then $\psi_T(x_0, z)$, $\psi_T(x_0, z + 1)$, $\psi_T(x_0, z + 2)$, ... all represent halting configurations. Since the Cantor-numbering function is monotonic in both its arguments, this means that there is an infinite sequence of increasing numbers $m_0, m_1, \ldots, m_i, \ldots$ such that $\psi_T(\sigma_1(m_i), \sigma_2(m_i))$ represents a halting configuration. And from this fact it follows that w is a total function, hence that f is a total function. Thus A is the range of a total recursive function and is therefore recursively enumerable. $\square$

The technique used in Theorem 6-11 to show that the range of every partial recursive function is recursively enumerable is referred to as *dovetailing*. Basically, dovetailing is a form of "timesharing" in which we perform a little bit of one computation, then a little bit of a second computation, then perhaps a little bit more of the first computation, and so on. In this way it is possible eventually to reach the end of any given terminating computation without getting permanently sidetracked by a nonterminating computation. Dovetailing is an important technique that finds many applications in the theory of recursive functions.

Theorem 6-11 provides a useful alternative to the definition of recursive enumerability, since it is often easier to show that a set is the range of a partial function than that it is the range of a total function. For a simple application, consider the set $\{x \mid \varphi_x(x)\!\downarrow\}$, which is commonly referred to as the set K. This set is not recursive, since its characteristic function is the diagonal domain function. However, it is recursively enumerable.

Theorem 6-12. *The set* $K = \{x \mid \varphi_x(x)\!\downarrow\}$ *is recursively enumerable.*

Proof. Let g be the one-variable function defined so that:

$$g(n) = \begin{cases} n & \text{if} & \varphi_n(n)\!\downarrow \\ \uparrow & \text{if} & \varphi_n(n)\!\uparrow \end{cases}$$

The function g is obviously recursive. Moreover, x belongs to the range of g iff $\varphi_x(x)\!\downarrow$; that is, iff x belongs to K. Thus K is the range of a one-variable recursive function and so is recursively enumerable by virtue of Theorem 6-11.

$\square$

Equally as useful as Theorem 6-11 is the fact that a set is recursively enumerable iff it is the *domain* of a partial recursive function. This fact can be established by a dovetailing argument closely related to that used in the proof of Theorem 6-11. It can also be established, as we now do, by building on Theorem 6-11.

Theorem 6-13. *The set A is recursively enumerable iff it is the domain of some one-variable recursive function.*

Proof. a) Suppose that A is recursively enumerable. If A is empty, it is the domain of the empty function $\Phi^{(1)}$. Otherwise A must by definition be the range of some total one-variable function f. Define the new function h so that

$$h(x) = \mu n[f(n) = x].$$

Since f is total, the function h is certainly recursive. And since h is defined for precisely those argument values that belong to the range of f, it follows that the domain of h is the set A.

b) Suppose that A is the domain of the one-variable recursive function h. Define the new function g so that:

$$g(n) = \begin{cases} n & \text{if} & h(n)\downarrow \\ \uparrow & \text{if} & h(n)\uparrow \end{cases}$$

The function g is certainly recursive, and its range coincides with the domain of h. Thus A is the range of a one-variable recursive function, whence Theorem 6-11 guarantees that A is recursively enumerable.   □

One byproduct of Theorem 6-13 is a naming scheme for recursively enumerable sets. Since the recursively enumerable sets are precisely the domains of the partial recursive functions, we can use the indices of the partial recursive functions to identify the recursively enumerable sets. In particular, we write W_i to denote the recursively enumerable set that is the domain of the function $\varphi_i^{(1)}$, and refer to i as an *index* of that set. Of course, a given recursively enumerable set may have many indices, since it may be the domain of many different recursive functions and each function may itself have many different indices.

Theorem 6-13 also provides a convenient way of showing that certain sets are *not* recursively enumerable. We now consider two such sets.

Theorem 6-14. *The set $\bar{K} = \{x \mid \varphi_x(x)\uparrow\}$ is not recursively enumerable.*

Proof. Assume that $\bar{K}$ is recursively enumerable. Then according to Theorem 6-13, $\bar{K}$ must be the domain of some one-variable recursive function. Suppose in particular that $\bar{K}$ is the domain of the recursive function h and that w is an index of h. By definition, w belongs to $\bar{K}$ iff $\varphi_w(w)$ is undefined—that is, iff $h(w)$ is undefined. Thus $w \in \bar{K}$ iff w is not in the domain of h. But this contradicts the assumption that $\bar{K}$ is the domain of h. From this contradiction we conclude that in fact $\bar{K}$ is not recursively enumerable.   □

Theorem 6-15. *The set* $\bar{T} = \{x \mid \varphi_x^{(1)}$ *is not total*$\}$ *is not recursively enumerable.*

Proof. Assume that $\bar{T}$ is recursively enumerable and, therefore, that it is the domain of some one-variable recursive function h. It is a routine matter to show[†] that there must exist a total one-variable recursive function r such that:

$$\varphi_{r(x)}^{(1)} = \begin{cases} C_0^{(1)} & \text{if} \qquad \varphi_x(x)\!\downarrow \\ \Phi^{(1)} & \text{if} \qquad \varphi_x(x)\!\uparrow \end{cases}$$

Now consider the composite function $g = h \circ (r)$. Note that $g(x)$ is defined iff $r(x)$ is in the domain of h. But since dom $h = \bar{T}$, $r(x) \in$ dom h iff $\varphi_{r(x)}^{(1)}$ is not total. And by definition of r, $\varphi_{r(x)}^{(1)}$ is not total iff $\varphi_x(x)$ is undefined; that is, iff $x \in \bar{K}$. Thus the domain of the function g is the set $\bar{K}$, which implies that $\bar{K}$ is recursively enumerable. Since this contradicts Theorem 6-14, we conclude that $\bar{T}$ cannot be recursively enumerable. $\qquad \square$

Theorems 6-8, 6-12, 6-14, and 6-15 imply the existence of a hierarchy of sets of natural numbers. The simplest sets are the recursive ones, for which it is possible to test for membership in an effective way. Next there are the sets (such as K) that are recursively enumerable but not recursive. Then come the sets (such as $\bar{K}$) that are not recursively enumerable but whose complements are. Finally, there are sets (such as T) that are not recursively enumerable and whose complements are not recursively enumerable.

Recursive and Recursively Enumerable Relations

Although the concepts of recursiveness and recursive enumerability have been defined for sets of natural numbers, they can be extended to apply to other kinds of sets as well. We consider now the extension of these concepts to number-theoretic relations—i.e., to sets of ordered n-tuples of natural numbers.

The case of recursiveness seems to be straightforward. Following the basic definitions for sets of natural numbers, we tentatively define a number-theoretic relation to be recursive iff its characteristic function is recursive. This definition is not only easy to use, but also consistent with the idea that a set should be recursive iff there is an effective procedure for determining whether a given object belongs to the set. Here the appropriate procedure amounts simply to evaluating the pertinent characteristic function for a given n-tuple of natural number arguments.

The case of recursive enumerability is more complicated. The basic definition—that a set is recursively enumerable iff it is the range of a total recursive function—cannot be directly applied to relations, since the range of a recursive function necessarily consists of individual natural numbers rather than n-tuples of natural numbers. One way around this problem would be to adopt the alternative criterion of Theorem 6-13 and define a relation to be

---

[†] See Lemma 1-1.

recursively enumerable iff it is the domain of some (partial) n-variable recursive function. But since this definition does not preserve the intuitive notion of enumerability, it is not immediately clear whether it is an appropriate and useful definition.

A better solution, and one that is also applicable to sets other than relations, is to agree on a convention for representing the elements of a number-theoretic relation by individual natural numbers. We will adopt the obvious Cantor-numbering scheme for this purpose, using the natural number $\pi^n(x_1, \ldots, x_n)$ to represent the n-tuple $(x_1, \ldots, x_n)$. With this scheme any n-ary number-theoretic relation R can be unambiguously described by the set of Cantor numbers of its elements. This set of Cantor numbers will be referred to as the *representing set* of R and denoted $\hat{R}$.

Example 6-11. Let R be the binary relation that holds between the natural numbers u and v iff $v = 2u$. That is, let R be the set $\{(0, 0), (1, 2), (2, 4), (3, 6), \ldots\}$ consisting of all ordered pairs of the form $(u, 2u)$. To obtain the associated representing set $\hat{R}$, we simply replace each member of R by its Cantor number. Thus $\hat{R}$ is the set $\{0, 7, 23, 48, \ldots\}$. Since $\pi^2(x, y) = (\frac{1}{2})(x^2 + 2xy + y^2 + 3x + y)$, we see that $\hat{R}$ must in fact be the set of all natural numbers of the form $(\frac{1}{2})(u^2 + 4u^2 + 4u^2 + 3u + 2u) = (u/2)(9u + 5)$.   □

Since representing sets provide unambiguous descriptions of n-ary number-theoretic relations, we may use them as a basis for defining both recursive and recursively enumerable relations. We now formally define the n-ary relation R to be *recursive* iff its representing set $\hat{R}$ is recursive, and *recursively enumerable* iff $\hat{R}$ is recursively enumerable.

Note that because these formal definitions are based on the corresponding definitions for sets of natural numbers, they do in fact preserve the appropriate intuitive notions. In particular, a relation R is considered to be recursive iff there exists a procedure for determining whether the n-tuple represented by a given natural number m belongs to R. The required procedure is, of course, that which determines whether m belongs to $\hat{R}$. Similarly, a relation R is considered to be recursively enumerable iff there exists a procedure for generating the natural numbers that represent the members of R—i.e., a procedure for generating the members of $\hat{R}$.

Example 6-12.

a) First consider the binary relation $R = \{(u, v) \mid v = 2u\}$ introduced in Example 6-11. As already noted, the representing set $\hat{R}$ for this relation consists of all the natural numbers of the form $(u/2)(9u + 5)$. Thus the characteristic function $\chi_{\hat{R}}$ of the set $\hat{R}$ has the form:

$$\chi_{\hat{R}}(x) = \begin{cases} 1 & \text{if} \quad (\exists u)[x = (u/2)(9u + 5)] \\ 0 & \text{otherwise} \end{cases}$$

Since $\chi_{\hat{R}}$ is obviously a recursive function, we conclude that $\hat{R}$ is a recursive set, and consequently that R is a recursive relation.

b) Now consider the binary relation R that holds between the natural numbers x and y iff $\varphi_x(x) = y$, so that $R = \{(x, y) \mid \varphi_x(x) = y\}$. The representing set of this relation is the set $\hat{R} = \{z \mid \varphi_{\sigma_1(z)}(\sigma_1(z)) = \sigma_2(z)\}$, and the characteristic function of $\hat{R}$ has the form:

$$\chi_{\hat{R}}(z) = \begin{cases} 1 & \text{if} \quad \varphi_{\sigma_1(z)}(\sigma_1(z)) = \sigma_2(z) \\ 0 & \text{otherwise} \end{cases}$$

Since this function can be shown to be nonrecursive, we conclude that the relation R is not recursive. On the other hand, we note that the set $\hat{R}$ is the domain of the function h, where:

$$h(z) = \begin{cases} 1 & \text{if} \quad \varphi_{\sigma_1(z)}(\sigma_1(z)) = \sigma_2(z) \\ \uparrow & \text{otherwise} \end{cases}$$

Since h is a recursive function, it follows from Theorem 6-13 that $\hat{R}$ is a recursively enumerable set. Therefore R is by definition a recursively enumerable relation. ☐

We are now ready to show that our formal definitions of recursiveness and recursive enumerability are equivalent to the tentative definitions suggested earlier. In the case of recursiveness, the argument is routine and is left to the reader. We simply state the result as:

Theorem 6-16. *An n-ary number-theoretic relation is recursive iff its characteristic function is recursive.*

Example 6-13. Again consider the binary relation R consisting of all ordered pairs of the form $(u, 2u)$. In Example 6-12(a) we established the recursiveness of R by showing that its representing set is recursive. Theorem 6-16 now provides a more direct way of doing the job. Using χ_R to denote the characteristic function of the relation R, we have:

$$\chi_R(x, y) = \begin{cases} 1 & \text{if} \quad y = 2x \\ 0 & \text{otherwise} \end{cases}$$

Since this function is obviously recursive, we conclude that R must also be recursive. ☐

The corresponding result for recursive enumerability is:

Theorem 6-17. *An n-ary number-theoretic relation is recursively enumerable iff it is the domain of some (partial) n-variable recursive function.*

Proof. a) Assume that the n-ary relation R is recursively enumerable. Then by definition the associated representing set $\hat{R}$ must be a recursively enumerable set. According to Theorem 6-13, this means that $\hat{R}$ must be the domain of some one-variable recursive function h. Now define the n-variable function θ so that

$$\theta(x_1, \ldots, x_n) = h(\pi^n(x_1, \ldots, x_n)).$$

This new function is obviously recursive. Moreover, it is easy to see that $(x_1, \ldots, x_n) \in \text{dom } \theta$ iff $\pi^n(x_1, \ldots, x_n) \in \hat{R}$. But since $\pi^n(x_1, \ldots, x_n) \in \hat{R}$ iff $(x_1, \ldots, x_n) \in R$, it follows that R is in fact the domain of θ.

   b) Assume that the n-ary relation R is the domain of a certain n-variable recursive function θ. Define the one-variable function h so that

$$h(z) = \theta(\sigma_1^n(z), \ldots, \sigma_n^n(z)).$$

This new function is certainly recursive. Moreover, $z \in \text{dom } h$ iff $(\sigma_1^n(z), \ldots, \sigma_n^n(z)) \in R$, which is to say $z \in \text{dom } h$ iff $z \in \hat{R}$. Thus $\hat{R}$ is the domain of h and is therefore a recursively enumerable set. This in turn means, by definition, that R is a recursively enumerable relation.   □

Example 6-14. Again consider the binary relation $R = \{(x, y) \mid \varphi_x(x) = y\}$. In Example 6-12(b) we showed that this relation is recursively enumerable by appealing to the formal definition. We now note that R is in fact the domain of the function θ, where:

$$\theta(x, y) = \begin{cases} 1 & \text{if} \quad \varphi_x(x) = y \\ \uparrow & \text{otherwise} \end{cases}$$

Since θ is easily shown to be recursive, Theorem 6-17 allows us to conclude directly that the relation R is recursively enumerable.   □

The result of Example 6-14 can be generalized in a natural way to yield:

Theorem 6-18. *Let φ be a (partial) n-variable function. If φ is recursive, the relation*

$$R = \{(x_1, \ldots, x_n, y) \mid \varphi(x_1, \ldots, x_n) = y\}$$

is recursively enumerable.

This theorem relates the functional interpretation of φ to the relational interpretation of φ: it says that if φ is recursive as a function, then φ must be recursively enumerable as a relation. It is left to the reader to determine whether the converse of Theorem 6-18 is also true.

Corollary 6-18.1. *For each positive integer n, the relation*

$$R = \{(w, x_1, \ldots, x_n, y) \mid \varphi_w^{(n)}(x_1, \ldots, x_n) = y\}$$

is recursively enumerable.

Proof. Apply Theorem 6-18 to the universal function $U^{(n+1)}$.   □

We conclude our introduction to recursive and recursively enumerable relations by extending Theorem 6-9 to apply to relations. If R is an n-ary number-theoretic relation, we define the *complement* of R to be the relation $\bar{R}$ that holds for a given n-tuple of natural numbers iff R does not. Thus $\bar{R}$ contains precisely those n-tuples that do not belong to R. It follows from this definition that the representing set of $\bar{R}$ is the complement (with respect to N) of the representing set of R. Coupling this fact with the result of Theorem 6-9 for sets of natural numbers, we have:

Theorem 6-19.   *The number-theoretic relation R is recursive iff both R and $\bar{R}$ are recursively enumerable.*

From the preceding results, we see that our formal definitions of recursive and recursively enumerable relations yield many of the same properties that hold for recursive and recursively enumerable sets. In particular, Theorems 6-16 and 6-17 provide us with useful working definitions of recursiveness and recursive enumerability for relations: A relation is recursive iff its characteristic function is, and a relation is recursively enumerable iff it is the domain of a recursive function. As usual, we will often rely on the Church-Turing Thesis to establish the recursiveness of the function in question.

Sections and Projections

We now consider two simple ways of constructing new relations from given relations—the operations of forming sections and projections. The importance of these operations lies in the fact that they preserve recursive enumerability and therefore provide a convenient way of showing that certain relations are recursively enumerable. Furthermore, we shall see that there is a "universal relation" from which it is possible to obtain any desired recursively enumerable set by forming appropriate sections and projections.

Suppose that R is an n-ary number-theoretic relation (with $n > 1$), that $1 \leq i \leq n$, and that u is any natural number. Then the $(n-1)$-ary relation

$$Q = \{(x_1, \ldots, x_{i-1}, x_{i+1}, \ldots, x_n) \mid (x_1, \ldots, x_{i-1}, u, x_{i+1}, \ldots, x_n) \in R\}$$

is called the *section of R at u along the ith coordinate*. Informally, Q is obtained from R by selecting those members of R that have u as their ith components and then "deleting" that common component. The name *section* reflects the obvious geometric interpretation of forming a cross-section of an n-dimensional space.

Example 6-15. Consider the ternary relation

$$R = \{(0, 1, 0), (0, 1, 2), (1, 2, 0), (2, 0, 1), (2, 2, 1)\}.$$

In this case the section of R at 0 along the first coordinate is the binary relation $\{(1, 0), (1, 2)\}$, the section at 2 along the second coordinate is the binary relation $\{(1, 0), (2, 1)\}$, and so on. Note that the section at, say, 5 along any coordinate is the empty binary relation.   □

The most important property of sections is that they preserve recursiveness and recursive enumerability.

Theorem 6-20. a) *Any section of a recursive relation is recursive, and* b) *any section of a recursively enumerable relation is recursively enumerable.*

Proof. a) The recursiveness of a section of a recursive relation is readily established through consideration of the appropriate characteristic functions. The details are entrusted to the reader.

b) Let R be a recursively enumerable n-ary relation, and let Q be the section of R at u along the ith coordinate. Since R is recursively enumerable it must be the domain of an n-variable recursive function θ. Define the $(n - 1)$-variable function θ' so that:

$$\theta'(x_1, \ldots, x_{i-1}, x_{i+1}, \ldots, x_n) = \begin{cases} 1 & \text{if} \quad \theta(x_1, \ldots, x_{i-1}, u, x_{i+1}, \ldots, x_n)\!\downarrow \\ \uparrow & \text{otherwise} \end{cases}$$

Since Q is the domain of θ', and since θ' is recursive, it follows that Q is recursively enumerable.   □

Theorem 6-20 is often useful in establishing that certain relations are recursive or recursively enumerable.

Example 6-16. Show that the relation $\{(w, y) \mid \varphi_w^{(2)}(5, y) = 7\}$ is recursively enumerable. First recall from Corollary 6-18.1 that the relation

$$R = \{(w, x, y, z) \mid \varphi_w^{(2)}(x, y) = z\}$$

is recursively enumerable. Forming the section at 5 along the second coordinate of R, we obtain the relation

$$Q = \{(w, y, z) \mid \varphi_w^{(2)}(5, y) = z\}$$

which must be recursively enumerable. Then forming the section at 7 along the third coordinate of Q, we obtain the desired relation

$$P = \{(w, y) \mid \varphi_w^{(2)}(5, y) = 7\}$$

which must also be recursively enumerable.   □

We turn now to the idea of a projection. Again suppose that R is an n-ary number-theoretic relation (with $n > 1$), and that $1 \le i \le n$. Then the $(n - 1)$-ary relation $R_{(i)}$, where

$$R_{(i)} = \{(x_1, \ldots, x_{i-1}, x_{i+1}, \ldots, x_n) \mid (\exists x_i)((x_1, \ldots, x_{i-1}, x_i, x_{i+1}, \ldots, x_n) \in R)\}$$

is called the *projection of R along the* ith *coordinate.* Informally, this relation consists of all the $(n-1)$-tuples of numbers that appear in positions $1, \ldots, i-1, i+1, \ldots, n$ among the members of R. Note that the projections of a binary relation along its first and second coordinates are respectively the range and domain of that relation.

Example 6-17

a) Again consider the ternary relation

$$R = \{(0,\ 1,\ 0),\ (0,\ 1,\ 2),\ (1,\ 2,\ 0),\ (2,\ 0,\ 1),\ (2,\ 2,\ 1)\}.$$

The projections of this relation along its first, second, and third coordinates are respectively the binary relations:

$$R_{(1)} = \{(1,\ 0),\ (1,\ 2),\ (2,\ 0),\ (0,\ 1),\ (2,\ 1)\}$$
$$R_{(2)} = \{(0,\ 0),\ (0,\ 2),\ (1,\ 0),\ (2,\ 1)\}$$
$$R_{(3)} = \{(0,\ 1),\ (1,\ 2),\ (2,\ 0),\ (2,\ 2)\}$$

b) Now consider the ternary relation $R = \{(x,\ y,\ z)\ |\ \varphi_x(y) = z\}$. In this case the projections of R along its first, second, and third coordinates are the relations:

$$R_{(1)} = \{(y,\ z)\ |\ \varphi_x(y) = z \quad \text{for some } x\} = N^2$$
$$R_{(2)} = \{(x,\ z)\ |\ z \in \text{ran } \varphi_x\}$$
$$R_{(3)} = \{(x,\ y)\ |\ \varphi_x(y)\!\downarrow\}$$

$\square$

Although projections do not necessarily preserve recursiveness, they do preserve recursive enumerability. This is easy enough to see intuitively. Suppose, for example, that R is a recursively enumerable binary relation. Then to enumerate, say, the range of R we need only enumerate the members of R and list the second component of each pair generated. This argument is formalized and generalized in:

Theorem 6-21. *Every projection of a recursively enumerable relation is recursively enumerable.*

Proof. Let R be a recursively enumerable n-ary relation and let $R_{(i)}$ be the projection of R along its ith coordinate. Since R is recursively enumerable, its representing set $\hat{R}$ must be the range of some recursive function f. Define the new function g so that

$$g(x) = \pi^{n-1}(\sigma_1^n(f(x)),\ \ldots,\ \sigma_{i-1}^n(f(x)),\ \sigma_{i+1}^n(f(x)),\ \ldots,\ \sigma_n^n(f(x))).$$

Certainly g is also a recursive function. Moreover, the range of g is just the set of Cantor numbers of the members of $R_{(i)}$; in other words, the range of g is the representing set $\hat{R}_{(i)}$ of the relation $R_{(i)}$. Thus $\hat{R}_{(i)}$ is a recursively enumerable set and $R_{(i)}$ is a recursively enumerable relation.   □

Since every recursive set is recursively enumerable, we immediately have:

Corollary 6-21.1. *Every projection of a recursive relation is recursively enumerable.*

It is important to remember, however, that projections of recursive relations need not be recursive.

Like Theorem 6-20, Theorem 6-21 can be useful in establishing the recursive enumerability of various relations.

Example 6-18. Consider the ternary relation R that holds for x, y, z iff the evaluation of $\varphi_x(y)$ terminates within z steps. Since there is clearly an effective procedure for determining whether R holds for a given triple of natural numbers, we conclude that R is a recursive relation. Now note that the projection of R along its third coordinate is the binary relation

$$Q = \{(x, y) \,|\, (\exists z)R(x, y, z)\} = \{(x, y) \,|\, \varphi_x(y)\!\downarrow\}.$$

Corollary 6-21.1 therefore ensures that Q is recursively enumerable. (Since Q is not recursive, we have verified that projection does not preserve recursiveness.) Next note that the projection of Q along its second coordinate is the unary relation

$$P = \{x \,|\, \varphi_x(y) \text{ is defined for some } y\} = \{x \,|\, \operatorname{ran} \varphi_x^{(1)} \text{ is nonempty}\}.$$

Theorem 6-21 therefore ensures that this relation is also recursively enumerable.   □

Of course, it is possible to use both sections and projections in establishing the recursive enumerability of a relation.

Example 6-19. Show that the unary relation $\{x \,|\, 7 \in \operatorname{ran} \varphi_x^{(1)}\}$ is recursively enumerable. First recall that the ternary relation

$$R = \{(x, y, z) \,|\, \varphi_x(y) = z\}$$

is recursively enumerable. Forming the section at 7 on the third coordinate yields the binary relation

$$Q = \{(x, y) \,|\, \varphi_x(y) = 7\}.$$

Forming the projection of Q along its second coordinate now yields the desired unary relation

$$P = \{x \,|\, (\exists y)(\varphi_x(y) = 7)\} = \{x \,|\, 7 \in \operatorname{ran} \varphi_x^{(1)}\}.$$

As a consequence of Theorems 6-20 and 6-21, this last relation must be recursively enumerable.   □

We now prove the converse of Corollary 6-21.1, namely that every recursively enumerable relation is a projection of some recursive relation. The germ of this result is contained in the definition of a recursively enumerable set. Every nonempty recursively enumerable set (unary relation) is the range of a total recursive function; but a total recursive function is in fact a recursive binary relation, and the range of the function is just the projection of the relation along its first coordinate.

Theorem 6-22. *Every recursively enumerable n-ary relation Q is a projection of some recursive $(n + 1)$-ary relation.*

Proof. If Q is empty, it is a projection of the empty $(n + 1)$-ary relation. If Q is nonempty, its representing set $\hat{Q}$ must be the range of a total recursive function f. Now consider the $(n + 1)$-ary relation

$$R = \{(w, x_1, \ldots, x_n) \mid f(w) = \pi^n(x_1, \ldots, x_n)\}.$$

Upon examination of the associated characteristic function, we find that R is recursive. But the given relation Q is the projection of R along its first coordinate, since:

$$Q = \{(x_1, \ldots, x_n) \mid \pi^n(x_1, \ldots, x_n) \in \hat{Q}\}$$
$$= \{(x_1, \ldots, x_n) \mid \pi^n(x_1, \ldots, x_n) \in \operatorname{ran} f\}$$
$$= \{(x_1, \ldots, x_n) \mid (\exists w)(f(w) = \pi^n(x_1, \ldots, x_n))\}$$
$$= \{(x_1, \ldots, x_n) \mid (\exists w)((w, x_1, \ldots, x_n) \in R)\}$$

Thus Q is in fact a projection of a recursive $(n + 1)$-ary relation. $\quad\square$

Two corollaries follow immediately from Theorems 6-21 and 6-22.

Corollary 6-22.1. *An n-ary number-theoretic relation is recursively enumerable iff it is a projection of a recursive $(n + 1)$-ary relation.*

Corollary 6-22.2. *An n-ary number-theoretic relation is recursively enumerable iff it is a projection of a recursively enumerable $(n + 1)$-ary relation.*

Thus the recursive relations serve as a source for all the recursively enumerable relations and sets. In fact, there is a single recursive relation from which all the recursively enumerable sets can be obtained by means of appropriate sections and projections.

Theorem 6-23. *There exists a recursive ternary relation R such that, for every recursively enumerable set W_a, $W_a = \{x \mid (\exists z)R(a, x, z)\}$.*

Proof. Let R be the relation that holds for i, x, z iff the evaluation of $\varphi_i(x)$ terminates with a defined value in z or fewer steps. As noted in Example 6-18, R is recursive. Now suppose that $W_a = \operatorname{dom} \varphi_a^{(1)}$ is any specified recursively enumerable set. The section of R at a on the first coordinate is the recursively enumerable relation Q that holds for x, z iff the evaluation of $\varphi_a(x)$ terminates with a defined value in z or fewer steps. The projection of Q along its second coordinate is therefore the recursively enumerable relation that holds for x iff $x \in \text{dom } \varphi_a$. Thus in fact $W_a = \{x \mid (\exists z)R(a, x, z)\}$. $\square$

6.3 DECISION PROBLEMS

We have by now seen many situations in which we wish to determine whether an arbitrarily specified object has a given property. As simple examples, we cite the problems of determining whether an arbitrary natural number is prime, whether an arbitrary Turing machine halts when started on blank tape, whether an arbitrary function is total, and so on. Such problems are known as decision problems. In this section we investigate decision problems in some detail, paying particular attention to the existence or nonexistence of algorithmic procedures for making the desired decisions. As might be expected, this subject is intimately related to the issues of computability discussed earlier in the chapter.

Basic Terminology

A *decision problem* is characterized by two things: a set of objects S, and an n-ary relation P on the set S. We refer to S as the *domain* of the decision problem and to P as the *property* of the decision problem. The decision problem itself is that of determining, for any arbitrary combination of n objects from S, whether or not the relation P holds for that combination. For the sake of explicitness, we sometimes refer to a decision problem whose relation is n-ary as an *n-ary decision problem*.

Example 6-20. The following are instances of decision problems.

a) The problem of determining, for any pair of nonzero natural numbers, whether those numbers are relatively prime. Here the domain S is the set of nonzero natural numbers and the property P is the binary relation that holds between two numbers iff they are relatively prime.

b) The problem of determining, for any finite string of symbols from a certain alphabet, whether that string is periodic—i.e., whether it consists of two or more repetitions of a basic subpattern. In this case the domain is the set of all finite strings on the given alphabet and the property is the unary relation that holds for a string iff that string is periodic.

c) The problem of determining, for any configuration of pieces on a chess board, whether it is possible for White to win from that configuration. Here, of course, the domain is the set of all possible configurations and the property is the unary relation that holds for a configuration iff White can win from that configuration. $\square$

An *n*-ary decision problem is said to be *effectively decidable* iff there exists an effective procedure that correctly determines, for every possible combination of *n* domain elements, whether or not the requisite property holds for that combination. Thus we see that the problem of Example 6-20(a) is effectively decidable; indeed, the familiar Euclidean algorithm provides an appropriate way of making the necessary decision. The string problem of Example 6-20(b) is also effectively decidable, since there are only a finite number of ways of dividing a given string into equal length substrings and each such division can be effectively checked to see whether its substrings are identical. It is left to the reader to determine whether the chess problem of Example 6-20(c) is effectively decidable.

A decision problem whose domain is the set of natural numbers will be referred to as a *number-theoretic decision problem*. We shall be concerned exclusively with decision problems that are either number-theoretic problems or that can be represented in a natural way by number-theoretic decision problems.

The problem of the relatively prime numbers obviously falls into this category, since it is a simple matter to extend the definition of the "relatively prime" relation to make it applicable to all natural numbers.[†] The string problem also lies within our purview, since every finite string on a specified finite alphabet can be represented by a single natural number via an appropriate Gödel-numbering scheme. Once that is done, the property to be tested for can be recast as a number-theoretic relation. In a similar way we find that any one of a number of coding schemes can be used to represent the chess problem as a number-theoretic decision problem.

Indeed, as long as a decision problem has a countable domain, it can be represented in one way or another as a number-theoretic decision problem. More to the point, in almost all cases of interest this representation can be effected in a natural and convenient way. By this we mean that there are almost always relatively simple algorithmic ways of converting back and forth between the members of the domain of the original problem and their natural number representations. Thus little generality will be lost by restricting our attention in the following discussion to number-theoretic decision problems.

We are now ready to formalize the notion of decidability for number-theoretic decision problems. Note that an effective procedure for determining whether a given relation holds really amounts to an effective procedure for evaluating the characteristic function of that relation. According to the Church-Turing Thesis, such a procedure exists iff the characteristic function is recursive—i.e., iff the relation itself is a recursive relation. We are therefore led to define a number-theoretic decision problem to be *recursively decidable* if its property is a recursive relation and to be *recursively undecidable* otherwise.

[†] For example, we can simply stipulate that 0 is not relatively prime to any natural number.

Since the Church-Turing Thesis implies that the concept of recursive decid-ability accurately represents the informal notion of effective decidability, it is common practice to use the single words *decidable* (or *solvable*) and *undecid-able* (or *unsolvable*) in place of recursively decidable and recursively undecid-able. These same terms are also applied to decision problems that, although not inherently number-theoretic problems, are easily recast as number-theoretic problems. Thus we speak of a nonnumber-theoretic problem as being decidable if, when expressed in number-theoretic terms, it is recursively decidable.

Example 6-21

a) Consider the problem of determining, for any natural number n, whether or not n is prime. This problem is recursively decidable since the unary relation that holds for a natural number n iff n is prime is in fact a recursive relation.

b) Consider the problem of determining, for any natural number n, whether or not $\varphi_n(n)$ is defined. This problem is recursively undecidable since the unary relation that holds for n iff $\varphi_n(n)\downarrow$ is not recursive. (This relation is just the set K defined in Section 6.2.)   □

It follows from the results of preceding sections that many decision prob-lems associated with the class of recursive functions are undecidable. Among these are:

1. the problem of deciding, for any x and y, whether $\varphi_x(y)$ is defined;
2. the problem of deciding, for any x, whether $\varphi_x^{(1)}$ is total;
3. the problem of deciding, for any x and y, whether $\varphi_x^{(1)} = \varphi_y^{(1)}$.

We now present a result, due to Rice, that automatically establishes the unde-cidability of a broad class of decision problems.

Theorem 6-24. *(Rice's Theorem) Let P be any proper, nonempty subset of the one-variable recursive functions. Then the set $A = \{x \mid \varphi_x^{(1)} \in P\}$ is not recursive.*

Proof. First consider the case in which the empty function $\Phi^{(1)}$ does *not* belong to P. Since P is nonempty, there must be some one-variable recursive function f that does belong to P. In the usual way we can establish the existence of a total recursive function r such that:

$$\varphi_{r(x)}(y) = \begin{cases} f(y) & \text{if} & \varphi_x(x)\downarrow \\ \uparrow & \text{if} & \varphi_x(x)\uparrow \end{cases}$$

Thus $\varphi_{r(x)}^{(1)}$ is the function f if $\varphi_x(x)$ is defined and the function $\Phi^{(1)}$ if $\varphi_x(x)$ is undefined. In other words, $\varphi_{r(x)}^{(1)} \in P$ iff $\varphi_x(x)$ is defined. Now, if the character-istic function χ_A of the set $A = \{x \mid \varphi_x^{(1)} \in P\}$ were recursive, the composite function $\chi_A \circ (r)$ would also be recursive. But $\chi_A \circ (r)$ is the diagonal domain function, since:

$$\chi_A(r(x)) = \begin{cases} 1 & \text{if} & \varphi_{r(x)}^{(1)} \in P \\ 0 & \text{if} & \varphi_{r(x)}^{(1)} \notin P \end{cases} = \begin{cases} 1 & \text{if} & \varphi_x(x)\downarrow \\ 0 & \text{if} & \varphi_x(x)\uparrow \end{cases}$$

Thus χ_A is not recursive, which in turn means that the set A is not recursive.

In case the empty function $\Phi^{(1)}$ does belong to P, we simply replace P by its complement and use the preceding argument to show that $\bar{A}$ is not recursive. It then follows from Lemma 6-1 that A itself is not recursive. $\square$

Invoking the definition of recursive decidability, we may recast Rice's Theorem in the following form.

Corollary 6-24.1. *Let P be any property that holds for at least one one-variable recursive function and fails to hold for at least one one-variable recursive function. Then the problem of determining, for any given index x, whether the property P holds for the function $\varphi_x^{(1)}$ is recursively undecidable.*

Stated less formally, this corollary says that there is no algorithmic way of determining, on the basis of an index alone, whether a function satisfies a given nontrivial property.

Rice's Theorem immediately establishes the undecidability of a large class of decision problems. Several of these problems are listed below:

1. the problem of determining, for any x, whether $\varphi_x^{(1)}$ is total;
2. the problem of determining, for any x, whether $\varphi_x^{(1)}$ is a constant function;
3. the problem of determining, for any x, whether the range of $\varphi_x^{(1)}$ is infinite;
4. the problem of determining, for any x, whether $\varphi_x^{(1)}$ includes 17 in its range.

Rice's Theorem can be extended in a rather natural way to properties that are not unary. However, it must not be thought that all undecidable problems having to do with recursive functions fall victim to Rice's Theorem or its extension. In particular, it is important to understand that the property referred to in Rice's Theorem must be a property of *functions* alone, not a property of indices or of functions and indices together. Thus, for example, Rice's Theorem does not apply to the problem of determining, for a specified total function f and an arbitrary choice of x, whether $\varphi_x^{(1)} = \varphi_{f(x)}^{(1)}$.

Many-One Reducibility

The most common way of showing that a decision problem is undecidable is to relate it in an appropriate way to another problem that is already known to be undecidable. This approach, which is referred to as *reducing* one decision problem to another, is based on the notion of relative computability introduced in Section 6.1. We now outline the general nature of the reduction process and investigate in detail its simplest and most useful form. We begin by considering the special case of unary decision problems.

Let A and B be sets of natural numbers. Extending the terminology of Section 6.1 in the obvious way, we define B to be *recursive in A* iff the characteristic function of B is recursive relative to the characteristic function of A. Thus B is recursive in A iff the ability to determine whether arbitrary numbers belong to A provides a procedure for determining whether a number belongs to B. Theorem 6-6, together with the definition of a recursive set, immediately gives us:

Theorem 6-25. *If the set B is recursive in the set A, and if A is a recursive set, then B must be a recursive set.*

Corollary 6-25.1. *If the set B is recursive in the set A, and if B is not recursive, then A cannot be recursive.*

Corollary 6-25.1 has the following application to decision problems. Suppose we wish to show that the problem of determining whether an arbitrary natural number belongs to a certain set A is undecidable. To this end, it suffices to find a set B such that (i) B is recursive in A, and (ii) the problem of determining membership in B is undecidable. For if the problem of determining membership in B is undecidable, B must not be recursive. Corollary 6-25.1 then ensures that the set A is not recursive, hence that the problem of determining membership in A is undecidable. In following this approach, we say that we have *reduced* the problem of determining membership in B to that of determining membership in A.

A simple way of showing that one set is recursive in another is to exhibit a recursive function that in effect maps the property of membership in the first set into the property of membership in the second set. To make this idea precise, again suppose that A and B are sets of natural numbers. We say that B is *many-one reducible* to A iff there exists a total recursive function f such that for every natural number x, $f(x) \in A$ iff $x \in B$. For the sake of simplicity, we will normally abbreviate "many-one reducible" to "m-reducible" or even, when the context permits, to "reducible". When B is m-reducible to A by virtue of the function f, we speak of f as (many-one) *reducing B to A*, or of B as being (many-one) *reduced* to A by f.

It is important to remember that in order for B to be m-reduced to A, the pertinent function f not only must map every member of B into some member of A, but also must map every member of $\bar{B}$ into some member of $\bar{A}$. Thus B is m-reduced to A by the function f iff $\bar{B}$ is m-reduced to $\bar{A}$ by the same function. On the other hand, the range of the function f need not include every member of A or every member of $\bar{A}$. Nor need f be a one-to-one function. (It is this last fact that provides the name "many-one" reducibility.)

Example 6-22

a) Let B be the set of odd natural numbers and A the set of perfect squares. Then B is m-reducible to A, since the function f, where

$$f(x) = \begin{cases} x^2 & \text{if} & x \text{ is odd} \\ 5 & \text{if} & x \text{ is even} \end{cases}$$

is a total recursive function having the property that $f(x) \in A$ iff $x \in B$. Note that the function g, where $g(x) = x^2$, does not m-reduce B to A; although g maps every member of B into some member of A, it fails to map every member of $\bar{B}$ into some member of $\bar{A}$.

b) Let B be the set $\{x \mid \varphi_x(x) \downarrow\}$ and let A be the set $\{x \mid \varphi_x^{(1)} \text{ is total}\}$. To see that B is m-reducible to A, we need only recall that there exists a total recursive function f such that:

$$\varphi_{f(x)}^{(1)} = \begin{cases} C_0^{(1)} & \text{if} & \varphi_x(x) \downarrow \\ \Phi^{(1)} & \text{if} & \varphi_x(x) \uparrow \end{cases}$$

Since $C_0^{(1)}$ is total and $\Phi^{(1)}$ is not, $f(x) \in A$ iff $x \in B$.   □

The importance of the concept of m-reducibility derives from the next theorem and its corollaries.

Theorem 6-26. *If the set B is m-reducible to A, then B is recursive in A.*

Proof. Suppose that B is m-reduced to A by the total recursive function f. Let χ_A and χ_B denote the characteristic functions of A and B. Noting that

$$\chi_A(f(x)) = \begin{cases} 1 & \text{if} & f(x) \in A \\ 0 & \text{if} & f(x) \notin A \end{cases} = \begin{cases} 1 & \text{if} & x \in B \\ 0 & \text{if} & x \notin B \end{cases}$$

we see that χ_B is the composite function $\chi_A \circ (f)$. Thus χ_B is recursive relative to χ_A and B is recursive in A.   □

Corollary 6-26.1. *If the set B is m-reducible to the set A, and if A is recursive, then B must be recursive.*

Corollary 6-26.2. *If the set B is m-reducible to the set A, and if B is not recursive, then A is not recursive.*

Thus to show that the problem of determining membership in a given set A is undecidable, it suffices to exhibit a set B that is m-reducible to A and for which the problem of determining membership is known to be undecidable.

The set most often used in the role of B in a reduction argument is the set $K = \{x \mid \varphi_x(x) \downarrow\}$. Proofs that such sets as $A_1 = \{x \mid \varphi_x^{(1)} \text{ is total}\}$, $A_2 = \{x \mid \varphi_x^{(1)} \text{ is monotonic}\}$, and $A_3 = \{x \mid \varphi_x^{(1)} \text{ is a constant function}\}$ are not recursive (and therefore that the associated membership problems are not decidable) can all be carried out by showing that K is reducible to the set in question. Indeed, Rice's Theorem provides a very general example of this approach, since the proof of this theorem really consists in showing that, whenever P is a nontrivial subset of the one-variable recursive functions, the set K is m-reducible to the set $A = \{x \mid \varphi_x^{(1)} \in P\}$.

In order to deal with nonunary decision problems, we must extend the concept of m-reducibility from sets of natural numbers to arbitrary number-theoretic relations. The basic idea is simple: all that is needed is an algorithmic procedure that maps each member of one relation into a member of the other, and each nonmember of the first relation into a nonmember of the second. This idea is made precise through the use of representing sets. Suppose that R and S are number-theoretic relations and that $\hat{R}$ and $\hat{S}$ are their representing sets. We say that S is *many-one reducible* to R iff the representing set $\hat{S}$ is many-one reducible to the representing set $\hat{R}$. Note that this definition does not require R and S to have the same arity.

As an illustration, consider the case in which R is the binary relation $\{(u, v) \mid \text{dom } \varphi_u^{(1)} = \text{dom } \varphi_v^{(1)}\}$ and S is the binary relation $\{(u, v) \mid \varphi_u^{(1)} = \varphi_v^{(1)}\}$. To show that S is m-reducible to R, we must find an algorithmic way of mapping members of S into members of R and nonmembers of S into nonmembers of R. In effect, we must find an algorithmic way of mapping each pair of one-variable recursive functions (φ_i, φ_j) into a new pair (ψ_i, ψ_j) in such a way that ψ_i and ψ_j have the same domain iff φ_i and φ_j are identical. One possibility is to choose the functions ψ_i and ψ_j so that

$$\psi_i(x) = \begin{cases} 0 & \text{if} \quad \varphi_i(x)\downarrow \text{ or } \varphi_j(x)\downarrow \\ \uparrow & \text{otherwise} \end{cases} \qquad (6\text{-}8a)$$

$$\psi_j(x) = \begin{cases} 0 & \text{if} \quad \varphi_i(x)\downarrow \text{ and } \varphi_j(x)\downarrow \text{ and } \varphi_i(x) = \varphi_j(x) \\ \uparrow & \text{otherwise} \end{cases} \qquad (6\text{-}8b)$$

On the basis of this definition, it is clear that ψ_i and ψ_j will have the same domain whenever φ_i and φ_j are identical. Conversely, a little thought shows that if ψ_i and ψ_j do have the same domain, then φ_i and φ_j must also have that same domain, and must in fact be identical. Thus ψ_i and ψ_j have the same domain iff φ_i and φ_j are the same function. And since the definitions of ψ_i and ψ_j can evidently be obtained in an effective way from those of φ_i and φ_j, we have—at least informally—reduced the relation S to the relation R.

To make the argument rigorous, we must recast the mapping process just outlined into the form of a function f that operates on Cantor numbers of ordered pairs of indices. We first note (and leave to the reader to show) that there exist total recursive functions r_1 and r_2 such that:

$$\varphi_{r_1(i,\,j)}(x) = \begin{cases} 0 & \text{if} \quad \varphi_i(x)\downarrow \text{ or } \varphi_j(x)\downarrow \\ \uparrow & \text{otherwise} \end{cases}$$

and

$$\varphi_{r_2(i,\,j)}(x) = \begin{cases} 0 & \text{if} \quad \varphi_i(x)\downarrow \text{ and } \varphi_j(x)\downarrow \text{ and } \varphi_i(x) = \varphi_j(x) \\ \uparrow & \text{otherwise} \end{cases}$$

These functions r_1 and r_2 serve to compute indices of ψ_i and ψ_j from those of φ_i and φ_j. It is the existence of such recursive index-computing functions that justifies the claim that ψ_i and ψ_j can be determined effectively from φ_i and φ_j.

The desired mapping function f can now be defined as follows:

$$f(z) = \pi^2(r_1(\sigma_1^2(z), \sigma_2^2(z)), r_2(\sigma_1^2(z), \sigma_2^2(z)))$$

This function is certainly total and recursive. And as a result of the choice of r_1 and r_2, f maps the Cantor number of the indices of φ_i and φ_j into the Cantor number of the indices of ψ_i and ψ_j. Consequently $f(z) \in \hat{R}$ iff $z \in \hat{S}$, which means that $\hat{S}$ is m-reducible to $\hat{R}$ and so, by definition, that S is m-reducible to R.

As should be expected, Corollaries 6-26.1 and 6-26.2 have counterparts for the case of relations.

Corollary 6-26.3. *If the number-theoretic relation S is m-reducible to the number-theoretic relation R, and if R is recursive, then S must also be recursive.*

Corollary 6-26.4. *If the number-theoretic relation S is m-reducible to the number-theoretic relation R, and if S is not recursive, then R is not recursive.*

The proofs are straightforward and will be omitted.

Corollary 6-26.4 provides a convenient means of showing that certain relations are not recursive, hence that certain decision problems are not decidable. For example, since we already know that the relation $S = \{(u, v) \mid \varphi_u^{(1)} = \varphi_v^{(1)}\}$ is not recursive, the fact that S is m-reducible to the relation $R = \{(u, v) \mid \operatorname{dom} \varphi_u^{(1)} = \operatorname{dom} \varphi_v^{(1)}\}$ allows us to conclude that R is not recursive either. And this means that the problem of determining whether two arbitrarily specified indices denote functions having the same domain is not decidable.[†]

While the concept of m-reducibility provides an important tool for establishing nonrecursiveness or undecidability, it is seldom applied in full detail. In practice, we normally supply only an informal description of the required function f, and rely on the Church-Turing Thesis to justify omission of a formal specification. This is especially true in the case of reductions from one relation to another, where we do not normally bother to phrase the argument in terms of representing sets. Thus in the example just discussed, we might simply define the new functions ψ_i and ψ_j as in Eqs. (6-8), observe that ψ_i and ψ_j will have identical domains iff φ_i and φ_j are identical, and omit the remainder of the argument. It must be kept in mind, however, that a complete formal argument requires a precise derivation of an appropriate mapping function f.

[†] It is worth noting that in this example each of the relations R and S is m-reducible to the other (as the reader may show). Thus the decision problems associated with these two relations are equivalent: the existence of an effective procedure for solving either problem would provide an effective procedure for solving the other.

The Correspondence Problem

It should not be thought that the only undecidable problems are those concerning the class of recursive functions. We now consider an important undecidable problem about lists of words. This problem was first formulated by the logician E. Post and is known as Post's Correspondence Problem. After describing the problem itself, we will use a reduction argument to show that the Correspondence Problem is undecidable. As we will later see, this result allows us to establish the undecidability of a variety of other problems.

Suppose that A is a finite alphabet of symbols. Let L be a list consisting of the k words $l_1, l_2, \ldots, l_k$ on the alphabet A, and let M be a list consisting of the k words $m_1, m_2, \ldots, m_k$ on the same alphabet. Then any finite sequence of numbers $i_1, i_2, \ldots, i_n$ in the range 1 to k serves to specify a concatenation $l_{i_1} l_{i_2} \cdots l_{i_n}$ of words from list L and a concatenation $m_{i_1} m_{i_2} \cdots m_{i_n}$ of words from the list M. We refer to the numbers $i_1, i_2, \ldots, i_n$ as *indices* and to the composite words $l_{i_1} l_{i_2} \cdots l_{i_n}$ and $m_{i_1} m_{i_2} \cdots m_{i_n}$ as the words *generated* from L and M by the sequence $i_1, i_2, \ldots, i_n$. If there exists a sequence of indices $i_1, i_2, \ldots, i_n$ (repetitions permitted) such that the generated words $l_{i_1} l_{i_2} \cdots l_{i_n}$ and $m_{i_1} m_{i_2} \cdots m_{i_n}$ are identical, we say that the lists L and M have a *match*.

Example 6-23

a) Let A be the alphabet $\{0, 1\}$; let L be the list consisting of the three words $l_1 = 0\,1$, $l_2 = 0$, $l_3 = 1\,0$; and let M be the list consisting of the three words $m_1 = 0$, $m_2 = 0\,0$, $m_3 = 1\,1$. Then, for example, the sequence of indices 1, 2, 1, 3 generates the word $l_1 l_2 l_1 l_3 = 0\,1\,0\,0\,1\,1\,0$ from list L and the word $m_1 m_2 m_1 m_3 = 0\,0\,0\,0\,1\,1$ from list M. The given lists L and M do have a match, since the sequence of indices 1, 3, 2 generates the same word $l_1 l_3 l_2 = m_1 m_3 m_2 = 0\,1\,1\,0\,0$ from each list.

b) Using the same alphabet $A = \{0, 1\}$, let L be the list consisting of the words $l_1 = 0\,1$, $l_2 = 1\,0$, $l_3 = 0$ and let M be the list consisting of the words $m_1 = 0$, $m_2 = 1\,1$, $m_3 = 0\,1$. In this case the lists L and M do not have a match, since no matter what sequence of indices is chosen, the words generated from L and M must end in different symbols. $\square$

The *Correspondence Problem* can now be stated quite simply. It is just the problem of determining, for any pair of equal-length lists on a common alphabet, whether or not that pair of lists has a match. We now propose to show that this problem is undecidable. In other words, we propose to show that when recast in number-theoretic form, the Correspondence Problem is recursively undecidable.

The proof consists in showing that the problem of determining membership in the set $K = \{x \mid \varphi_x(x)\downarrow\}$ is reducible to the Correspondence Problem. We will describe the reduction informally, showing how to construct, for any given natural number x, a pair of lists that have a match iff $x \in K$. Once this construction is understood, it is a routine matter to recast the argument in the number-theoretic form needed to establish formal m-reducibility.

The argument is formulated in terms of Turing-machine computations. Recall that each step in a computation is characterized by a machine configuration consisting of a tape pattern, an internal state, and a head position. For our present purposes, such a configuration is conveniently described by a string of symbols consisting of the nonblank portion of the tape pattern, with a special symbol denoting the internal state inserted immediately to the left of the currently scanned tape symbol. Thus the configuration in which a given machine is in state q_3 and is scanning the second 1 in the pattern $1\,0\,2\,1\,1\,0\,3$ will be described by the string $1\,0\,2\,q_3\,1\,1\,0\,3$.

Now suppose that **T** is a Turing machine whose internal states are q_0, q_1, ..., q_h and whose tape symbols are 0, 1, ..., r. Without loss of generality, we assume that **T** performs its number-theoretic computations in standard form. Further suppose that w_1 is a configuration of the form $q_0 01^{n_1}$ and that w_2 is a configuration of the form $q_h 01^{n_2}$. Our first job is to show how to construct lists L and M that have a match iff the machine **T** eventually reaches configuration w_2 after being started in configuration w_1. Later we will apply this construction to a particular choice of **T**, w_1, and w_2 in order to establish the undecidability of the Correspondence Problem.

The alphabet from which the desired lists L and M are formed consists of the symbols 0, 1, ..., r, q_0, ..., q_h, $*$, $\square$, and $\triangle$. The individual words of L and M are determined by the quintuples that define the behavior of **T**, together with the configurations w_1 and w_2. These words are described in Table 6-1. Note that this table does not explicitly specify the indices of the words of L and M. Instead, the members of the two lists are presented in pairs, each member of L being paired with the member of M having the same index.

Now we must argue that the lists defined in Table 6-1 have a match iff the machine **T** eventually reaches configuration w_2 after being started in configuration w_1. The key to the argument lies in noting that if the lists are to have a match, the matching strings must have the form:

$$\triangle\square \longleftarrow c_1 \longrightarrow \square \longleftarrow c_2 \longrightarrow \square \longleftarrow \cdots \longrightarrow \square \longleftarrow c_p \longrightarrow \square\triangle \qquad (6\text{-}9)$$

Here the segments denoted c_1, c_2, ..., c_p represent successive configurations in the computation performed by **T**, starting with w_1 and ending with w_2. The convention for representing configurations is the same as that introduced above, except that the special symbol $*$ appears before and after each tape or state symbol. Thus the configuration $1\,0\,2\,q_3\,1\,1\,0\,3$ is to be represented by the segment $*1*0*2*q_3*1*1*0*3*$.

To see why the matching strings must have the form indicated in (6-9), let us consider an illustrative example. Suppose that w_1 is the configuration $q_0\,0\,1\,1$ and that w_2 is the configuration $q_h\,0\,1$. Then the pair of words

Table 6-1

	List L	List M
a) Corresponding to the initial configuration $w_1 = q_0 01^{n_1}$, the pair:	$\triangle$	$\overbrace{\triangle\square *q_0 *0*1* \cdots *1}^{n_1}$
b) Corresponding to the final configuration $w_2 = q_h 01^{n_2}$, the pair:	$\overbrace{q_h *0*1* \cdots *1*\square\triangle}^{n_2}$	$\triangle$
c) For each tape symbol s, the pair:	$s*$	$*s$
Plus the pair:	$\square *$	$*\square$
d) i) For each quintuple of the form (q, s, s', R, q') and each choice of tape symbol v, the pair:	$q*s*v*$	$*s' *q' *v$
ii) For each quintuple of the form (q, s, s', R, q'), the pair:	$q*s*\square *$	$*s' *q' *0*\square$
e) i) For each quintuple of the form (q, s, s', L, q') and each choice of tape symbols u and v, the pair:	$u*q*s*v*$	$*q' *u*s' *v$
ii) For each quintuple of the form $(q, s, s' \neq 0, L, q')$ and each choice of tape symbol u, the pair:	$u*q*s*\square *$	$*q' *u*s' *\square$
iii) For each quintuple of the form $(q, s, 0, L, q')$ and each choice of tape symbol u, the pair:	$u*q*s*\square *$	$*q' *u*\square$

specified in line (a) of Table 6-1 will consist of the word $\triangle$ in list L and the word $\triangle\square *q_0 *0*1*1$ in list M, while the pair specified in line (b) will consist of the word $q_h *0*1*\square\triangle$ in list L and the word $\triangle$ in list M. Suppose further that the description of $\mathbf{T}$ contains the quintuple $(q_0, 0, 2, R, q_1)$, so that when $\mathbf{T}$ is started in the configuration $q_0 011$ its first move will take it to the configuration $2q_1 11$. Then among the pairs specified by line $(d$-$i)$ of the table will be one consisting of the word $q_0 *0*1*$ in list L and the word $*2*q_1 *1$ in list M.

Now, if the lists L and M are to have a match, the strings that constitute that match must begin with the pair of words specified in line (a) of the table, since this is the only pair whose members have the same first symbol. Figure 6-2(a) shows the beginning of the match created by these two words. In order for the match to be continued, the next word from list L must begin with the symbol $\square$. The only such word is the word $\square *$ specified in the second part of line (c). Appending this word and its counterpart $*\square$ from list M gives the strings shown in Fig. 6-2(b). Next we need a member of L that is an initial segment of the string $q_0 *0*1*1* \cdots$. As already noted, the presence of the quintuple $(q_0, 0, 2, R, q_1)$ in the description of $\mathbf{T}$ implies the inclusion of

(a) List L $\triangle$
     List M $\triangle \square * q_0 * 0 * 1 * 1$

(b) List L $\triangle \square *$
     List M $\triangle \square * q_0 * 0 * 1 * 1 * \square$

(c) List L $\triangle \square * q_0 * 0 * 1 *$
     List M $\triangle \square * q_0 * 0 * 1 * 1 * \square * 2 * q_1 * 1$

(d) List L $\triangle \square * q_0 * 0 * 1 * 1 *$
     List M $\triangle \square * q_0 * 0 * 1 * 1 * \square * 2 * q_1 * 1 * 1$

(e) List L $\triangle \square * q_0 * 0 * 1 * 1 * \square *$
     List M $\triangle \square * q_0 * 0 * 1 * 1 * \square * 2 * q_1 * 1 * 1 * \square$

Figure 6-2

the word $q_0 * 0 * 1 *$ in list L. Moreover, inspection of the table reveals that this is the only member of L that is an initial segment of $q_0 * 0 * 1 * 1 * \cdots$. Adding this word, together with its counterpart $* 2 * q_1 * 1$ from list M, yields the strings shown in Fig. 6-2(c). To continue the match, we must next add the words $1 *$ and $* 1$ specified in line (c), and then the words $\square *$ and $* \square$, so as to obtain the strings shown in Fig. 6-2(d) and (e).

At this point we note that the strings shown in part (b) of Fig. 6-2 have the form

$$\triangle \square *$$
$$\triangle \square \longleftarrow c_1 \longrightarrow \square$$

where c_1 represents the initial configuration w_1. We also see that the strings shown in part (e) of the figure have the form

$$\triangle \square \longleftarrow c_1 \longrightarrow \square *$$
$$\triangle \square \longleftarrow c_1 \longrightarrow \square \longleftarrow c_2 \longrightarrow \square$$

where c_2 represents the configuration assumed by **T** after its first step. The situation may therefore be summarized as follows. In order to create a pair of matching strings, we were forced to begin with a string from list M that represents the initial configuration w_1; and in order to match that string with words from list L, we were forced to extend the M-string to include a representation of the next configuration assumed by **T**.

Similar reasoning applies from this point on. In order to continue the match, we must extend the M-string to include a representation of the third configuration assumed by **T**, then the fourth, and so on. In each case the extension is accomplished by "copying" the bulk of the current configuration by means of pairs of words of the form $s *$ and $* s$, and "updating" the state and

symbol representations by means of the pairs of words defined by the quintuples of **T**. In most cases this updating involves words specified in lines (d-i),
(e-i), and (e-ii) of Table 6-1. The words specified in line (d-ii) provide for the
addition of extra 0's when it is necessary to lengthen a tape-pattern representation, and the words specified in line (e-iii) provide for the removal of such extra
0's as soon as the reading head moves to their left.[†]

Finally, we observe that the potential match can be terminated only by the
addition of the pair of words $q_h*0*1*\square\triangle$ and $\triangle$ specified in line (b) of the
table, since this is the only pair whose members have the same last symbol. But
these words can be added to achieve a complete match only if the strings
eventually assume the form:

$$\cdots *\square *$$

$$\cdots *\square *q_h*0*1*\square$$

Evidently this will be the case only if **T** eventually assumes the configuration
$q_h\,0\,1$. Thus the lists L and M have a match iff the given machine **T** eventually
reaches the configuration w_2 after being started in configuration w_1.

By studying the way in which the lists L and M are defined, and by
experimenting with other examples, the reader should convince himself that in
each case a match will exist in precisely the desired circumstances. We state this
conclusion as:

Lemma 6-5. *Let* **T** *be any specified Turing machine that carries out its number-
theoretic computations in standard form, let* w_1 *be a machine configuration of the
form* $q_0\,0\,1\cdots 1$, *and let* w_2 *be a configuration of the form* $q_h\,0\,1\cdots 1$. *Then there
exists a pair of lists* L *and* M *that have a match iff* **T** *eventually assumes
configuration* w_2 *after being started in configuration* w_1. *Moreover, there is an
effective procedure for constructing* L *and* M *from a description of* **T**, w_1, *and* w_2.

Lemma 6-5 provides the foundation for our major result.

Theorem 6-27. *The Correspondence Problem—i.e., the problem of determining
whether two arbitrary lists have a match—is undecidable.*

Proof. We propose to show that the problem of determining membership in
the set $K = \{x\,|\,\varphi_x(x)\downarrow\}$ is reducible to the Correspondence Problem. To do this,
we must show how to construct, for each natural number x, a pair of lists that
have a match iff $x \in K$.

Note that the function g, where

$$g(x) = \begin{cases} 0 & \text{if} & \varphi_x(x)\downarrow \\ \uparrow & \text{if} & \varphi_x(x)\uparrow \end{cases}$$

[†] Since **T** performs its computations in standard form, it is never necessary to extend a tape pattern
to the left.

is certainly recursive. Let $\mathbf{T}_0$ be a Turing machine that evaluates g in standard form. Suppose that the starting state of $\mathbf{T}_0$ is q_0 and that the sole halting state of $\mathbf{T}_0$ is q_h. Thus $\mathbf{T}_0$ will eventually reach the configuration $q_h 0\, 1$ after starting in the initial configuration $q_0 0\bar{x}$ iff $g(x) = 0$; that is, iff $\varphi_x(x)$ is defined.

Now to each natural number x we assign the pair of lists L_x and M_x associated with the machine $\mathbf{T}_0$, the initial configuration $w_1 = q_0 0\bar{x}$, and the final configuration $w_2 = q_h 0\, 1$. According to Lemma 6-5, there is an effective procedure for constructing these lists from the given number x. In addition, Lemma 6-5 ensures that L_x and M_x will have a match iff $\mathbf{T}_0$ eventually reaches configuration w_2 after starting in configuration w_1. But as just noted, this will be the case iff $\varphi_x(x)$ is defined or, equivalently, iff $x \in K$.

Thus the problem of determining membership in K is reducible to the Correspondence Problem, and the latter problem is therefore undecidable. □

The preceding proof represents an informal reduction argument. To formalize the argument, we must first recast the Correspondence Problem in number-theoretic terms. Adopting any one of a variety of simple coding schemes will enable us to represent each possible pair of lists by a single natural number. Then if C denotes the set of numbers representing lists that have matches, the Correspondence Problem becomes that of determining membership in C. Using the proof given for Theorem 6-27 as a guide, we can derive a total recursive function f such that $f(x)$ is the representation of the pair of lists associated with the machine $\mathbf{T}_0$, the initial configuration $w_1 = q_0 0\bar{x}$, and the final configuration $w_2 = q_h 0\, 1$. Since f maps members of K into members of C and nonmembers of K into nonmembers of C, it follows that K is m-reducible to C and that C is nonrecursive. The details are routine and will not be given.

Applications to Context-Free Grammars

Among the tools used in the study of computer programming are various formal languages and the grammars by which they may be defined. And associated with such languages and grammars are a number of important decision problems. Since some of these problems can be shown to be undecidable with the help of the Correspondence Problem, it is appropriate to consider them briefly here. We begin by reviewing the basic definitions underlying the use of formal grammars in general and context-free grammars in particular.[†]

A *formal grammar* is characterized by four entities: a finite set of *variables* V_N, a finite set of *terminals* V_T (disjoint from V_N), a finite set of *productions* P, and a single *sentence symbol* S drawn from V_N. Thus a formal grammar can be represented as a quadruple of the form (V_N, V_T, P, S). Variables are commonly

---

[†] Readers with some prior exposure to context-free grammars and languages should be able to follow the discussion of this section without difficulty. Others may wish to consult an introductory text on formal languages for background.

denoted by upper case latin letters and terminals by lower case latin letters, although in the case of terminals it is often convenient to use other symbols as well. The productions of a formal grammar are expressions of the form $\alpha \to \beta$, where α and β are specified strings of variable and terminal letters. The sentence symbol will almost always be denoted S.

The productions of a formal grammar provide rules for generating new strings of symbols from given strings. Suppose that u and v are strings formed from the variables and terminals of the grammar G. Then the production $\alpha \to \beta$ is said to *apply to u to yield v* iff there exist substrings x and y (possibly null) such that $u = x\alpha y$ and $v = x\beta y$. We say that a string u *directly generates* a string v (in G) iff there is some production in G that applies to u to yield v. And we say that a string w_1 *generates* a string w_n if w_1 directly generates w_n or if there exist intermediate strings $w_2, \ldots, w_{n-1}$ such that w_1 directly generates w_2, w_2 directly generates w_3, $\ldots$, and w_{n-1} directly generates w_n. Finally, we define the *language generated by* the formal grammar $G = (V_N, V_T, P, S)$ to be the set of all those strings of *terminals* that are generated by the sentence symbol S.

Formal grammars may be classified according to the nature of their productions. Of special interest are grammars whose productions all have the form $A \to \beta$, where A is a single variable symbol and β is a string of terminals and variables. Such grammars are called *context-free grammars* and the languages they generate are called *context-free languages*. Thus a set of strings constitutes a context-free language iff there is some context-free grammar that generates that set of strings. Context-free languages are important because they can be generated by simple rules and yet can be used to represent large segments of practical programming languages.

Example 6-24

a) Let G_1 be the formal grammar whose variables are the letters S and A; whose terminals are the letters a, b, and c; whose sentence symbol is S; and whose productions are:

$$P_1: S \to aSa \qquad P_3: S \to A$$
$$P_2: A \to bAb \qquad P_4: A \to c$$

Here we find, for example, that the string $aaSaa$ generates the string $aaabcbaaa$, since production P_1 applies to $aaSaa$ to yield $aaaSaaa$, P_3 applies to $aaaSaaa$ to yield $aaaAaaa$, P_2 applies to $aaaAaaa$ to yield $aaabAbaaa$, and P_4 applies to $aaabAbaaa$ to yield $aaabcbaaa$. A little thought shows that the language generated by G consists of all the strings of the form $a^n b^m c b^m a^n$, since these are the strings of terminals that can be obtained from the sentence symbol S through successive applications of various productions. Since G_1 is a context-free grammar, we conclude that the strings of the form $a^n b^m c b^m a^n$ constitute a context-free language.

b) Let G_2 be the context-free grammar whose variables are S and A; whose terminals are the symbols x, y, (,), and $+$; whose·sentence symbol is S; and whose productions are:

$$P_1: S \to S + S \qquad P_5: A \to x$$
$$P_2: S \to AA \qquad P_6: A \to y$$
$$P_3: A \to (S + S) \qquad P_7: S \to x$$
$$P_4: A \to AA \qquad P_8: S \to y$$

Among the members of the language generated by this grammar are $x + xy$, which is obtained by applying productions P_1, P_7, P_2, P_5, and P_6 in the obvious way; and $x(x + (y + x)y)$, which is obtained by applying productions P_2, P_5, P_3, P_7, P_2, P_3, P_8, P_7, and P_6. Indeed, as the reader may convince himself, the language generated by this grammar consists of all the "arithmetic expressions" involving x and y—that is, all the expressions conventionally used to denote sums and products involving x and y.   □

The way in which a given string is generated by a context-free grammar can be conveniently represented by a *parse tree*. Figure 6-3 shows the parse tree for the generation of the string $x(x + yx)$ in the grammar of Example 6-24(b). Here the sentence symbol S at the top of the tree represents the starting point of the generation, and each set of edges descending from a single node represents the application of a single production. Thus the initial application of the production $S \to AA$ is represented by the edges leading from the topmost S to the symbols A and A in the second row, the application of the production $A \to x$ is represented by the edge leading from A to x, and so on. As a consequence of this construction, the symbols of the generated string can be read off the terminal points of the tree, working from left to right.

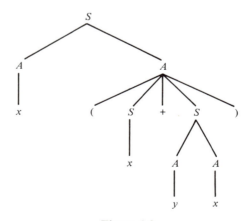

Figure 6-3

Parse trees are helpful in assigning "meanings" to the strings generated by a context-free grammar and are commonly used by compilers to translate algebraic expressions into sequences of operations for evaluating such expressions. For this purpose it is usually desirable for the parse tree of a given string to be uniquely determined by the grammar, since otherwise it might not be possible to determine what meaning to assign. A context-free grammar is accordingly defined to be *unambiguous* if every string in the associated language has only one possible parse tree, and *ambiguous* otherwise. Informally, then, a grammar is ambiguous if it provides two essentially different ways of generating some string.

Example 6-25

a) Consider the context-free grammar whose only variable is S, whose terminals are a and b, and whose productions are

$$P_1: S \rightarrow SS \qquad P_2: S \rightarrow SaS \qquad P_3: S \rightarrow b$$

This grammar is ambiguous, since as the parse trees below indicate, there are two distinct ways of generating the string $babb$ (among others).

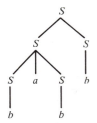   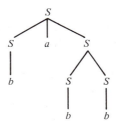

b) The grammars G_1 and G_2 of Example 6-24 are unambiguous, although in the case of G_2 this fact is not so easy to demonstrate.   □

A number of important decision problems are associated with context-free grammars and their languages. Among these are:

1. the problem of determining whether an arbitrary string belongs to the language generated by a given context-free grammar;

2. the problem of determining whether a context-free grammar generates a nonempty language;

3. the problem of determining whether the languages generated by two context-free grammars have any strings in common;

4. the problem of determining whether a context-free grammar is ambiguous;

5. the problem of determining whether two context-free grammars generate the same language.

The first two of these problems turn out to be decidable, although we cannot undertake to provide proofs here. The remaining three problems are undecidable, and two of them can be shown to be undecidable quite easily. We begin by showing that the Correspondence Problem is reducible to problem (3), hence that the latter problem is undecidable.

What we must do is show how to associate a pair of context-free grammars with every pair of lists in such a way that the resulting languages will have some string in common iff the given lists have a match. To this end, suppose that L is the list $l_1, \ldots, l_k$ on the alphabet $\{a_1, \ldots, a_q\}$ and that M is the list $m_1, \ldots, m_k$ on the same alphabet. We now define the context-free grammars G_L and G_M as follows:

G_L Variables: S_L
       Terminals: $1, 2, \ldots, k, a_1, \ldots, a_q, *$
       Productions: $S_L \to iS_L l_i$ for each i, $1 \le i \le k$
                 $S_L \to i*l_i$ for each i, $1 \le i \le k$

G_M Variables: S_M
       Terminals: $1, 2, \ldots, k, a_1, \ldots, a_q, *$
       Productions: $S_M \to iS_M m_i$ for each i, $1 \le i \le k$
                 $S_M \to i*m_i$ for each i, $1 \le i \le k$

The sentence symbols of the two grammars are, of course, S_L and S_M. Note that the process of constructing G_L and G_M from L and M is an effective one.

Example 6-26. Let L be the list $l_1 = ab$, $l_2 = baa$, $l_3 = b$ on the alphabet $\{a, b\}$, and let M be the list $m_1 = a$, $m_2 = bb$, $m_3 = ba$ on the same alphabet. Then the associated grammars G_L and G_M each have as their terminals the symbols 1, 2, 3, a, b, *. The sole variable of G_L is S_L, the sole variable of G_M is S_M, and the productions of the two grammars are those listed below.

$$G_L: \; S_L \to 1S_L ab \qquad\qquad G_M: \; S_M \to 1S_M a$$
$$S_L \to 2S_L baa \qquad\qquad\qquad S_M \to 2S_M bb$$
$$S_L \to 3S_L b \qquad\qquad\qquad\;\; S_M \to 3S_M ba$$
$$S_L \to 1*ab \qquad\qquad\qquad\;\; S_M \to 1*a$$
$$S_L \to 2*baa \qquad\qquad\qquad S_M \to 2*bb$$
$$S_L \to 3*b \qquad\qquad\qquad\quad\; S_M \to 3*ba$$

It is not hard to see that the language generated by the grammar G_L consists of all those strings of the form $i_1 i_2 \cdots i_n * l_{i_n} \cdots l_{i_2} l_{i_1}$, where the indices $i_1, i_2, \ldots, i_n$ are chosen arbitrarily from among $1, 2, \ldots, k$. Similarly, the language generated by G_M consists of all those strings of the form $i_1 i_2 \cdots i_n * m_{i_n} \cdots m_{i_2} m_{i_1}$. From these observations it follows that the two languages will have some string in common iff there exists a sequence of indices $i_1, i_2, \ldots, i_n$ such that $l_{i_1} l_{i_2} \cdots l_{i_n} = m_{i_1} m_{i_2} \cdots m_{i_n}$. In other words, the languages will have a string in common iff the lists L and M have a match.

Thus we have succeeded, at least informally, in reducing the Correspondence Problem to the problem of determining whether the languages generated by two context-free grammars have any strings in common. As a consequence, we may conclude that the latter problem is undecidable. Of course, to make the argument rigorous, we must recast both problems in number-theoretic form and exhibit a recursive function f that m-reduces one to the other. Since there is little to be learned by going through these formalities, we shall not bother to do so.

Essentially the same argument can be used to establish the undecidability of problem (4), that of determining whether an arbitrary context-free grammar is ambiguous. Again let L and M be lists on the common alphabet $\{a_1, \ldots, a_q\}$ and let the grammars G_L and G_M be specified as above. We now define the new grammar G from G_L and G_M as follows. The variables of G are the symbols S, S_L, and S_M, with S as the sentence symbol. The terminals of G are the same as those of G_L and G_M, namely 1, 2, ..., k, a_1, ..., a_q, and $*$. The productions of G consist of all the productions of G_L and G_M, together with the two additional productions $S \rightarrow S_L$ and $S \rightarrow S_M$.

Since the only productions that can be applied to the new sentence symbol S are $S \rightarrow S_L$ and $S \rightarrow S_M$, it follows that the language generated by G is just the union of the sets of strings that can be generated by S_L and S_M—that is, the union of the languages generated by G_L and G_M. Next note that both G_L and G_M are unambiguous; in each case there is only one way of generating any given string in the language. Thus a string in the new language can be generated in two different ways iff that string belongs to both the language of G_L and the language of G_M. This means that G is ambiguous iff the languages generated by G_L and G_M have some member in common—i.e., iff the lists L and M have a match. We have therefore succeeded in reducing the Correspondence Problem to the problem of determining whether a context-free grammar is ambiguous and can conclude that the latter problem is undecidable.

The grammars G_L and G_M also play a role in a third important reduction argument. Although the details are too complicated to present here, the conclusion can be stated quite simply: There is an effective procedure for constructing, for any pair of lists L and M, a context-free grammar G' whose language consists of all possible strings on a certain alphabet iff L and M do not have a match. Now, it is easy to establish the existence of a context-free grammar G'' whose language is in fact the set of all possible strings on the alphabet in question. The grammars G' and G'' therefore generate the same language iff L and M do not have a match. Thus the Correspondence Problem is reducible to the equivalence problem for context-free grammars,[†] from which we conclude that the latter problem is undecidable.

---

† Strictly speaking, the *in*equivalence problem.

The arguments outlined above illustrate the role of the Correspondence Problem in establishing the undecidability of various questions about context-free grammars. These results in turn provide more or less practical applications of the notion of undecidability, for they imply that certain questions arising naturally in the study of programming languages cannot be settled by algorithmic means.

Other Undecidable Problems

We close this chapter with a brief discussion of several additional examples of undecidable problems. The intent is simply to suggest the range of problems that turn out to be effectively undecidable, and we will not attempt to provide proofs of the results cited.

Our first example is the so-called word problem for semigroups. We begin with a review of terminology. A *semigroup* is a mathematical system composed of a set of elements on which an associative binary operation has been defined. Thus the set of natural numbers forms a semigroup under the operation of addition, the set of all finite strings of symbols from a given alphabet forms a semigroup under the operation of concatenation, and so on. It is common to think of the operation of a semigroup as "multiplication" and to represent its application by the infix symbol · or by simple juxtaposition of the elements involved.

A subset of the elements of a semigroup is said to *generate* the semigroup if every other element can be expressed as a product of the given elements (repetitions allowed). Such products are commonly referred to as *words* on the generating elements. Of course, it is quite possible for two words to represent the same element of the semigroup. That is, there will in general exist *constraints* among the generating elements. These constraints often provide a convenient way of characterizing the semigroup. In particular, when a semigroup is generated by a finite set of elements, it is common to describe that semigroup by specifying enough constraints on the generating elements to uniquely determine the structure of the semigroup.

Example 6-27.   A certain semigroup is generated by the two elements a and b, and these elements are subject to the following constraints:

$$aaa = a \qquad aab = b$$
$$bb = b \qquad ba = b$$

It is not difficult to show that the operation table for this semigroup must be that shown in Fig. 6-4, where the elements c and d represent the products aa and ab, respectively. To illustrate the formation of this table, consider the product $b \cdot d$. Making use of the given constraints, the associativity of the semigroup operation, and the fact that $d = ab$, we find that

$$b \cdot d = b \cdot ab = ba \cdot b = b \cdot b = b$$

Similar arguments apply to the other entries.   □

$\cdot$	a	b	c	d
a	c	d	a	b
b	b	b	b	b
$aa = c$	a	b	c	d
$ab = d$	d	d	d	d

Figure 6-4

Now suppose that a certain semigroup S is generated by the elements $a_1, \ldots, a_k$, and that the structure of S is specified by a finite number of constraints on these generators, each constraint being denoted by an equality between two products of generators. Then two words (products) of the form $a_{i_1} a_{i_2} \cdots a_{i_m}$ and $a_{j_1} a_{j_2} \cdots a_{j_n}$ will represent the same element of S iff it is possible to obtain one from the other by successive applications of the defining constraints. In terms of the generators $a_1, \ldots, a_k$ the *word problem* for S is the problem of determining, for any two words $a_{i_1} a_{i_2} \cdots a_{i_m}$ and $a_{j_1} a_{j_2} \cdots a_{j_n}$, whether those words represent the same element of S.

For some choices of a semigroup and a set of defining generator constraints, the associated word problem is decidable. The semigroup of Example 6-27 is a case in point, since to determine whether two products of a's and b's denote the same element, we need only make use of the operation table of Fig. 6-4. On the other hand, there are (infinite) semigroups whose word problems are not decidable. Although we shall not attempt to supply a detailed proof of this fact, it is perhaps appropriate to indicate how the argument goes.

In essence, the proof consists in showing that the problem of determining membership in the set $K = \{x \mid \varphi_x(x)\!\downarrow\}$ is reducible to the word problem for an appropriately defined semigroup. The first step is to establish a scheme whereby a semigroup can be thought of as "simulating" the computations of a given Turing machine. To this end, we may choose the generating elements of the semigroup to represent the tape symbols and internal states of the machine in question. Words formed from the generators can then be used to represent machine configurations according to the convention in which the internal state symbol appears to the left of the currently scanned tape symbol.

With this representation scheme in mind, constraints on the generators can be chosen so as to permit the transformation of each configuration that the machine might assume into the next succeeding one. This can be accomplished in much the same manner as was used in dealing with the Correspondence Problem. For instance, if the machine writes the symbol s', moves right, and enters state q' after scanning the symbol s in state q, we include the identity $qs = s'q'$ among the generator constraints. When the details are properly arranged, we find that the word denoting configuration v represents the same semigroup element as that denoting configuration u iff the machine in question eventually reaches configuration v after being started in configuration u, or vice versa.

Now let **T** be a Turing machine that evaluates the function g, where

$$g(x) = \begin{cases} 0 & \text{if} & \varphi_x(x)\downarrow \\ \uparrow & \text{if} & \varphi_x(x)\uparrow \end{cases}$$

and let S be the corresponding semigroup, defined as outlined above. Assume that q_0 is the starting state of **T**, and that q_h is the sole halting state. Then with each natural number x we may associate the words u_x and v_x on the generators of S that represent the initial configuration $q_0\,0\,\bar{x}$ and the halting configuration $q_h\,0\,1$. Because of the way in which the semigroup S is defined, u_x and v_x will represent the same element of S iff $g(x) = 0$; that is, iff $x \in K$. Thus the problem of determining membership in K is reducible to the word problem for S, and the latter problem is undecidable.

For our next example, we consider a more general type of decision problem arising in the study of mathematics and logic. It is common in mathematics to investigate various classes of structures—such as semigroups, groups, fields, etc.—with the intent of determining what statements hold true for all members of a given class. A convenient vehicle for expressing such statements is the notion of a first-order language, which we now review.

A *first-order language* is based on a set of constant letters (standing for certain fixed elements); a set of function letters (standing for operations); a set of predicate letters (standing for predicates or properties); an infinite set of variable letters (whose values range over the set of elements to be considered); the logical connectives $\wedge$ (and), $\vee$ (or), $\neg$ (not), and $\rightarrow$ (implies); and the universal and existential quantifiers $\forall$ (for all) and $\exists$ (there exists). Sentences in the language are built up from variable, constant, function, and predicate letters using connectives and quantifiers in a more or less natural way. We shall not bother to set forth in detail the rules for constructing legal sentences; for these rules the interested reader may consult any introductory text on mathematical logic.

Example 6-28. Recall that a group is a mathematical structure consisting of a set of elements together with an associative binary operation for which there is an identity element and with respect to which every element has an inverse. The first-order language of groups has a single constant letter 1 standing for the identity element, a single function letter f standing for the operation, and the single predicate letter $=$ standing for the binary relation of equality between elements. Among the sentences in this language are "$\forall x \exists y(f(x, y) = 1 \wedge f(y, x) = 1)$," which represents the (true) statement that for every group element x there exists some element y that serves as an inverse for x; and

"$\forall x(\neg (x = 1) \rightarrow \neg (f(x, x) = 1))$," which represents the (false) statement that if x is an element other than the identity, then applying the group operation to x and x does not yield the identity element.   □

The set of first-order sentences that hold for all members of a given class of mathematical systems constitutes what is called the *first-order theory* of that class. Thus the first-order theory of groups, for example, consists of all the sentences in the first-order language of groups that are true of all groups.

It is obviously of interest in the study of a given class of systems to be able to determine what things are true of the members of that class. We are therefore faced with the fundamental decision problem for a first-order theory—that of determining, for an arbitrary sentence in the appropriate language, whether or not that sentence belongs to the theory. The first-order theory of a given class of systems is said to be *decidable* or *undecidable* according as this fundamental question is or is not decidable.

As might be expected, there are many classes of systems whose first-order theories are decidable. Moreover, it can be shown that as long as the defining characteristics of a class of structures can be described by a computable set of sentences, the theory of that class must be effectively enumerable. Thus there are effective (although not necessarily practical) procedures for generating the first-order sentences that are true of groups, of semigroups, of fields, or of many other structures.

But the ability to generate the members of a set is not the same as the ability to decide the question of membership in a set, and there are in fact many important classes of structures whose theories are not decidable. Among these are the first-order theories of groups, rings, and fields. Also undecidable is the first-order theory of arithmetic—i.e., the theory of the structure formed by the natural numbers under the operations of ordinary addition and multiplication. The arguments leading to these conclusions are complex and we shall not attempt to present them here. We only observe that in each case the argument consists of a reduction or chain of reductions, with a problem such as that of membership in K serving as an ultimate starting point.

Results such as the undecidability of certain first-order theories or the undecidability of the word problem for certain semigroups tell us something about the nature of mathematics. While it is often possible—indeed highly desirable—to provide simple characterizations of complex mathematical structures, such characterizations may not help us settle certain natural questions about the structures they define. In particular, when dealing with mathematical systems of reasonable complexity, it may not be possible to devise algorithms for resolving questions about the structure of a system given only a finite description of that system.

The realization that some decision problems cannot be settled algorithmically is relatively new to mathematics. In this regard it is appropriate to consider briefly what is commonly referred to as "Hilbert's tenth problem."

In 1900 the mathematician David Hilbert presented a list of what he considered to be the most important problems that would occupy mathematicians during the coming century. Among these were several decision problems, of which number ten has become the best known. It is the problem of determining, for an arbitrary (multivariable) polynomial equation with integer coefficients, whether that equation has an integer solution. This problem is not as artificial as it may at first seem, since many practical applications are governed by polynomial constraints for which integer solutions are required.

When Hilbert's tenth problem was first posed it was generally assumed that the "solution" of the problem simply amounted to finding an appropriate algorithm for making the desired decision. It was obvious that it might be very difficult, perhaps impractically so, to obtain such an algorithm; what was not obvious was that such an algorithm might not even exist. Only after certain questions concerning the behavior of Turing machines were discovered to be undecidable did mathematicians begin to suspect that Hilbert's tenth problem might also be undecidable in the formal sense discussed here. That this is in fact the case was finally established in 1970. Although the proof is quite complicated, it basically amounts to showing that the problem of determining membership in K is reducible to that of determining whether a polynomial with integer coefficients has an integer solution.

The history of Hilbert's tenth problem illustrates the way in which the philosophy of mathematics can change. An attitude that is taken for granted in one era may later prove to be quite inappropriate. As a consequence of the work of Turing, Church, Gödel, Post, and others, we now realize that problems—even very simple sounding problems—do not necessarily have algorithmic solutions. Thus when we are faced with the task of finding a general procedure for solving a certain class of problem, we may find it appropriate to ask first whether or not such a procedure exists. And as Hilbert's problem also illustrates, answering this question may not be easy.

6.4 SUMMARY

The notion of effective computability can be formalized in many ways. But the resulting formal concepts, which include Turing computability, μ-recursiveness, and general recursiveness, prove to be equivalent. The common class of functions defined by the various formal approaches is known as the class of recursive functions. The comprehensiveness of this class, and of the techniques used to establish membership in it, lead us to conjecture that the effectively computable functions are just the recursive functions. Similar conclusions are reached in the case of relative computability. Informally, one function is computable relative to another if a procedure for evaluating the second provides a procedure for evaluating the first. This notion can also be modeled in terms of Turing machines, μ-recursive functions, and general recursive functions. The various formalizations again turn out to be equivalent, and yield the formal concept of relative recursiveness. The hypothesis that recursiveness and relative recursiveness accurately reflect the intuitive ideas of computability and relative computability is known as the Church-Turing Thesis.

The recursive functions can be indexed quite simply by assigning to each function the indices of all the Turing machines that compute it. This indexing scheme, like those that can be obtained from it via recursive translation functions, is one for which the five basic properties of Chapter 1 hold. The recursive functions therefore represent a standard family of algorithms and share all the characteristics of such families. Moreover, the Recursion Theorem can be used to show that any family in which the five basic properties hold must be closed under primitive recursion and minimalization, and so must include all the recursive functions. This means that the recursive functions represent the smallest family for which the basic properties hold, and that these five properties completely characterize the set of functions that can be effectively computed.

There are two basic computational ideas associated with sets. Informally, a set is computable if there is a procedure for determining whether an object belongs to the set, and enumerable if there is a procedure for generating the members of the set. These ideas are formalized by defining a set of natural numbers to be recursive if its characteristic function is recursive, and recursively enumerable if it is the range of a total recursive function. The two concepts are related by the fact that a set is recursive iff both it and its complement are recursively enumerable. The process of dovetailing, or interleaving two or more computations, provides a way of establishing the recursive enumerability of many sets, such as the set $K = \{x \mid \varphi_x(x)\downarrow\}$ which is recursively enumerable but not recursive. Dovetailing arguments also reveal that a set is recursively enumerable iff it is the range of a partial recursive function or, alternatively, iff it is the domain of a partial recursive function.

The use of a Cantor-numbering scheme for representing n-ary relations by sets of natural numbers permits the definitions of recursiveness and recursive enumerability to be extended to number-theoretic relations. As in the case of simple sets, a relation is recursive iff both it and its complement are recursively enumerable and recursively enumerable iff it is the domain of a recursive function. The operation of projecting a relation along a given coordinate preserves recursive enumerability, while the operation of forming a cross section at a given coordinate value preserves both recursive enumerability and recursiveness. These facts make it possible to establish the recursive enumerability of many relations. They also enable us to show that every recursively enumerable n-ary relation can be derived from a single recursive $(n + 1)$-ary relation by means of suitable projections and sections.

The concept of computability is closely related to that of decidability. Informally, the problem of determining whether a designated property holds for arbitrarily chosen objects is said to be decidable if there is an effective procedure for making the required decision. Formally, such a decision problem is defined to be recursively decidable if the number-theoretic relation corresponding to the property in question is recursive. Many important problems turn out to be undecidable in this sense. Undecidability is most commonly established by the method of reduction, which consists in showing that a decision procedure for the problem at hand would provide a decision procedure for some problem already known to be undecidable. Rice's Theorem uses this approach to show that many simple problems concerning recursive functions are undecidable. The import of Rice's Theorem is that there can be no algorithmic procedure for determining, on the basis of an index alone, whether a recursive function satisfies a given nontrivial property.

One of the most useful reduction techniques is that known as many-one reduction. A set B is said to be many-one reducible to a set A if there is a total recursive function that maps members of B into members of A and nonmembers of B into nonmembers of A. The existence of such a function implies that the problem of determining membership in A must be undecidable whenever the problem of determining membership in B is undecidable. This fact provides a way of establishing the undecidability of many interesting problems. Among these is the so-called Correspondence Problem—that of determining whether identical strings can be formed by concatenating corresponding members of two given lists. And the undecidability of the Correspondence Problem can in turn be used to establish the undecidability of a number of problems associated with formal grammars and the languages these grammars define.

A great many problems have by now been shown to be undecidable, in most cases by some form of reduction from the problem of determining membership in the set K. In fact, so many fundamental problems have been found to be undecidable that the questions of computability and decidability must be raised whenever a general-purpose algorithm is sought for a complex problem.

6.5 PROBLEMS

Section 6.1

6.1.1   Indicate which of the functions described below are recursive and which are not. Proofs are not required.

a) $f_1(x) = \begin{cases} 1 & \text{if} \quad \text{dom } \varphi_x^{(1)} = \phi \\ 0 & \text{otherwise} \end{cases}$

b) $f_2(x, n) = $ the smallest number y such that $\mathbf{T}_y$ halts in exactly n steps when started on the pattern $\underline{0}\ \bar{x}\ 0 \cdots$

c) $f_3(x) = \begin{cases} 1 & \text{if} \quad 5 \in \text{ran } \varphi_x^{(1)} \\ 0 & \text{otherwise} \end{cases}$

d) $f_4(x, y) = \begin{cases} 1 & \text{if} \quad y \in \text{ran } \varphi_x^{(1)} \\ \uparrow & \text{otherwise} \end{cases}$

e) $f_5(x) = \begin{cases} 1 & \text{if} \quad \varphi_x^{(1)} \text{ is total} \\ \uparrow & \text{otherwise} \end{cases}$

6.1.2 Indicate which of the predicates described below are recursive and which are not. Proofs are not required.

a) $P_1(x) \quad \Leftrightarrow \quad \text{dom } \varphi_x^{(1)} \neq \phi$

b) $P_2(x, y, n) \quad \Leftrightarrow \quad \mathbf{T}_x$ halts within n steps for some initial
$\qquad\qquad\qquad\qquad$ tape pattern of the form $\underline{0} \; \bar{w} \; 0 \ldots$, where $w \leq y$

c) $P_3(x, y, z) \quad \Leftrightarrow \quad \varphi_x(y) = z$

d) $P_4(x, y) \quad \Leftrightarrow \quad$ there exists an initial tape pattern of the form
$\qquad\qquad\qquad\qquad \underline{0} \; \bar{w} \; 0 \cdots$ for which $\mathbf{T}_x$ and $\mathbf{T}_y$ both halt

e) $P_5(x, n) \quad \Leftrightarrow \quad \mathbf{T}_x$, after starting at the left end of a *finite*
$\qquad\qquad\qquad\qquad$ blank tape of length n, eventually runs off
$\qquad\qquad\qquad\qquad$ one end of the tape

6.1.3 Let f_1 and f_2 be specified one-variable recursive functions. Consider the new function f defined as follows.

$$f(x) = \begin{cases} f_1(x) & \text{if} \quad f_1(x)\downarrow \\ f_2(x) & \text{if} \quad f_1(x)\uparrow \text{ and } f_2(x)\downarrow \\ \uparrow & \text{otherwise} \end{cases}$$

Assume that dom $f_1 \cap$ dom $f_2 = \phi$.

a) Describe an informal algorithm for evaluating f.

b) Formalize the algorithm of part (a) and thereby show that the function f is recursive.

c) Explain why f might fail to be recursive if dom $f_1 \cap$ dom $f_2 \neq \phi$.

6.1.4 Give a brief criticism of each of the following arguments.

a) Consider the function f such that:

$$f(x, y) = \begin{cases} 1 & \text{if} \quad f_0(x) = y \qquad \text{(where } f_0 \text{ is a designated} \\ 0 & \text{otherwise} \qquad\qquad \text{recursive function)} \end{cases}$$

Evaluating $f(x, y)$ amounts to determining whether $f_0(x)\downarrow$. But there is no effective procedure for determining whether a function is defined for a given argument value. Therefore f is not recursive.

b) Consider the function f such that:

$$f(x, y, z) = \begin{cases} 1 & \text{if} \quad \varphi_x(z) = \varphi_y(z) \\ 0 & \text{otherwise} \end{cases}$$

To evaluate $f(x, y, z)$, it is first necessary to determine whether $\varphi_x(z)$ and $\varphi_y(z)$ are defined. Since there is no effective procedure for doing this, f is not recursive.

6.1.5 With each real number r we associate the number-theoretic function γ_r defined as follows: If $n = 0$, $\gamma_r(n)$ is the integer part of r; if $n > 0$, $\gamma_r(n)$ is the nth digit in the decimal representation of the fractional part of r. Thus $\gamma_\pi(0) = 3$, $\gamma_\pi(1) = 1$, $\gamma_\pi(2) = 4$, $\gamma_\pi(3) = 1$, $\gamma_\pi(4) = 5$, etc. For the purposes of this problem we consider a real number r to be *computable* iff the associated function γ_r is recursive.

   a) Describe a real number that is not computable.

   b) Show that every rational number is computable.

   c) Show that $\sqrt{2}$ is computable.

   d) (Hard) Show that the sum of two computable real numbers is also computable.

6.1.6 Let P be any specified property that a one-variable number-theoretic function might have. For each method of indexing the recursive functions, let h be the function such that

$$h(x) = \begin{cases} 1 & \text{if} \quad \varphi_x^{(1)} \text{ has property } P \\ 0 & \text{otherwise} \end{cases}$$

Prove that if h is not recursive for the standard indexing scheme, then it is not recursive for any acceptable indexing scheme.

6.1.7 The one-variable function f is said to be an *extension* of the one-variable function g if $f(x)$ is defined and equal to $g(x)$ whenever the latter is defined. Show that there exists a partial one-variable recursive function φ that has no total recursive extension.

6.1.8 Let $\mathcal{F}$ be a class of functions satisfying Properties 1 through 5. Let g and h be respectively n- and one-variable members of $\mathcal{F}$, and let f be defined from g and h by iteration, thus:

$$f(x_1, \ldots, x_n, 0) = g(x_1, \ldots, x_n)$$
$$f(x_1, \ldots, x_n, y^+) = h(f(x_1, \ldots, x_n, y))$$

Use a simplified version of the proof of Theorem 6-1 to show that f must be a member of $\mathcal{F}$. (The auxiliary variable w is not needed in this case.)

6.1.9 Let $\mathcal{F}$ be a class of functions satisfying Properties 1 through 5.

   a) Give a specialized version of the proof of Theorem 6-1 showing that $\mathcal{F}$ contains the predecessor function.

b) Using the result of part (a), give a proof of Theorem 6-1 that does not require the auxiliary variables z and w.

6.1.10   Let f_0 be a given total one-variable number-theoretic function. Explain why there are one-variable number-theoretic functions that are not recursive relative to f_0. Describe such a function.

6.1.11   Show that each of the functions described below is recursive relative to the designated total function.

  a) $f_1(x, y) =$ the number of values of z in the range $0 \leq z \leq y$ for which
  $$\varphi_x(z){\downarrow}$$
  Recursive relative to the domain function $D^{(2)}$

  b) $f_2(x) =$ the smallest value of z such that $\varphi_x(z) = 0$, or undefined if
  there is no such z
  Recursive relative to the domain function $D^{(2)}$

  c) $f_3(x, y) = D^{(2)}(x, y) = \begin{cases} 1 & \text{if} & \varphi_x(y){\downarrow} \\ 0 & \text{if} & \varphi_x(y){\uparrow} \end{cases}$

  Recursive relative to the function f_1 of part (a)

  d) $f_4(x, y) = \begin{cases} 1 & \text{if} & \varphi_x^{(1)} \neq \varphi_y^{(1)} \\ {\uparrow} & \text{otherwise} \end{cases}$

  Recursive relative to the domain function $D^{(2)}$

  e) $f_5(x) = t(x) = \begin{cases} 1 & \text{if} & \varphi_x^{(1)} \text{ is total} \\ 0 & \text{otherwise} \end{cases}$

  Recursive relative to the equivalence function e, where

  $$e(x, y) = \begin{cases} 1 & \text{if} & \varphi_x^{(1)} = \varphi_y^{(1)} \\ 0 & \text{otherwise} \end{cases}$$

6.1.12   a) Give an appropriate definition of relative computability for general recursive functions.

  b) Show that the function f of Example 6-5 is general recursive relative to the domain function $D^{(2)}$.

  c) Outline an approach that could be used to show that relative μ-recursiveness and relative general recursiveness are equivalent.

Section 6.2

6.2.1   Describe an effective procedure for enumerating each of the following sets.

  a) $\{x \mid \text{the evaluation of } \varphi_x(x) \text{ terminates within } 7{,}128 \text{ steps}\}$
  b) $\{x \mid x \text{ is prime}\}$ *(continued)*

c) $\{n \mid x^n + y^n = z^n$ for some choice of positive integers $x, y, z\}$

d) $\{x \mid x \in \operatorname{ran} f_1 \quad \text{or} \quad x \in \operatorname{ran} f_2\}$ where $\operatorname{dom} f_1 \cup \operatorname{dom} f_2 = N$ and $\operatorname{dom} f_1 \cap \operatorname{dom} f_2 = \phi$.

e) $\{x \mid \text{Turing machine } \mathbf{T}_x$ halts after starting on blank tape$\}$

6.2.2 Determine, to the best of your ability, which of the following sets are recursive and which are recursively enumerable.

a) $S_1 = \{x \mid$ the xth digit in the decimal expansion of π is a 7$\}$

b) $S_2 = \{x \mid$ the decimal expansion of π contains a run of at least x consecutive 7's$\}$

c) $S_3 = \{x \mid$ the decimal expansion of π contains a run of exactly x consecutive 7's$\}$

6.2.3 Use Theorem 6-10 to show that every infinite recursively enumerable set has an infinite recursive subset.

6.2.4 Let A_1 and A_2 be any two recursively enumerable sets. Prove or disprove each of the following claims.

a) $A_1 \cup A_2$ is necessarily recursively enumerable.

b) $A_1 \cap A_2$ is necessarily recursively enumerable.

c) $A_1 - A_2$ is necessarily recursively enumerable.

6.2.5 Prove that neither of the functions described below is recursive.

a) $f(x) = \begin{cases} 0 & \text{if} \quad \varphi_x^{(1)} \text{ is total} \\ \uparrow & \text{otherwise} \end{cases}$

b) $f(x, y) = \mu z[z \geq y \wedge \varphi_x(z)\downarrow]$ \qquad *Hint:* Use Theorem 6-10.

6.2.6 Use dovetailing to establish the recursiveness of each of the following functions.

a) $g(x) = \begin{cases} 1 & \text{if} \quad \operatorname{dom} \varphi_x^{(1)} \neq \phi \\ \uparrow & \text{otherwise} \end{cases}$

b) $g(x) = \begin{cases} 1 & \text{if} \quad 0 \in \operatorname{ran} \varphi_x^{(1)} \\ \uparrow & \text{otherwise} \end{cases}$

c) $g(x, y) = \begin{cases} 1 & \text{if there exists a number } z \text{ such that } \varphi_x(z)\downarrow, \varphi_y(z)\downarrow, \\ & \text{and } \varphi_x(z) = \varphi_y(z) \\ \uparrow & \text{otherwise} \end{cases}$

6.2.7 Let A be any recursively enumerable set. Prove that the set $V = \bigcup_{x \in A} W_x$ is also recursively enumerable. *Hint:* Use Theorem 6-11 and an appropriate dovetailing strategy.

6.2.8 a) Prove that the Cartesian product of any two recursive sets is a recursive relation.

b) Prove that the Cartesian product of any two recursively enumerable sets is a recursively enumerable relation.

6.2.9 Use Theorems 6-20 and 6-21 to establish the recursive enumerability of each of the following sets.

a) $S_1 = \{x \mid \varphi_x(700)\downarrow\}$
b) $S_2 = \{x \mid \text{dom } \varphi_x^{(1)} \neq \phi\}$
c) $S_3 = \{(x, y) \mid \text{the evaluation of } \varphi_x(y) \text{ terminates in an odd number of}$
$$\text{steps}\}$$
d) $S_4 = \{(x, y, z) \mid \varphi_x(z) = \varphi_y(z)\}$
e) $S_5 = \{(x, y) \mid \text{ran } \varphi_x^{(1)} \cap \text{ran } \varphi_y^{(1)} \neq \phi\}$

6.2.10 a) Show that the composition of two recursively enumerable binary relations is necessarily recursively enumerable.

b) Show that the composition of two recursive binary relations need not be recursive.

6.2.11 a) Establish the following generalization of Theorem 6-20.

Theorem. *Let R be a given n-ary number-theoretic relation, and let $f_1, \ldots, f_n$ be total m-variable recursive functions, where $m \leq n$. Define R_* to be the m-ary relation such that $(x_1, \ldots, x_m) \in R_*$ iff $(f_1(x_1, \ldots, x_m), \ldots, f_n(x_1, \ldots, x_m)) \in R$. Then R_* is recursive whenever R is recursive and recursively enumerable whenever R is recursively enumerable.*

b) Use the theorem of part (a) to prove that:
   i) $R_1 = \{x \mid \varphi_x(x) = 3\}$ is recursively enumerable.
   ii) $R_2 = \{(x, y) \mid \varphi_x(y)\uparrow\}$ is not recursively enumerable.
   iii) $R_3 = \{x \mid \varphi_x^{(1)} = C_7^{(1)}\}$ is not recursively enumerable.

6.2.12 Show that none of the following sets are recursively enumerable. You may use the results of Problem 6.2.9.

a) $\{x \mid \text{dom } \varphi_x^{(1)} = \phi\}$
b) $\{x \mid \varphi_x(700)\uparrow\}$

(*continued*)

c) $\{(x, y) \mid \varphi_x^{(1)} = \varphi_y^{(1)}\}$

d) $\{(x, y, z) \mid \varphi_x(z) \downarrow$ and $\varphi_y(z) \uparrow$ or vice versa$\}$

e) $\{(x, y) \mid \text{dom } \varphi_x^{(1)} \cap \text{dom } \varphi_y^{(1)} = \phi\}$

Section 6.3

6.3.1 Which of the following decision problems are effectively decidable?

a) Given x and n, determine whether the machine $\mathbf{T}_x$ visits more than n squares after being started on blank tape.

b) Given x, determine whether the machine $\mathbf{T}_x$ ever writes the same symbol during two consecutive moves after being started on blank tape.

c) Given x, determine whether $x \in S$, where the set S is defined inductively as follows: $0 \in S$; if $u \in S$, then $u^2 + 1$, $3u + 2$, and $u!$ are all members of S.

d) Given x, determine whether there exists a Turing machine with fewer states than $\mathbf{T}_x$ that computes the same one-variable function as $\mathbf{T}_x$.

e) Given x_1, x_2, x_3, determine whether $f(x_1) = \pi^2(x_2, x_3)$, where f is a fixed, nontotal recursive function.

6.3.2 Consider the problem of determining, for an arbitrary choice of integers $c_0, \ldots, c_n$ (with $c_n \neq 0$), whether the equation

$$c_0 + c_1 x + c_2 x^2 + \cdots + c_n x^n = 0$$

has an integer solution. Show that this problem is decidable.

6.3.3 Which of the following decision problems succumb to Rice's Theorem?

a) Given x, determine whether $\varphi_x^{(1)}$ is a one-to-one function.

b) Given x, determine whether x is the smallest index of $\varphi_x^{(1)}$.

c) Given x, determine whether $\varphi_x^{(1)}$ can be evaluated by a twenty-three-state Turing machine.

d) Given x, determine whether $\varphi_x(3x) = \varphi_{3x}(x)$.

e) Given x, determine whether $\varphi_x(x) \downarrow$.

6.3.4 a) Prove the following generalization of Rice's Theorem:

Theorem. *Let R be an n-ary relation on the set of one-variable recursive functions. Assume that R holds for at least one n-tuple of functions and fails to hold for at least one n-tuple of functions. Then the set $A = \{(x_1, \ldots, x_n) \mid R \text{ holds for } \varphi_{x_1}^{(1)}, \ldots, \varphi_{x_n}^{(1)}\}$ is not recursive.*

b) Use the theorem of part (a) to establish the undecidability of each of the following decision problems.

    i) Given x and y, determine whether $\varphi_x^{(1)} = \varphi_y^{(1)}$.

    ii) Given x and y, determine whether $\varphi_y(z)$ is defined and greater than $\varphi_x(z)$ whenever the latter is defined.

    iii) Given x, y, and z, determine whether $\varphi_x^{(1)} = \varphi_y^{(1)} \circ (\varphi_z^{(1)})$.

6.3.5   In each of the following parts, show that the set S_1 is m-reducible to the set S_2.

    a) $S_1 = \{x \mid 7 \in \text{ran } \varphi_x^{(1)}\};$ $S_2 = \{x \mid 13 \in \text{ran } \varphi_x^{(1)}\}$

    b) $S_1 = \{x \mid \text{dom } \varphi_x^{(1)} \text{ is infinite}\};$ $S_2 = \{x \mid \text{ran } \varphi_x^{(1)} \text{ is infinite}\}$

    c) $S_1 = \{x \mid \varphi_x^{(1)} \text{ is total}\};$ $S_2 = \{(x, y) \mid \varphi_x^{(1)} = \varphi_y^{(1)}\}$

    d) $S_1 = \{(x, y) \mid y \in \text{ran } \varphi_x^{(1)}\};$ $S_2 = \{x \mid 7 \in \text{ran } \varphi_x^{(1)}\}$

    e) $S_1 = \{x \mid \text{ran } \varphi_x^{(1)} \text{ is infinite}\};$ $S_2 = \{x \mid \text{dom } \varphi_x^{(1)} \text{ is infinite}\}$

6.3.6   a) Prove that if the set B is m-reducible to the set A, and if A is recursively enumerable, then B must be recursively enumerable.

    b) Exhibit a set that is not m-reducible to K.

    c) Exhibit a set to which K is not m-reducible.

6.3.7   Use the result of Problem 6.3.6(a) to show that none of the following sets are recursively enumerable.

    a) $S_1 = \{(x, y) \mid \varphi_x(y)\!\uparrow\}$

    b) $S_2 = \{x \mid \varphi_x^{(1)} \neq C_7^{(1)}\}$

    c) $S_3 = \{(x, y) \mid \varphi_x^{(1)} = \varphi_y^{(1)}\}$

6.3.8   Consider the problem of determining, for any natural number x, whether the recursively enumerable set W_x is a recursive set. Prove that this problem is not recursively decidable. *Hint:* Find a function f such that the domain of $\varphi_{f(x)}^{(1)}$ is cofinite if $\varphi_x(x)\!\downarrow$ and is the set K if $\varphi_x(x)\!\uparrow$.

6.3.9   Consider the problem of determining whether a given set of formal equations, together with a specified principal function letter, actually defines a (general recursive) function. Is this problem decidable? Justify your answer briefly.

6.3.10   It is desired to find a pair of lists whose shortest match is very long. One possibility is to design the lists to "simulate" the operation of a binary counter. Assume that the low-order digit of the counter is on the left, and that two different signals, C and N, can be propagated through the counter from left to right. The signal C represents a carry, and causes each digit that it encounters to be complemented. The signal N represents no carry and has no effect on the digits it encounters. A typical sequence of steps in a four-bit counter might be:

$$C1\ 0\ 1\ 0 \rightarrow 0C0\ 1\ 0 \rightarrow 0\ 1N1\ 0 \rightarrow 0\ 1\ 1N0 \rightarrow 0\ 1\ 1\ 0N$$

When either of the signals C or N emerges from the right end of the counter, a new carry signal C is entered at the left. Adapt the technique used in the proof of Lemma 6-5 to the design of a pair of lists that "simulate" the behavior of a three-bit counter, starting with the configuration $N0\ 0\ 0$ and ending with the configuration $0\ 0\ 0C$. Exhibit the lists and determine the length of their shortest match.

6.3.11   The *free semigroup* on the alphabet A is the semigroup formed by the finite strings of symbols from A under the operation of concatenation. Use the undecidability of the Correspondence Problem to establish the undecidability of the following problem.

> Given any two free semigroups S_1 and S_2 on finite alphabets A_1 and A_2, and any two homomorphisms φ_1 and φ_2 from S_1 to S_2, determine whether there is an element in S_1 that has the same image under both φ_1 and φ_2.

Note that a homomorphism from S_1 to S_2 is completely determined by its effect on the members of S_1 that consist of a single symbol.

6.3.12   In this problem we show that the equivalence problem for primitive recursive functions is undecidable.

a) Let $L = \{l_1, \ldots, l_k\}$ be a list of words on the alphabet $\{a_1, \ldots, a_q\}$, and let C be an arbitrary constant. Show that there exists a primitive recursive function $\psi_{L,C}$ such that: (i) if z is the Gödel number of a sequence of indices $i_1, i_2, \ldots, i_n$ (where $1 \le i_j \le k$), then $\psi_{L,C}(z)$ is the Gödel number of the string $l_{i_1}l_{i_2} \cdots l_{i_n}$; and (ii) if z is not the Gödel number of a sequence of indices, then $\psi_{L,C}(z) = C$.

b) Use the undecidability of the Correspondence Problem to show that there is no effective procedure for determining, for two arbitrarily specified one-variable primitive recursive functions g_1 and g_2, whether there exists a number z such that $g_1(z) = g_2(z)$.

c) Show that there is no effective procedure for determining, for any two given primitive recursive derivations, whether those derivations define the same primitive recursive function.

Postscript

Now that we understand the basic attributes of the computable functions, it is appropriate to comment on the relationship between our introductory presentation and the general development of the theory of computation. The following remarks provide some additional background, suggest supplementary reading, and indicate possible areas for further study. The bibliography includes many of the most directly relevant references, but it is by no means exhaustive. More extensive bibliographies appear in Rogers [1967], Kleene [1952], and Minsky [1967].

The first attempts to devise models for algorithmic computation were made during the 1930's by Church, Turing, and Kleene, among others. A brief account of their work is given in Chapter 1 of Hermes [1965]. The basic concepts of Turing machines, recursive functions, computability, and decidability that grew out of this work are now standard concepts. Most of the results and proofs presented in this book are widely known and readily available in the literature. In particular, most of the material discussed here can be found in such texts as Minsky [1967], Hermes [1965], and Rogers [1967]. Thus the basic content of this book is not new.

There are, however, two ways in which our development differs from other introductions to the theory of computation. The first is that we have begun with an abstract view of computation rather than with a specific computational model. This approach was motivated in part by the work of Blum [1967]. It has the disadvantage of lacking the concrete foundation that would be provided by beginning with, say, Turing machines. But it does allow us to get a quick overview of all the major results to be developed in concrete form later. Even more important, it emphasizes the fact that the fundamental characteristics of the theory of computability do not depend on the details of any particular implementation.

The second difference lies in the treatment of the five basic properties introduced in Chapter 1. Since it may not be initially clear why these properties are of particular interest, we have tried to justify them by reference to the capabilities of practical programming languages. But the primary reason for emphasizing these properties is that they completely characterize the class of computable functions. This fact, which follows from Theorem 6-3, is perhaps the one key idea not commonly presented in introductory texts. Theorem 6-3 and its proof are derived from the work of Wagner [1969] and Strong [1968] at IBM.

As indicated earlier, there are a number of sources of additional reading for the material presented in this book. In the area of Turing machines, the interested reader may consult Minsky [1967], Hermes [1965], or Arbib [1969], among others. Chapter 14 of Minsky describes an elegant seven-state universal Turing machine. Chapter 4 of Arbib discusses techniques for simulating multi-head, multitape, and multidimensional machines with ordinary machines. This same chapter also presents a method, due to Shannon [1956], for converting any ordinary Turing machine into an equivalent two-state machine. The basis for the Busy Beaver function introduced in Problem 3.3.9 is to be found in the paper by Rado [1962].

The basic references for the various classes of recursive functions include Péter [1967], Hermes [1965], and Kleene [1952]. Péter's book emphasizes primitive recursive functions. It examines several types of recursive definitions, shows which ones yield primitive recusive functions, and develops some simple bases for the class of primitive recursive functions. Hermes and Kleene are broader in scope and somewhat easier to use. They both treat primitive recursive, μ-recursive, and general recursive functions, among other topics. An interesting relationship between the primitive recursive functions and a restricted class of programming languages is presented in the paper by Meyer and Ritchie [1967].

The basic ideas of computability and the Church-Turing Thesis are developed in many texts, among them Kleene [1952], Hermes [1965], Minsky [1967], and Davis [1958]. Those interested in reading some of the early work in this area may consult the papers of Turing [1936, 1954], Post [1944], and Markov [1960]. Additional background on decision problems will be found in Davis [1965], Kleene [1952], or Hermes [1965]. In particular, the book by Davis is a collection of papers on undecidable problems. The original proof of the undecidability of the Correspondence Problem appears in Post [1946]. The proof given here is patterned after an unpublished proof by J. B. Dennis and P. Denning.

Several areas of further study are open to the reader who wishes to build on the introduction provided here. Among those concerned primarily with functions and algorithms, recursive function theory is the oldest and most firmly established. It deals with the properties of the recursive functions, with hierarchies of sets and functions, and with reductions among decision problems. The study of computational complexity, on the other hand, attempts to assess and compare the numbers of steps or amounts of storage space needed to evaluate various recursive functions. A recent outgrowth of this study forms the beginning of what might be called a theory of algorithms. The goals of this theory are to devise efficient algorithms for common computational tasks, to establish lower bounds on the numbers of basic operations needed to perform various computations, and to classify computational problems according to their difficulty.

The primary reference for recursive function theory is the text by Rogers [1967]. Several aspects of the theory of computational complexity, including work of Blum [1967], are discussed in Chapter 7 of Arbib [1969]. Interested readers should also see Hartmanis and Stearns [1965]. An introduction to the analysis of algorithms is to be found in Aho, Hopcroft, and Ullman [1974], along with an extensive bibliography of papers on algorithms and complexity.

An understanding of computability also provides a foundation for the study of formal systems, in which the notion of computation is replaced by that of derivation. Among the most important formal systems are those treated by mathematical logic. The basic goals of mathematical logic are to formalize the notion of proof and to determine what statements are provable. Provability is related to computability, and some of the key results of logic are reminiscent of results presented here. The study of formal systems also includes formal grammars and the languages they define. Here the intent is to compare various types of grammars, investigate their decision problems, and find machine models for their associated languages. The study of formal grammars is motivated in part by their relationships to programming languages.

There are many good texts on mathematical logic. The reader who wants an introduction to the basic ideas and results should find Margaris [1967] useful. Those who want a more advanced treatment may consult Boolos and Jeffrey [1974], Kleene [1967], or Shoenfield [1967], among others. An introduction to grammars and languages will be found in Hopcroft and Ullman [1969], while Ginsburg [1966] provides a detailed treatment of context-free grammars and languages.

References

1. Aho, A. V., J. E. Hopcroft, and J. D. Ullman, *The Design and Analysis of Computer Algorithms*, Reading, Mass.: Addison-Wesley, 1974.
2. Arbib, M. A., *Theories of Abstract Automata*, Englewood Cliffs, N.J.: Prentice Hall, 1969.
3. Blum, M., "A machine independent theory of the complexity of recursive functions," *JACM* **14**, pp. 322–336, 1967.
4. Boolos, G. S., and R. C. Jeffrey, *Computability and Logic*, London: Cambridge University Press, 1974.
5. Davis, M., *Computability and Unsolvability*, New York: McGraw-Hill, 1958.
6. Davis, M. (Editor), *The Undecidable*, Hewlett, N.Y.: Raven Press, 1965.
7. Ginsburg, S., *The Mathematical Theory of Context Free Languages*, New York: McGraw-Hill, 1966.
8. Hartmanis, J., and R. E. Stearns, "On the computational complexity of algorithms," *Trans. Amer. Math. Soc.*, **117**, pp. 285–306, 1965.
9. Hermes, H., *Enumerability, Decidability, Computability*, New York: Academic Press, 1965.
10. Hopcroft, J. E., and J. D. Ullman, *Formal Languages and Their Relation to Automata*, Reading, Mass.: Addison-Wesley, 1969.
11. Kleene, S. C., *Introduction to Metamathematics*, Princeton, N.J.: Van Nostrand, 1952.
12. Kleene, S. C., *Mathematical Logic*, New York: Wiley, 1967.
13. Margaris, A., *First Order Mathematical Logic*, Waltham, Mass.: Blaisdell, 1967.
14. Markov, A. A., *Theory of Algorithms*, (in) *Trans. Amer. Math. Soc.*, Series 2, **15**, 1960.
15. Mendelson, E., *Introduction to Mathematical Logic*, Princeton, N.J.: Van Nostrand, 1964.

16. Meyer, A. R., and D. M. Ritchie, "The complexity of loop programs," *Proceedings of 23rd National Conference of the ACM*, 1967.

17. Minsky, M., *Computation: Finite and Infinite Machines*, Englewood Cliffs, N.J.: Prentice Hall, 1967.

18. Péter, R., *Recursive Functions*, New York: Academic Press, 1967.

19. Post, E. L., "Recursively enumerable sets of positive integers," *Bulletin American Mathematical Society*, **50**, pp. 284–316, 1944.

20. Post, E. L., "A variant of a recursively undecidable problem," *Bulletin American Mathematical Society*, **52**, pp. 264–268, 1946.

21. Rado, T., "On non-computable functions," *Bell System Technical Journal*, **41**, pp. 877–884, 1962.

22. Rogers, H., *Theory of Recursive Functions and Effective Computability*, New York: McGraw-Hill, 1967.

23. Shannon, C. E., "A universal Turing machine with two internal states," *Automata Studies* (C. Shannon and J. McCarthy, Eds.) *Annals of Mathematical Studies*, **34**, Princeton, N.J., 1956.

24. Shoenfield, J. R., *Mathematical Logic*, Reading, Mass.: Addison-Wesley, 1967.

25. Strong, H. R., "Algebraically generalized recursive function theory," *IBM Journal of Research and Development*, **12**, pp. 465–475, 1968.

26. Turing, A. M., "On computable numbers, with an application to the entscheidungsproblem," *Proceedings London Mathematical Society, Series 2*, **42**, pp. 230–265, 1936.

27. Turing, A. M., "Computing machinery and intelligence," *Mind*, **59**, pp. 433–460, 1950. (Reprinted in) *The World of Mathematics*, **4**, New York: Simon and Schuster, 1954.

28. Wagner, E. G., "Uniformly reflexive structures: On the nature of Gödelizations and relative computability," *Trans. Amer. Math. Soc.*, **144**, 1969.

Index